CSR, Sustainability, Ethics & Governance

Series Editors

Samuel O. Idowu, London Metropolitan University, Calcutta House, London, UK
René Schmidpeter, Cologne Business School, Cologne, Germany

In recent years the discussion concerning the relation between business and society has made immense strides. This has in turn led to a broad academic and practical discussion on innovative management concepts, such as Corporate Social Responsibility, Corporate Governance and Sustainability Management. This series offers a comprehensive overview of the latest theoretical and empirical research and provides sound concepts for sustainable business strategies. In order to do so, it combines the insights of leading researchers and thinkers in the fields of management theory and the social sciences – and from all over the world, thus contributing to the interdisciplinary and intercultural discussion on the role of business in society. The underlying intention of this series is to help solve the world's most challenging problems by developing new management concepts that create value for business and society alike. In order to support those managers, researchers and students who are pursuing sustainable business approaches for our common future, the series offers them access to cutting-edge management approaches.

CSR, Sustainability, Ethics & Governance is accepted by the Norwegian Register for Scientific Journals, Series and Publishers, maintained and operated by the Norwegian Social Science Data Services (NSD)

Elisabeth Fröhlich • Kristina Steinbiß •
Maud H. Schmiedeknecht •
Cristian R. Loza Adaui
Editors

Circular Economy in Sustainable Supply Chains

A Global Perspective on Challenges, Concepts and Cases

Springer

Editors
Elisabeth Fröhlich
ispira Think Tank for Sustainable Supply Chains
Hürth, Germany

Kristina Steinbiß
ESB Business School, Reutlingen University
Reutlingen, Baden-Württemberg, Germany

Maud H. Schmiedeknecht
ESB Business School, Reutlingen University
Reutlingen, Germany

Cristian R. Loza Adaui
THI Business School, Technische Hochschule Ingolstadt
Ingolstadt, Germany

ISSN 2196-7075 ISSN 2196-7083 (electronic)
CSR, Sustainability, Ethics & Governance
ISBN 978-3-031-70751-3 ISBN 978-3-031-70749-0 (eBook)
https://doi.org/10.1007/978-3-031-70749-0

This Springer imprint is published by the registered company Springer Nature Switzerland AG
The registered company address is: Gewerbestrasse 11, 6330 Cham, Switzerland

Contents

About the Editors

Elisabeth Fröhlich is the founder of ispira Think Tank for Sustainable Supply Chains. She invented an innovative business coach concept to prepare companies for the challenges of the different European sustainability laws. Her major aim is to bridge the gap between research and practice in this field to enable companies to implement the sustainable transformation of global supply chains successfully. Prof. Dr. habil. Elisabeth Fröhlich was president of the CBS International Business School in Germany until 2022. She holds a full professorship in Sustainable Procurement and Supply Chains and is an internationally recognised expert in sustainable supply chain management. She is also highly engaged in the field of responsible management education. She serves as a board member of PRME, chairs the PRME Nomination and Governance Committee and is the PRME DACH Chapter Chair. Additionally, she leads the AOM Management Education Division Ambassador Program as vice-chair, leads the Board of JARO Institute and offers online training in the field of sustainable procurement. Her research focuses on sustainable supply chain management and green procurement, qualification in purchasing, Procurement 4.0 and strategic supplier relationship management. Innovative teaching formats of responsible management education are further main areas of her research. She has published several books and articles on the above-mentioned topics and supports several journals as an external reviewer.

Kristina Steinbiß holds a professorship in Management with a focus on Marketing at the ESB Business School of the Reutlingen University in Germany, where she teaches primarily in the Industrial Engineering programmes. Her fundamental teaching goal is to enable her students to think and act sustainably while adopting an integrative perspective combining business and technology. Kristina's current research focuses on implementing new business models to accomplish a company's sustainability goals, as well as on influencing and nudging consumers towards sustainable behaviour. She is the author of numerous publications in the above-mentioned fields, including the textbook *Marketing*, which is already available in its second edition. Next to her research activities, she is also enthusiastic about the Theory of Inventive Problem Solving, the TIPS innovation methodology. As a TIPS trainer and coach, she likes to combine this methodology with her marketing approach and adapt it to implement sustainable solutions in day-to-day business.

Maud H. Schmiedeknecht is a Professor of Corporate Governance and Corporate Social Responsibility at the ESB Business School of the Reutlingen University in Germany and a Business Design Coach. What drives her is supporting and empowering business leaders and business school students to develop skills and learn methods to (re)design sustainable, circular and regenerative business models. Maud previously worked as a consultant at the global management consulting firm Kearney in Germany and abroad. As a research assistant at the Konstanz Institute for Intercultural Management, Values and Communication of Prof. Dr. Josef Wieland, she undertook consulting and research projects in the fields of corporate social responsibility and compliance and participated in national and international committees, e.g. the German CSR-Forum and ISO 26000 Working Group on Social Responsibility. She was an associate of the Stiftung Neue Verantwortung, a think tank for digital technologies, politics and society, on the topic of global economic ethic. Her dissertation on the governance of multistakeholder dialogues was awarded the Max Weber Prize for Business Ethics. Maud's current teaching, research and consulting focus is on corporate sustainability management, (social) entrepreneurship and sustainable business model innovation, business transformation and circular economy.

Cristian R. Loza Adaui is a Professor of Management and Corporate Social Responsibility at the THI Business School of the Technische Hochschule Ingolstadt in Germany. He previously worked for the CBS International Business School in Cologne, Germany, where he also served as Deputy Director of the Centre for Advanced Sustainability Management (CASM). He has held teaching and research positions at the Institute of Economics of the Friedrich-Alexander University Erlangen-Nürnberg in Germany and the Universidad Católica San Pablo in Peru. Prof. Loza Adaui received his Doctoral Degree in Management from the Ingolstadt School of Management at the Catholic University of Eichstätt-Ingolstadt in Germany. His research and consulting activities focus on how businesses become more sustainable, responsible, human-centred and purpose-driven through strategic management, corporate social responsibility, sustainability management, non-financial reporting, circular economy and corporate purpose. His research has appeared in outlets such as *Business Strategy and the Environment*, *Journal of Business Research*, *Long Range Planning*, *Voluntas: International Journal of Voluntary and Nonprofit Organizations* and *Journal of Management Development*, among others. He serves on the editorial boards of journals such as *Technological Sustainability* and *Frontiers in Sustainability*.

Circular Economy in Sustainable Supply Chains: Presenting a Global Perspective on Challenges, Concepts and Cases

Cristian R. Loza Adaui, Maud H. Schmiedeknecht, Kristina Steinbiß, and Elisabeth Fröhlich

1 Introduction

The term "circular economy", central to this book, has gained significant and increasing attention in recent decades. Scientists, managers, policymakers, corporations and governmental bodies are among those who use the term as an alternative to a linear model of production, consumption and resource usage (Kirchherr et al., 2023a). The coining of the term "circular economy" is not credited to one single person or institution. There are at least five persons who are recurrently linked to the origin of this idea: John Lyle, Willian McDonough, Michael Braungart, Walter Stahel and Genvieve Reday (Winans et al., 2017). All these scholars, in one way or another, shared a concern about resource depletion and waste management.

Similarly, the decades of the 1960s and 1970s witnessed the rise of a modern environmental movement, with thinkers like Kenneth Boulding (1966) advocating for a "spaceship economy" and raising awareness for the physical limits of the earth and the necessity to strive towards a "closed-loop" economic system that mimicked nature's cycles. In parallel, the concept of industrial ecology gained momentum. Pioneered by scientists like Frosch and Gallopoulos (1989), industrial systems were envisioned as closed-loop ecosystems where waste from one process becomes the input for another. Early examples included industrial symbiosis, where companies collaborated to exchange by-products, minimising virgin resource use.

C. R. Loza Adaui (✉)
THI Business School, Technische Hochschule Ingolstadt, Ingolstadt, Germany
e-mail: Cristian.LozaAdaui@thi.de

M. H. Schmiedeknecht · K. Steinbiß
ESB Business School, Reutlingen University, Reutlingen, Germany

E. Fröhlich
ispira Think Tank for Sustainable Supply Chains, Hürth, Germany

E. Fröhlich et al. (eds.), *Circular Economy in Sustainable Supply Chains*, CSR, Sustainability, Ethics & Governance,
https://doi.org/10.1007/978-3-031-70749-0_1

Other influential concepts that contributed to the development of a circular economy include Walter Stahel's (2010) idea of "product-service systems", highlighting the possibility of providing access to the functionality of a product (e.g. mobility) instead of car ownership, fostering a shift towards product life extension and reuse (Baines et al., 2007). Similarly, Geoffrey Boothroyd's (1992) concept of "design for disassembly" advocated for products designed for easy disassembly and material recovery at the end of their useful life or Michael Braungart and William McDonough's (2002) concept of "cradle-to-cradle" design.

1.1 A Journey of Convergence Among Circular Economy and Supply Chain Management

During the 1990s, the circular economy gained recognition. The formal introduction of the concept into European Union policy frameworks certainly played a role, together with the emergence of organisations like the Ellen MacArthur Foundation, that prominently promoted and advocated for this concept. Especially the influential "Butterfly Model" developed by the Ellen MacArthur Foundation Circular Economy Team emphasises a range of strategies to be implemented into biological and technical cycles (Aggeri, 2020; Kirchherr et al., 2023a).

The early twenty-first century witnessed a significant convergence of the circular economy and the supply chain management literature, driven by factors like resource scarcity, rising waste management costs and growing consumer demand for sustainable products. Businesses began to recognise the potential of circular economy principles, such as enhancing resource efficiency and reducing environmental impact, that potentially generated benefits at a company level, including cost reductions and improved brand value and reputation, besides the impact at the systemic level.

Several key milestones marked this period of convergence. For instance, the European Commission's (2008) Sustainable Consumption and Production and Industrial Policy Action Plan explicitly linked the concept of circular economy to supply chains, advocating for "greener" supply chains that minimise waste, maximise resource recovery and promote smarter consumption (Nash, 2009). The development of standards like ISO 14001:2015 on Environmental Management Systems and ISO 26000 on Social Responsibility further emphasised the integration of sustainability principles into supply chain practices. This integration process is also present in the ISO 59000 family of standards explicitly dedicated to circular economy (Arana-Landin et al., 2024; Heras-Saizarbitoria et al., 2024).

The convergence of these two concepts—circular economy and supply chain management—has been characterised by an evolutionary shift towards a more systemic perspective. Early approaches often focused on individual elements of the supply chain, such as recycling or energy efficiency. However, a growing understanding of the complex interactions within supply networks has led to the development of more holistic approaches that consider the entire life cycle of products and services, as well as the development of circular business models and ecosystems.

1.2 *Rationale for Circular Practices in the Supply Chain*

The integration of circular economy principles into supply chain management represents a compelling rationale for businesses, encompassing environmental, economic and social dimensions. For each one of these dimensions, the implementation of a circular economy could be beneficial; therefore, there are environmental, social and economic arguments for circular supply chains.

1.2.1 The Environmental Case for Circular Supply Chains

The traditional linear model has a significant environmental footprint. Extracting virgin resources, energy consumption during production and waste disposal contribute to greenhouse gas emissions, pollution and ecosystem destruction (Lacy et al., 2020; Stahel, 2016). In opposition, circular practices offer a solution, presenting a framework for keeping resources in use for as long as possible and extracting the maximum value from them before they are ultimately returned to a biological or technical cycle (Kirchherr et al., 2017; Kirchherr et al., 2023b). Companies can significantly reduce resource consumption and waste generation by designing products for durability, repairability and recyclability. Remanufacturing and refurbishment programmes extend product lifespans, further reducing the need for virgin materials. Additionally, closed-loop recycling systems ensure materials stay in the loop, minimising environmental impact (MahmoumGonbadi et al., 2021; Souza, 2013).

According to a study by the Ellen MacArthur Foundation (2019), it was estimated that a fully circular economy implemented within the cement, steel, plastic and aluminium industries could contribute to reducing global greenhouse gas emissions from these key industry materials by 40% by 2050. This is not a distant goal but a pressing need. Moreover, potential benefits for the environment include the reduction of natural resource use pace, the reduction of landscape utilisation, the minimisation of landscape disruption and the limitation of biodiversity loss. Additionally, the extension of the lifespan of products would contribute to the reduction in the demand for raw materials and the consequent reduction of waste (Knäble et al., 2022; Velenturf & Purnell, 2021). These potential benefits underscore the urgent need for circular practices to mitigate environmental damage.

1.2.2 The Economic Case for Circular Supply Chains

Beyond the environmental case, a strong economic case for embracing circularity in supply chains exists. In 2019, a study by the World Economic Forum (2019, p. 1) estimated that the circular economy could unlock $4.5 trillion in economic growth by 2030. Similarly, a report produced jointly by the Ellen McArthur Foundation and McKinsey (2016, p. 12) suggested that capturing the full potential of circularity could generate a net benefit of €1.8 trillion by 2030 in Europe. These figures demonstrate the substantial economic opportunities associated with circular supply chain practices.

The potential economic benefits are related to several issues. Among them are virgin resource scarcity and fluctuating prices. Shifting to a circular model could reduce dependence on virgin materials, leading to cost savings and more resilient supply chains. Additionally, by extending product lifespans, companies can create new revenue streams through repair, refurbishment and take-back programmes (Lieder & Rashid, 2016).

Furthermore, circular practices often lead to increased innovation. Designing products for disassembly and remanufacturing necessitates exploring new design principles and technologies. This can lead to developing more efficient and sustainable products, enhancing a company's competitive advantage (Bakker et al., 2018; Bocken et al., 2016; Potting et al., 2017).

1.2.3 The Social Case for Circular Supply Chains

The social dimension of circularity is also gaining recognition (Mies & Gold, 2021; Vázquez-Brust & de Souza Campos, 2023). Linear models often contribute to social inequalities through resource extraction practices that can impact local communities negatively (Neves & Marques, 2022). Circular approaches, on the other hand, can promote local job creation through activities like repair, refurbishment and remanufacturing (Sulich & Sołoducho-Pelc, 2022).

Additionally, circularity can extend product lifespans, making them more accessible to individuals who might not be able to afford new products. This can contribute to improved living standards and social equity (Preston et al., 2019; Schröder et al., 2020) as well as social inclusion (Souza Piao et al., 2023).

For example, according to He et al. (2024), a circular economy for electronics could create between 0.3 and 1.5 million job opportunities globally. These figures highlight the potential of circularity to contribute to social well-being and create new opportunities. However, it must be mentioned that tensions are also generated by the implementation of a circular economy in the labour market due to precarious work in some areas of the circular economy or the rapid growth of circular jobs that can contribute to the generation of poor working conditions (Barford & Ahmad, 2021; Dufourmont & Brown, 2017; Hartmann et al., 2022).

Efficient supply chains are characterised by the fact that all processes are coordinated to the minute and interlock in the best possible way. Methods have been developed to further optimise these value creation networks by, for example, reducing inventories or increasing globalisation to fully exhaust price differences in terms of labour, energy or raw material costs. Nevertheless, the focus is on customer satisfaction. Unfortunately, the goals of further optimising and streamlining supply chains are not necessarily in harmony with the goal of meeting customer needs in the best possible manner.

2 Rationale for the Structure of the Book

This book delves into the multifaceted and evolving landscape of the circular economy, particularly within the context of sustainable supply chains. Our intention has been to address global challenges, thereby elucidating concepts and presenting implementation cases. We organised the contributions of this book into three parts: "Part I: The R-Principles of Circular Economy", "Part II: Theoretical Perspectives of Circular Economy in Global Supply Chains" and "Part III: Practical Implementations of Circular Economy in Global Supply Chains". In the following sections of this introductory chapter, we present the contributions in detail.

2.1 The R-Principles of Circular Economy (Part I of the Book)

Principles can be defined as fundamental rules or starting points that guide various systems and disciplines ranging from philosophy to science (McDonald, 2009). They serve as normative regulators that operate across time and space, ensuring regularities and guiding change (McDonald, 2009). Principles are not merely conventional but hold significant truth and methodological value, acting as both the beginning and end of understanding. Aristotle's perspective, for example, adds that principles are the origins from which knowledge and actions stem, forming the basis of sciences and practical reasoning. Principles can also be action-oriented statements guiding behaviour, and for some scholars, they build a bridge between theory and practice (Skaar et al., 2020).

This understanding of principles as a bridge between theory and practice lies at the core of our decision to dedicate the first part of this book to the so-called "R-principles" of circular economy. As opposed to hard laws that obligate agents to comply with regulations in a "one-size-fits-all" way, principles, especially in social sciences, have the advantage of informing decision-making processes and thus allow for differentiated implementation in terms of time and space. In other words, principles applied to different situations can generate different actions because, despite their fundamental value, the practical implementation of them could be different for different agents.

Consequently, in the first part of this book, we explore the foundational principles of the circular economy encapsulated by the R-principles. These principles are addressed in microchapters, in which each principle is meticulously examined, offering an overview of the principle and its implementation in supply chain management. Furthermore, each chapter illustrates, with the help of a real business example or case study, how a particular organisation or organisations implemented a principle within its managerial practices.

The R-principles analysed in this first part of the book include respect, responsibility, regeneration, rethink, redesign, reduce, reuse, repair, refurbish, remanufacture, recycling and recovery. The rationale for the order of the R-principles is presented in Fig. 1.

The wheel diagram is a comprehensive visualisation of these principles, structured to illustrate how they interrelate and contribute to increasing circularity within economic systems. We considered it important to initially address a set of four R-principles that provide awareness of systemic interdependence and guidelines for action at a business level. These principles are respect, responsibility, restore and regenerate. They serve as a precondition for the implementation of circular

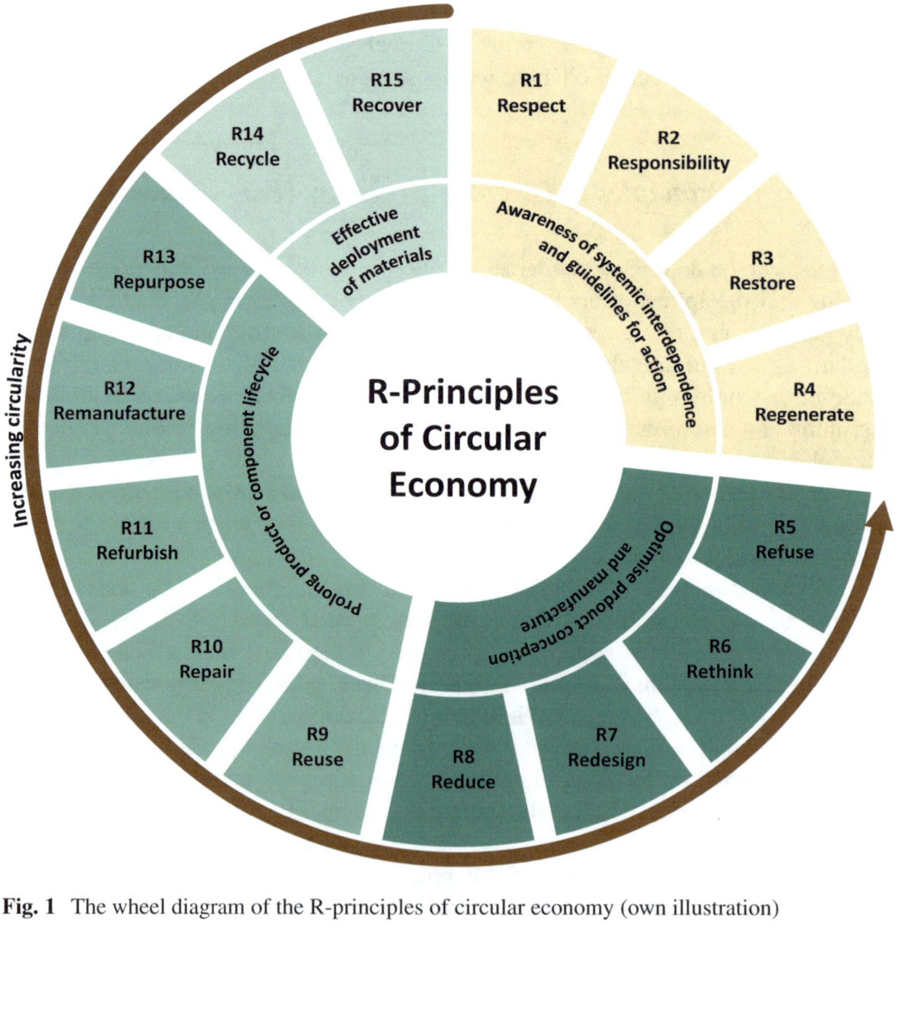

Fig. 1 The wheel diagram of the R-principles of circular economy (own illustration)

economy in supply chain management because they answer more normative questions related to the moral reasons for a circular economy. Respect (R1) acknowledges planetary boundaries, ensuring that economic activities do not exceed ecological limits. Responsibility (R2) underscores the duty of individuals and organisations to make sustainable decisions. Restore (R3) involves activities oriented to bring natural and human-made systems to their original or improved states. Regenerate (R4) focuses on designing processes that revitalise and replenish ecosystems.

Further along, the wheel addresses principles that optimise resource usage and product life cycle. Refuse (R5) encourages the rejection of materials, unnecessary products and packaging. Rethink (R6) challenges traditional business models and promotes innovative solutions. Redesign (R7) aims to optimise products for minimal resource use and maximal longevity. Reduce (R8) emphasises cutting down on resource consumption and waste generation. Reuse (R9) promotes using products multiple times in their original form or with minor adjustments.

Beyond this, the principles focus on extending the life cycle of products in different ways. Repair (R10) involves fixing products to prolong their usability. Refurbish (R11) restores used products to a like-new condition. Remanufacture (R12) goes a step further, bringing products back to a state that meets original specifications. Repurpose (R13) finds new applications for products or materials, extending their functional life. Finally, two R-principles are dedicated to the effective deployment of materials: recycle (R14) transforms waste materials into new products, closing the resource loop, and recover (R15) focuses on extracting valuable materials or energy from waste that cannot be reused or recycled.

Despite our efforts, it was not possible to gather a microchapter for each R-principle. Specifically, the principles of restore, repurpose and refuse are not included. The omission of repurpose is justifiable, as the principles of remanufacture and refurbish already encompass many overlapping strategies, thus making repurpose somewhat redundant. As for refuse, the strategy is often challenging to implement in economic practice. Therefore, we chose to focus on redesign and rethink, which offer more pragmatic and actionable approaches for businesses. Finally, the principle of restore was not included due to the overlapping nature of its implementation with other principles like repair and regenerate.

Additionally, Table 1 briefly defines all the R-principles and lists the main academic sources and, if available, the reference to the microchapter in this book related to that R-principle.

This first section serves as a crucial introduction, grounding readers in the essential concepts and practices that drive circular economic systems.

Table 1 Principles of circular economy (based on Kirchherr et al., 2017, p. 224; Kirchherr et al., 2023b, p. 8; Morseletto, 2020b, p. 4; Potting et al., 2017, p. 5)

Target	R	Principle	Rationale	Sources
Awareness of systemic interdependence and guidelines for action	R1	Respect	Recognise and honour planetary boundaries as catalysts for business activities.	Desing et al. (2020) and Loza Adaui (2024b)
	R2	Responsibility	Embrace the duty to make decisions and take actions that promote sustainable practices.	Jonas (1979), Purvis et al. (2023) and Loza Adaui (2024c)
	R3	Restore	Renew natural and human-made systems to return them to their original state or to a previous defined state.	Morseletto (2020a)
	R4	Regenerate	Design systems and processes that replenish natural resources and ecosystems, aiming to sustain and enhance their health and productivity over time.	Morseletto (2020a) and Loza Adaui (2024a)

(continued)

Table 1 (continued)

Target	R	Principle	Rationale	Sources
Optimise product conception and manufacture	R5	Refuse	Render a product redundant by either discontinuing its function or by providing the same function using a radically different product.	Potting et al. (2017), Morseletto (2020b) and Kirchherr et al. (2023b)
	R6	Rethink	Propose new ideas and solutions on product/service, business model and ecosystem level.	Potting et al. (2017), Morseletto (2020b), Kirchherr et al. (2023b) and Schmiedeknecht (2024c)
	R7	Redesign	Optimise product or process design to embrace circularity.	Potting et al. (2017), Morseletto (2020b), Kirchherr et al. (2023b) and Fröhlich (2024a)
	R8	Reduce	Increase efficiency of resource consumption in product manufacture or use and implement conservation practices.	Potting et al. (2017), Morseletto (2020b), Kirchherr et al. (2023b) and Fröhlich (2024b)
Prolong product or component life cycle	R9	Reuse	Use discarded products or components repeatedly for their original function, extending their lifespan.	Potting et al. (2017), Morseletto (2020a, 2020b), Kirchherr et al. (2023b) and Fröhlich (2024c)
	R10	Repair	Repair or provide maintenance to a defective product to enable its continued use with its original functionality.	Potting et al. (2017), Morseletto (2020a, 2020b), Kirchherr et al. (2023b) and Steinbiß (2024c)
	R11	Refurbish	Restore and update old products to like-new conditions extending their usability and life cycle.	Potting et al. (2017), Morseletto (2020a, 2020b), Kirchherr et al. (2023b) and Steinbiß (2024a)
	R12	Remanufacture	Use parts of discarded products in a new product with the same function.	Potting et al. (2017), Morseletto (2020a, 2020b), Kirchherr et al. (2023b) and Steinbiß (2024b)
	R13	Repurpose	Use discarded products or their components in a new product with a different functionality.	Potting et al. (2017), Morseletto (2020a, 2020b) and Kirchherr et al. (2023b)

(continued)

Table 1 (continued)

Target	R	Principle	Rationale	Sources
Effective deployment of materials	R14	Recycle	Transform materials from used or discarded products to obtain either the same quality (high grade) or lower quality (low grade) material (excluding energy recovery and the use of waste as a fuel).	Potting et al. (2017), Morseletto (2020a, 2020b), Kirchherr et al. (2023b) and Schmiedeknecht (2024b)
	R15	Recover	Use waste as a source of energy, e.g. by incineration of material with energy recovery.	Potting et al. (2017), Morseletto (2020a, 2020b), Kirchherr et al. (2023b) and Schmiedeknecht (2024a)

2.2 Theoretical Perspectives of Circular Economy in Global Supply Chains (Part II of the Book)

The second part of the book transitions to theoretical perspectives on circular economy in global supply chains, providing a robust analytical framework. The chapter "Circular Economy and Sustainable Supply Chain Management" by Elisabeth Fröhlich and Kristina Steinbiß explores how circular economy principles intersect with sustainable supply chain management, highlighting the challenges and opportunities inherent in this synergy.

The chapter "Circular Procurement in Global Supply Chains" by Chiara Bernd elucidates the pivotal role of circular procurement in the sustainable transformation of companies within global supply chains. It discusses how the adoption of circular economy principles in procurement practices can contribute significantly to achieving the Sustainable Development Goals (SDGs) by integrating the 7R model to enhance circularity efforts. This exploration emphasises the shift needed from linear business models to circular strategies to promote sustainability and environmental responsibility in procurement within the context of global supply chains.

A chapter dedicated to discussing the definition of circular economy is provided by Roelof Vogel, Tharaka de Vass, Glen Croy and Amrik Sohal: "Circular Economy: Contested Definitions and an Evolving Concept". The chapter delves into the nature of circular economy definitions, highlighting two distinct streams of thought—one advocating for circular economy's capacity to achieve all sustainability pillars and the other constraining its contribution to resource efficiency and effectiveness. Despite this contention, the chapter underscores a consistent conceptualisation of circular economy revolving around several key frames. These frames include the Earth's finite resources, value creation, regenerative systems, closed-loop systems, material flows, and value capture and re-creation. Additionally, the chapter contributes by tracing the evolution of the circular economy concept through three significant phases, shedding light on the progressive shift in focus and priorities within the circular economy framework. The initial phase primarily addressed waste management, while the subsequent phase emphasised waste reduction opportunities across supply chains. The latest phase, however, places a paramount emphasis on retaining the value of materials, components and products, showing a clear evolution towards a more holistic and value-centric approach to circularity.

In the chapter "The Economic Geography of the Moral Supply Chain in a Circular Economy", Guli-Sanam Karimova and Stephen LeMay offer a critical examination of the ethical and moral facets of the circular economy paradigm, presenting valuable insights that can guide practitioners, policymakers and researchers in formulating more comprehensive and ethically sound strategies for advancing sustainability in supply chain management. The chapter introduces the concept of the moral supply chain (TMSC) and underscores its significance in understanding and addressing the ethical challenges that accompany the transition to circularity. Firstly, the authors highlight the current limitations of existing frameworks, such as corporate social responsibility (CSR), in fully encapsulating the ethical implications of circular economy and circular supply chains. While CSR traditionally focuses on sustainability, it tends to overlook crucial ethical considerations like labour practices, corruption and broader social issues that are essential components of a sustainable supply chain. By emphasising the gaps in CSR with regard to these ethical aspects, the paper offers a fresh perspective on the moral responsibilities that extend beyond traditional corporate boundaries.

Moreover, the chapter of Karimova and LeMay delves into the complexities that arise when supply chains intersect national and cultural borders. It elucidates how the ethical and moral considerations in supply chains cannot be disregarded in the pursuit of circularity. Highlighting that circularity itself does not serve as a panacea for ethical dilemmas, the research sheds light on the necessity of incorporating TMSC to identify and address the barriers that hinder the ethical implementation of circular economy and circular supply chains.

2.3 *Practical Implementations of Circular Economy in Global Supply Chains (Part III of the Book)*

Finally, the book presents practical implementations of circular economy principles in global supply chains. This third part features case studies from various industries and regions, illustrating real-world applications and successes. Examples include the creation of a circular value ecosystem by a multinational brewery in Latin America, regional transformation through organisational educational networks and innovative solutions like replacing cardboard boxes with circular alternatives in Austria. By showcasing these practical implementations, the book demonstrates how theory can be effectively translated into practice, offering valuable insights for businesses and policymakers striving to foster sustainable and circular supply chains.

In the pursuit of business models incorporating sustainability within their core activities, life cycle sustainability assessment (LCSA) plays a crucial role. The chapter "Life Cycle Sustainability Assessment as a Tool for Businesses in a Circular Economy: Business Examples from Egypt" by Ola Mamdouh explains how businesses can identify areas where they can reduce inputs and outputs, such as energy, water, greenhouse gas emissions and waste by evaluating the different phases of a product's life cycle, from design to end-of-life. It examines the implementation of LCSA as a tool for businesses of different sizes and types from a variety of industry sectors in Egypt to apply circular economy principles and adopt sustainable practices in their business models and supply chains.

In order to address the challenges posed by the linear economy model and achieve the SDGs, the chapter "Structural Framework Development for Smooth Circular Supply Chain Management Adoption" by Monika Vyas and Gunjan Yadav focuses on developing a structural framework for the adoption of circular supply chain management. To identify the drivers that promote circularity in the supply chain and determine their mutual relationship, they apply an interpretative structural modelling method as a multi-criteria decision-making technique with expert input. The analysis showcases that higher authority, government policy and finance possess the highest driving power.

For many years, the logistics industry has relied on a "per unit" pricing model for selling services like containers, air freight and road transport. However, this traditional pricing model has numerous disadvantages when it comes to selling technically advanced and green logistics products, such as not being conducive to accurately reflecting the value that these products provide to customers. The chapter "Value-Based Selling of Green Products: A Radical New Approach to Commercial Growth" by Alexander Nowroth proposes a new approach to pricing these green products—shifting from a "per unit" model to a "value-based" model. This new pricing model allows logistics companies to showcase the value they bring to their customers better and enables them to compete effectively in the market.

In response to supply chain disruptions, political tensions and cost considerations, many North American companies have turned to nearshoring in Mexico, leading to a boom in the U.S.-Mexico cross-border logistics market. This market, valued at over 80 billion U.S. dollars, has created a demand for efficient and sustainable logistics solutions. In the chapter "Green and Collaborative Logistics in the U.S-Mexico Border: The Blue5PL Case" by Olivia Hernandez-Pozas and Rodrigo Martinazioli, North America's logistics industry is observed. Furthermore, a case study is conducted with Blue5PL, a FreightTech company, to showcase green logistics practices with a focus on the reduce principle of circular economy.

How can manufacturing industries create circular value networks in developing countries through an industrial symbiosis approach? The experience of a multinational brewery in northern Mexico serves as a compelling example of establishing a circular value ecosystem in a developing country, as outlined in the chapter "Creating the First Circular Value Ecosystem in Latin America: The Experience of a Multinational Brewery" by Eduardo Aguiñaga and Ana R. Leal. Based on insider action research, this case study provides valuable insights for successfully implementing a circular economy at a regional level in developing countries.

The interest in regional circular economies as a sustainable alternative to linear economic models has grown, as they can enhance self-sufficiency and create value, especially in structurally weak regions. In the chapter "Supporting Regional Transformation Through an Organisational Educational Network Consulting Approach: An Empirical Example Regarding the Implementation of a Regional Circular Economy", Marc-André Heidelmann and Tobias Klös discuss the potential of organisational educational network consulting in establishing sustainable circular economies, using a piloted approach of regional network consulting as a case analysis. This approach can facilitate knowledge sharing, capacity building and the development of collaborative initiatives among organisations, leading to the successful implementation of a regional circular economy.

The status quo of the courier, express and package (CEP) industry involves the predominant use of disposable cardboard boxes and, in some cases, plastic polybags. However, with the increasing energy and fuel costs, regulatory pressures and heightened competition, there is a growing need for companies to invest in circular economy solutions. One such solution is the adoption of returnable transport items to replace disposable packaging. Implementing new technologies and sustainable approaches may bring unexpected consequences, making the support of key stakeholders crucial for success. The chapter "Replacing Cardboard-Based Boxes With a Circular Economy Solution in Austria: A Framework for Stakeholder Assessment and Impact Evaluation of Social, Ecological and Economic Factors" by Nils Kruse, Melanie Rainer, Gerald Schneikart and Walter Mayrhofer focuses on the Austrian CEP industry and presents a framework for identifying and prioritising relevant stakeholders. It argues that stakeholders should be evaluated based on their resources for implementing circular economy solutions and the impact these solutions have on social, ecological and economic aspects.

Applying the principles of circular economy to global supply chains is a complex and multifaceted subject that involves various stakeholders and processes. Get inspired by exploring global perspectives on challenges, concepts and cases.

Acknowledgements We would like to acknowledge the help of all the people involved in this project to jointly create a global perspective on challenges, concepts and cases of Circular Economy in sustainable supply chains. We extend our heartfelt thanks to each author for their expertise in the field and their invaluable contributions to this book. This book would not have been possible without the tireless dedication of Jana Fröhlich in organising and proofreading the manuscript. Our deepest thanks go to her. We are also grateful to the publishing team at Springer for their support and professionalism. We are profoundly grateful to our respective families for their unwavering support. Special thanks also go to our colleagues and friends who have supported us throughout the process of bringing this publication to fruition. Finally, we apologise for any errors or omissions that may be found in this book. They were unintentional and not meant to cause any harm.

References

Aggeri, F. (2020). The circular economy: Historical perspective and contemporary issues. In K. Delchet-Cochet (Ed.), *Modern finance, management innovation and economic growth SET: Vol. 3. Circular economy: From waste reduction to value creation* (pp. 3–12). ISTE Ltd / John Wiley and Sons Inc.

Arana-Landin, G., Sigüenza, W., Landeta-Manzano, B., & Laskurain-Iturbe, I. (2024). Circular economy: On the road to ISO 59000 family of standards. *Corporate Social Responsibility and Environmental Management, 31*(3), 1977–2009. https://doi.org/10.1002/csr.2665

Baines, T. S., Lightfoot, H. W., Evans, S., Neely, A., Greenough, R., Peppard, J., Roy, R., Shehab, E., Braganza, A., Tiwari, A., Alcock, J. R., Angus, J. P., Bastl, M., Cousens, A., Irving, P., Johnson, M., Kingston, J., Lockett, H., Martinez, V., Michele, P., Tranfield, D., Walton, I. M., & Wilson, H. (2007). State-of-the-art in product-service systems. *Proceedings of the Institution of Mechanical Engineers, Part B: Journal of Engineering Manufacture, 221*(10), 1543–1552. https://doi.org/10.1243/09544054JEM858

Bakker, C., Balkenende, R., & Poppelaars, F. (2018). Design for product integrity in a circular economy. In M. Charter (Ed.), *Designing for the circular economy* (pp. 148–156). Routledge.

Barford, A., & Ahmad, S. R. (2021). A call for a socially restorative circular economy: Waste pickers in the recycled plastics supply chain. *Circular Economy and Sustainability, 1*(2), 761–782. https://doi.org/10.1007/s43615-021-00056-7

Bocken, N. M. P., de Pauw, I., Bakker, C., & van der Grinten, B. (2016). Product design and business model strategies for a circular economy. *Journal of Industrial and Production Engineering, 33*(5), 308–320. https://doi.org/10.1080/21681015.2016.1172124

Boothroyd, G., & Alting, L. (1992). Design for assembly and disassembly. *CIRP Annals, 41*(2), 625–636. https://doi.org/10.1016/S0007-8506(07)63249-1

Boulding, K. E. (1966). *The economics of the coming spaceship earth*. Prentice-Hall.

Desing, H., Brunner, D., Takacs, F., Nahrath, S., Frankenberger, K., & Hischier, R. (2020). A circular economy within the planetary boundaries: Towards a resource-based, systemic approach. *Resources, Conservation and Recycling, 155*, Article 104673. https://doi.org/10.1016/J.RESCONREC.2019.104673

Dufourmont, J., & Brown, E. G. (2017). *Jobs & skills in the circular economy: State of play and future pathways*. Circle Economy. https://www.circle-economy.com/resources/jobs-skills-in-the-circular-economy-state-of-play-and-future-pathways

Ellen McArthur Foundation, & McKinsey. (2016). *Growth within: A circular economy vision for a competitive Europe*. https://www.mckinsey.com/capabilities/sustainability/our-insights/growth-within-a-circular-economy-vision-for-a-competitive-europe#/

Fröhlich, E. (2024a). The principle of redesign in circular economy: At the heart of circular economy. In E. Fröhlich, K. Steinbiß, M. H. Schmiedeknecht, & C. R. Loza Adaui (Eds.), *Circular economy in sustainable supply chains: A global perspective on challenges, concepts and cases*. Springer Nature.

Fröhlich, E. (2024b). The principle of reduce in circular economy: Simply using less? In E. Fröhlich, K. Steinbiß, M. H. Schmiedeknecht, & C. R. Loza Adaui (Eds.), *Circular economy in sustainable supply chains: A global perspective on challenges, concepts and cases*. Springer Nature.

Fröhlich, E. (2024c). The principle of reuse in circular economy: Trash is treasure. In E. Fröhlich, K. Steinbiß, M. H. Schmiedeknecht, & C. R. Loza Adaui (Eds.), *Circular economy in sustainable supply chains: A global perspective on challenges, concepts and cases*. Springer Nature.

Frosch, R. A., & Gallopoulos, N. E. (1989). Strategies for manufacturing. *Scientific American, 261*(3), 144–153. http://www.jstor.org/stable/24987406

Hartmann, C., Hegel, C., & Boampong, O. (2022). The forgotten essential workers in the circular economy? Waste picker precarity and resilience amidst the COVID-19 pandemic. *Local Environment, 27*(10-11), 1272–1286. https://doi.org/10.1080/13549839.2022.2040464

He, Y., Kiehbadroudinezhad, M., Hosseinzadeh-Bandbafha, H., Gupta, V. K., Peng, W., Lam, S. S., Tabatabaei, M., & Aghbashlo, M. (2024). Driving sustainable circular economy in electronics: A comprehensive review on environmental life cycle assessment of e-waste recycling. *Environmental Pollution, 342*, Article 123081. https://doi.org/10.1016/j.envpol.2023.123081

Heras-Saizarbitoria, I., Boiral, O., & Testa, F. (2024). ISO 59000 standards for the circular economy: A call for accuracy. *Circular Economy and Sustainability*, 1–7. https://doi.org/10.1007/s43615-024-00370-w

Jonas, H. (1979). *Das Prinzip Verantwortung: Versuch einer Ethik für die technologische Zivilisation* (1st ed.). Insel-Verlag.

Kirchherr, J., Reike, D., & Hekkert, M. (2017). Conceptualizing the circular economy: An analysis of 114 definitions. *Resources, Conservation and Recycling, 127*, 221–232. https://doi.org/10.1016/j.resconrec.2017.09.005

Kirchherr, J., Urbinati, A., & Hartley, K. (2023a). Circular economy: A new research field? *Journal of Industrial Ecology, 27*(5), 1239–1251. https://doi.org/10.1111/jiec.13426

Kirchherr, J., Yang, N.-H. N., Schulze-Spüntrup, F., Heerink, M. J., & Hartley, K. (2023b). Conceptualizing the circular economy (revisited): An analysis of 221 definitions. *Resources, Conservation and Recycling, 194*(1–32), Article 107001. https://doi.org/10.1016/j.resconrec.2023.107001

Knäble, D., de Quevedo Puente, E., Pérez-Cornejo, C., & Baumgärtler, T. (2022). The impact of the circular economy on sustainable development: A European panel data approach. *Sustainable Production and Consumption, 34*, 233–243. https://doi.org/10.1016/j.spc.2022.09.016

Lacy, P., Long, J., & Spindler, W. (2020). *The circular economy handbook*. Palgrave Macmillan UK.

Lieder, M., & Rashid, A. (2016). Towards circular economy implementation: A comprehensive review in context of manufacturing industry. *Journal of Cleaner Production, 115*, 36–51. https://doi.org/10.1016/j.jclepro.2015.12.042

Loza Adaui, C. R. (2024a). The principle of regeneration in circular economy: Revitalising for resilience. In E. Fröhlich, K. Steinbiß, M. H. Schmiedeknecht, & C. R. Loza Adaui (Eds.), *Circular economy in sustainable supply chains: A global perspective on challenges, concepts and cases*. Springer Nature.

Loza Adaui, C. R. (2024b). The principle of respect in circular economy: Recognising worth and boundaries. In E. Fröhlich, K. Steinbiß, M. H. Schmiedeknecht, & C. R. Loza Adaui (Eds.), *Circular economy in sustainable supply chains: A global perspective on challenges, concepts and cases*. Springer Nature.

Loza Adaui, C. R. (2024c). The principle of responsibility in circular economy: Between freedom and duty. In E. Fröhlich, K. Steinbiß, M. H. Schmiedeknecht, & C. R. Loza Adaui (Eds.), *Circular economy in sustainable supply chains: A global perspective on challenges, concepts and cases*. Springer Nature.

MahmoumGonbadi, A., Genovese, A., & Sgalambro, A. (2021). Closed-loop supply chain design for the transition towards a circular economy: A systematic literature review of methods, applications and current gaps. *Journal of Cleaner Production, 323*, Article 129101. https://doi.org/10.1016/j.jclepro.2021.129101

McDonald, H. P. (2009). Principles: The principles of principles. *The Pluralist, 4*(3), 98–126. https://doi.org/10.1353/plu.0.0028

McDonough, W., & Braungart, M. (2002). *Cradle to cradle: Remaking the way we make things* (1st ed.). North Point Press.

Mies, A., & Gold, S. (2021). Mapping the social dimension of the circular economy. *Journal of Cleaner Production, 321*, Article 128960. https://doi.org/10.1016/j.jclepro.2021.128960

Morseletto, P. (2020a). Restorative and regenerative: Exploring the concepts in the circular economy. *Journal of Industrial Ecology, 24*(4), 763–773. https://doi.org/10.1111/jiec.12987

Morseletto, P. (2020b). Targets for a circular economy. *Resources, Conservation and Recycling, 153*, Article 104553. https://doi.org/10.1016/j.resconrec.2019.104553

Nash, H. A. (2009). The European Commission's sustainable consumption and production and sustainable industrial policy action plan. *Journal of Cleaner Production, 17*(4), 496–498. https://doi.org/10.1016/j.jclepro.2008.08.020

Neves, S. A., & Marques, A. C. (2022). Drivers and barriers in the transition from a linear economy to a circular economy. *Journal of Cleaner Production, 341*, Article 130865. https://doi.org/10.1016/j.jclepro.2022.130865
Potting, J., Hekkert, M., Worrell, E., & Hanemaaijer, A. (2017). *Circular economy: Measuring innovation in the product chain.* PBL Netherlands Environmental Assessment Agency. https://www.pbl.nl/uploads/default/downloads/pbl-2016-circular-economy-measuring-innovation-in-product-chains-2544.pdf
Preston, F., Lehne, J., & Wellesley, L. (2019). *An inclusive circular economy: Priorities for developing countries.* Chatham House.
Purvis, B., Celebi, D., & Pansera, M. (2023). A framework for a responsible circular economy. *Journal of Cleaner Production, 400*, Article 136679. https://doi.org/10.1016/j.jclepro.2023.136679
Schmiedeknecht, M. H. (2024a). The principle of recovery in circular economy: Generating energy. In E. Fröhlich, K. Steinbiß, M. H. Schmiedeknecht, & C. R. Loza Adaui (Eds.), *Circular economy in sustainable supply chains: A global perspective on challenges, concepts and cases.* Springer Nature.
Schmiedeknecht, M. H. (2024b). The principle of recycling in circular economy: Transforming waste. In E. Fröhlich, K. Steinbiß, M. H. Schmiedeknecht, & C. R. Loza Adaui (Eds.), *Circular economy in sustainable supply chains: A global perspective on challenges, concepts and cases.* Springer Nature.
Schmiedeknecht, M. H. (2024c). The principle of rethink in circular economy: Redefining business solutions. In E. Fröhlich, K. Steinbiß, M. H. Schmiedeknecht, & C. R. Loza Adaui (Eds.), *Circular economy in sustainable supply chains: A global perspective on challenges, concepts and cases.* Springer Nature.
Schröder, P., Lemille, A., & Desmond, P. (2020). Making the circular economy work for human development. *Resources, Conservation and Recycling, 156*, Article 104686. https://doi.org/10.1016/j.resconrec.2020.104686
Skaar, J., Bølviken, T., Koskela, L., & Kalsaas, B. T. (2020). Principles as a bridge between theory and practice. In I. D. Tommelein & E. Daniel (Eds.), *Proceedings of the 28th Annual Conference of the International Group for Lean Construction (IGLC28)* (pp. 1–12). https://doi.org/10.24928/2020/0065
Souza, G. C. (2013). Closed-loop supply chains: A critical review, and future research. *Decision Sciences, 44*(1), 7–38. https://doi.org/10.1111/j.1540-5915.2012.00394.x
Souza Piao, R., de Vincenzi, T. B., Da Silva, A. L. F., de Oliveira, M. C. C., Vazquez-Brust, D., & Monteiro Carvalho, M. (2023). How is the circular economy embracing social inclusion? *Journal of Cleaner Production, 411*, Article 137340. https://doi.org/10.1016/j.jclepro.2023.137340
Stahel, W. R. (2010). *The performance economy* (2nd ed.). Palgrave Macmillan.
Stahel, W. R. (2016). The circular economy. *Nature, 531*(7595), 435–438. https://doi.org/10.1038/531435a
Steinbiß, K. (2024a). The principle of refurbish in circular economy: Products better than new. In E. Fröhlich, K. Steinbiß, M. H. Schmiedeknecht, & C. R. Loza Adaui (Eds.), *Circular economy in sustainable supply chains: A global perspective on challenges, concepts and cases.* Springer Nature.
Steinbiß, K. (2024b). The principle of remanufacture in circular economy: Same as new product. In E. Fröhlich, K. Steinbiß, M. H. Schmiedeknecht, & C. R. Loza Adaui (Eds.), *Circular economy in sustainable supply chains: A global perspective on challenges, concepts and cases.* Springer Nature.
Steinbiß, K. (2024c). The principle of repair in circular economy: Extending the lifespan of products. In E. Fröhlich, K. Steinbiß, M. H. Schmiedeknecht, & C. R. Loza Adaui (Eds.), *Circular economy in sustainable supply chains: A global perspective on challenges, concepts and cases.* Springer Nature.

Sulich, A., & Sołoducho-Pelc, L. (2022). The circular economy and the Green Jobs creation. *Environmental Science and Pollution Research, 29*(10), 14231–14247. https://doi.org/10.1007/s11356-021-16562-y

Vázquez-Brust, D., & de Souza Campos, L. M. (2023). Introduction – Social dimension of circular economy: Step forward or step back? *The social dimensions of the circular economy, 10*, 1–25. https://doi.org/10.1007/978-3-031-25436-9_1

Velenturf, A. P., & Purnell, P. (2021). Principles for a sustainable circular economy. *Sustainable Production and Consumption, 27*, 1437–1457. https://doi.org/10.1016/j.spc.2021.02.018

Winans, K., Kendall, A., & Deng, H. (2017). The history and current applications of the circular economy concept. *Renewable and Sustainable Energy Reviews, 68*, 825–833. https://doi.org/10.1016/j.rser.2016.09.123

World Economic Forum. (2019). *Platform for accelerating the circular economy: A global public-private collaboration platform and project accelerator*. https://www3.weforum.org/docs/WEF_PACE_Platform_for_Accelerating_the_Circular_Economy.pdf

Cristian R. Loza Adaui is a Professor of Management and Corporate Social Responsibility at the THI Business School of the Technische Hochschule Ingolstadt in Germany. He previously worked for the CBS International Business School in Cologne, Germany, where he also served as Deputy Director of the Centre for Advanced Sustainability Management (CASM). He has held teaching and research positions at the Institute of Economics of the Friedrich-Alexander University Erlangen-Nürnberg in Germany and the Universidad Católica San Pablo in Peru. Prof. Loza Adaui received his Doctoral Degree in Management from the Ingolstadt School of Management at the Catholic University of Eichstätt-Ingolstadt in Germany. His research and consulting activities focus on how businesses become more sustainable, responsible, human-centred and purpose-driven through strategic management, corporate social responsibility, sustainability management, non-financial reporting, circular economy and corporate purpose. His research has appeared in outlets such as *Business Strategy and the Environment*, *Journal of Business Research*, *Long Range Planning*, *Voluntas: International Journal of Voluntary and Nonprofit Organizations* and *Journal of Management Development*, among others. He serves on the editorial boards of journals such as *Technological Sustainability* and *Frontiers in Sustainability*.

Maud H. Schmiedeknecht is a Professor of Corporate Governance and Corporate Social Responsibility at the ESB Business School of the Reutlingen University in Germany and a Business Design Coach. What drives her is supporting and empowering business leaders and business school students to develop skills and learn methods to (re)design sustainable, circular and regenerative business models. Maud previously worked as a consultant at the global management consulting firm Kearney in Germany and abroad. As a research assistant at the Konstanz Institute for Intercultural Management, Values and Communication of Prof. Dr. Josef Wieland, she undertook consulting and research projects in the fields of corporate social responsibility and compliance and participated in national and international committees, e.g. the German CSR-Forum and ISO 26000 Working Group on Social Responsibility. She was an associate of the Stiftung Neue Verantwortung, a think tank for digital technologies, politics and society, on the topic of global economic ethic. Her dissertation on the governance of multistakeholder dialogues was awarded the Max Weber Prize for Business Ethics. Maud's current teaching, research and consulting focus is on corporate sustainability management, (social) entrepreneurship and sustainable business model innovation, business transformation and circular economy.

Kristina Steinbiß holds a professorship in Management with a focus on Marketing at the ESB Business School of the Reutlingen University in Germany, where she teaches primarily in the Industrial Engineering programmes. Her fundamental teaching goal is to enable her students to think and act sustainably while adopting an integrative perspective combining business and technology. Kristina's current research focuses on implementing new business models to accomplish a company's sustainability goals, as well as on influencing and nudging consumers towards sustainable behaviour. She is the author of numerous publications in the above-mentioned fields, including the textbook *Marketing*, which is already available in its second edition. Next to her research activities, she is also enthusiastic about the Theory of Inventive Problem Solving, the TIPS innovation methodology. As a TIPS trainer and coach, she likes to combine this methodology with her marketing approach and adapt it to implement sustainable solutions in day-to-day business.

Elisabeth Fröhlich is the founder of ispira Think Tank for Sustainable Supply Chains. She invented an innovative business coach concept to prepare companies for the challenges of the different European sustainability laws. Her major aim is to bridge the gap between research and practice in this field to enable companies to implement the sustainable transformation of global supply chains successfully. Prof. Dr. habil. Elisabeth Fröhlich was president of the CBS International Business School in Germany until 2022. She holds a full professorship in Sustainable Procurement and Supply Chains and is an internationally recognised expert in sustainable supply chain management. She is also highly engaged in the field of responsible management education. She serves as a board member of PRME, chairs the PRME Nomination and Governance Committee and is the PRME DACH Chapter Chair. Additionally, she leads the AOM Management Education Division Ambassador Program as vice-chair, leads the Board of JARO Institute and offers online training in the field of sustainable procurement. Her research focuses on sustainable supply chain management and green procurement, qualification in purchasing, Procurement 4.0 and strategic supplier relationship management. Innovative teaching formats of responsible management education are further main areas of her research. She has published several books and articles on the above-mentioned topics and supports several journals as an external reviewer.

Part I
The R-Principles of Circular Economy

The Principle of Respect in Circular Economy: Recognising Worth and Boundaries

Cristian R. Loza Adaui

1 Introduction

The circular economy has gained significant attention as a means to build a sustainable future, but its scope must be expanded to incorporate ethical considerations beyond just material efficiency (Kopnina et al., 2018). The circular economy aims to change the whole economic system through the transformation of business processes into closed-loop resource systems. This can happen, for example, through the integration of a new rationale for decision-making guided by principles oriented to one of the following purposes: first, the optimisation of product use or manufacturing (refuse, rethink, reduce); second, the extension of use and lifespan of products or its components (reuse, repair, refurbish, remanufacture, repurpose); or third, the useful deployment of materials (recycle and recover) (Kirchherr et al., 2023; Morseletto, 2020; Potting et al., 2017). However, the focus on micro-level implementation, i.e. at the business or organisational level, has overlooked the need for a more holistic, i.e. systemic or macro-level approach that recognises and respects planetary boundaries and the inherent worth of human and non-human entities.

Planetary boundaries refer to the limits within which humanity can safely operate to avoid irreversible environmental damage. These boundaries encompass various critical Earth system processes, such as climate change, biodiversity loss and nitrogen and phosphorus cycles (Röckstrom et al., 2009; Steffen et al., 2015). Respecting these boundaries is crucial for maintaining the planet's health and ensuring long-term sustainability.

Incorporating planetary boundaries into the discussions that shape business strategies, particularly in supply chain management, represents, on the one hand, a

C. R. Loza Adaui (✉)
THI Business School, Technische Hochschule Ingolstadt, Ingolstadt, Germany
e-mail: Cristian.LozaAdaui@thi.de

E. Fröhlich et al. (eds.), *Circular Economy in Sustainable Supply Chains*, CSR, Sustainability, Ethics & Governance,
https://doi.org/10.1007/978-3-031-70749-0_2

sustainability duty and, on the other hand, a significant challenge (Clift et al., 2017; Muñoz-Torres et al., 2018; Nilsen, 2024; Schneider & Murray, 2024; Whiteman et al., 2013). Accordingly, companies must strive for material efficiency and ensure that their operations do not exceed the Earth's ecological limits. This requires a comprehensive understanding of the interconnectedness of global supply chains and the environmental impacts at each stage of production and distribution and, at the same time, a deep understanding of interdependences and the impact of global supply chains on the issues related to planetary boundaries (Clift et al., 2017).

Moreover, in addition to the planetary boundaries, which focus more on environmental issues, the social dimension of the circular economy cannot be overlooked (Zavos et al., 2024). For example, the informal sector, which often comprises marginalised communities, plays a significant role in waste disposal and recycling. Inclusive approaches that empower these communities are essential for ensuring sustainable and equitable waste management practices. Developing strategies that integrate the informal sector into formal waste management systems can help minimise negative environmental impacts while promoting social equity (Ezeah et al., 2013; Linzner & Lange, 2013; Sembiring & Nitivattananon, 2010; Wilson et al., 2006).

Consequently, social boundaries emerge as complementary to planetary boundaries (Leach et al., 2013). The work of Kate Raworth (Leach et al., 2013; Raworth, 2012, 2017) has been influential in understanding the interdependence and complementarity of social and environmental issues and is better known as the doughnut economy. Accordingly, there is a "safe and just operating space for humanity", a wording that has also been adopted within the planetary boundaries literature (Rockström et al., 2023) and considers two types of constraints that operate simultaneously. First, an economic system where society can satisfy the basic needs of its members, and second, respect for environmental limits as established in planetary boundaries literature (Raworth, 2012, 2017). Similarly, as in the case of the planetary boundaries, the doughnut economy literature has mainly focused on macro-level and systemic challenges of measurement and analysis (O'Neill et al., 2018; Schlesier et al., 2024), and the business and organisational integration has only recently received scientific attention (Hausdorf & Timm, 2023; Sahan et al., 2022).

This chapter frames the concepts of planetary and social boundaries within the circular economy discussion. To that aim, it introduces the principle of respect within the circular economy, postulating that recognising planetary and social boundaries mediates the respect of the inherent value of both human and non-human entities. The following research question is addressed: How can companies integrate the principle of respect to planetary and social boundaries into their circular economy strategies and supply chain management practices? To illustrate the practical application of the principle of respect, the chapter analyses Unilever's approach to circular economy and its efforts to respect planetary and social boundaries.

2 The Principle of Respect and Its Typology

The principle of respect is a fundamental tenet that underpins moral deliberations, guides interactions and shapes how human and non-human entities are perceived and treated. This principle is rooted in various religious, cultural and philosophical traditions. According to Dillon (2007), there are different types of respect. In this regard, Hudson (1980) categorises respect into four types: evaluative respect, directive respect, institutional respect and obstacle respect.

2.1 Evaluative Respect

Evaluative respect encompasses the appreciation or admiration for an entity's merit, qualities or achievements. It involves recognising the inherent worth or value of an object, person or concept based on subjective evaluation criteria. Evaluative respect is an internal acknowledgement of value or significance, often contingent on subjective perceptions and personal judgments. While evaluative respect may vary among individuals based on diverse evaluative standards or cultural backgrounds, it generally implies a positive valuation of the respected entity (Hudson, 1980). For example, a company might show evaluative respect by recognising and valuing its employees' efforts in innovating sustainable products through redesign for modularity. This appreciation encourages further creativity and commitment to sustainability and circular economy goals.

2.2 Directive Respect

Directive respect focuses on honouring moral directives, obligations or rights as guiding principles for behaviour. It involves a sense of duty or obligation to adhere to prescribed norms, rules or ethical standards, regardless of personal feelings or preferences. Directive respect manifests in actions that align with ethical guidelines, laws or social norms, demonstrating a commitment to upholding moral imperatives or fulfilling obligations towards others or the broader community. This form of respect is characterised by a sense of accountability and responsibility towards following directives or norms that govern individual or organisational conduct or interactions (Hudson, 1980). For example, a business might demonstrate directive respect by implementing rigorous environmental policies that comply or exceed legal requirements, thus honouring its ethical commitment to environmental stewardship and sustainability.

2.3 Institutional Respect

Institutional respect is attributed to individuals or entities based on their roles, positions or status within institutional frameworks, social practices or organisational structures. This type of respect acknowledges and defers to the authority, responsibilities or functions associated with specific institutional roles or positions. Institutional respect is grounded in recognising the significance of maintaining hierarchical relationships and adhering to established norms or procedures within institutional settings. It emphasises upholding the integrity and legitimacy of institutional frameworks (Hudson, 1980). For example, institutional respect is shown when a company adheres to the guidelines and standards set by international environmental organisations, recognising their authority and the importance of maintaining global environmental standards.

2.4 Obstacle Respect

Finally, obstacle respect pertains to a distinct form of respect that arises from recognising obstacles or hindrances that possess inherent significance, challenge or influence within a given context. It involves acknowledging and esteeming barriers, challenges or constraints that impact individuals, organisations, societies or systems. Obstacle respect signifies an appreciation for the complexity, adversity or transformative potential embodied by obstacles, highlighting the need to approach them with mindfulness, care or strategic deliberation in various spheres of human endeavours (Hudson, 1980). For example, a business may exhibit obstacle respect by recognising the challenge of reducing its carbon footprint due to existing planetary boundaries. This respect leads to strategic efforts to innovate and find feasible solutions within the limits of the Earth's capacity to sustain life.

3 Respect for the Rights of Future Generations in Alignment with Planetary and Social Boundaries

Regardless of the different types of respect, there is always an object towards which respect is directed (Ege, 2019). Additionally, respect is based on recognising inherent value or rights. For example, respect towards individuals could be based on recognising their human dignity, respect towards communities on their rights and respect for the natural environment and non-human entities on their inherent value.

Within the sustainability discussion, which is closely linked to the circular economy (Geissdoerfer et al., 2017; Schöggl et al., 2020), the principle of respect plays a role in at least two different ways: first, respecting future generations' rights to satisfy their needs, and second, respecting the constraints imposed by planetary and social boundaries on humanity.

3.1 Respect for Future Generations' Rights

Respect for the rights of future generations is a complex ethical consideration, intertwining intergenerational justice, sustainability and responsibility towards preserving resources and ecosystems for posterity (Hendlin, 2014; Li et al., 2021). In this context, directive respect emerges as the most relevant type of respect, aligning closely with principles related to environmental stewardship and sustainable practices.

Directive respect extends beyond immediate interpersonal interactions to broader ethical responsibilities towards safeguarding the well-being and interests of future generations. By acknowledging and respecting these rights, individuals and societies adopt a stance rooted in the moral obligation to act in ways that promote sustainability, address environmental challenges and ensure the long-term viability of ecosystems and resources.

Directive respect for the rights of future generations involves recognising the ethical imperative to consider the consequences of present actions on the well-being of individuals who will inhabit the planet in the future. This respect signifies a commitment to intergenerational justice and environmental sustainability by adhering to moral imperatives that transcend temporal boundaries and prioritise preserving resources and ecosystems for the benefit of future societies (Campos, 2018).

3.2 Respect for Planetary and Social Boundaries

Directive respect for planetary boundaries emphasises the moral obligation to recognise and adhere to principles that safeguard the delicate balance of environmental systems and limit human impact on the planet. This respect involves acknowledging the environment's inherent rights to exist and flourish, underscoring the ethical duty to act in ways that promote sustainability, biodiversity and the longevity of ecosystems (Heikkurinen et al., 2021).

By embracing directive respect in the context of planetary boundaries, individuals and societies acknowledge the inherent worth of environmental systems and engage in behaviours demonstrating conscientious regard for the finite resources and ecological boundaries that sustain life on Earth (Steffen et al., 2015). This proactive dimension of directive respect underscores the importance of recognising the fragility of planetary systems and taking tangible steps to ensure their preservation and regeneration for present and future generations (Nilsen, 2024).

Moreover, respect for planetary boundaries can be understood as a form of obstacle respect, recognising the inherent limitations and challenges that the Earth's ecological systems impose on human activities. This perspective involves acknowledging the significant barriers that planetary boundaries represent in maintaining the delicate balance necessary for a sustainable environment. Obstacle respect in this context means appreciating the complexities and constraints of these boundaries and understanding that they are not merely hindrances but essential conditions for the continued health and stability of the planet. By recognising planetary boundaries as obstacles, individuals and organisations are called to approach them with strategic deliberation, mindful of the need to operate within these limits to prevent irreversible environmental damage and ensure the long-term viability of the Earth's ecosystems (Schaltegger et al., 2018).

Approaching planetary boundaries with obstacle respect necessitates a proactive and respectful stance towards environmental stewardship. It involves understanding that these boundaries represent critical thresholds that, if crossed, could lead to catastrophic ecological consequences (Rockström & Gaffney, 2021). This respect compels businesses and policymakers to innovate and adopt sustainable practices that align with the finite nature of the Earth's resources. For example, recognising the carbon budget as an obstacle requires significant changes in energy consumption and production patterns, promoting renewable energy sources and reducing greenhouse gas emissions. By treating planetary boundaries with obstacle respect, society acknowledges the transformative potential of these challenges, driving efforts to develop solutions that respect ecological limits and promote resilience and sustainability in the face of environmental constraints (Sjåfjell, 2020).

4 Implementing the Principle of Respect Within Planetary Boundaries in Global Supply Chains

The academic literature on global supply chains already explored the possibilities of integrating planetary boundaries (Muñoz-Torres et al., 2018). For example, Mwangi et al. (2022) explored the relationship between planetary boundaries in the Kenyan tea supply chain. Ehrenstein et al. (2020) use the concept of planetary boundaries to evaluate the sustainability of hydrogen fuel supply chains, aiming to minimise the risk of exceeding the Earth's ecological limits by optimising production, storage and transport methods. They conclude by affirming that transitioning to hydrogen fuel can substantially reduce the transgression levels of planetary boundaries, but it is still not enough to operate entirely within all planetary boundaries simultaneously, indicating the need for further improvements and innovations (Ehrenstein et al., 2020). The work of Wheeler et al. (2021) focuses on biomass supply chains; they concentrate on the analysis of the bioethanol supply chain in Argentina and showcase various optimisation models used to design biomass supply chains that convert biomass into fuels, considering factors like demand, capacity, biomass availability and economic data. Their results show how different fuel blends and technologies can significantly reduce environmental impacts; however, Wheeler et al. (2021) conclude that meeting all planetary boundaries simultaneously may require additional measures like higher ethanol content in fuel blends or cleaner transport means.

Despite the efforts done in the previous studies, implementing planetary boundaries in global supply chains presents several challenges that need to be addressed for effective strategic decision-making.

- The primary challenge lies in translating the high-level, global planetary boundary parameters into practical metrics that can be applied at regional or local scales. This translation, which has been called downscaling, is crucial for operationalising the planetary boundaries framework within the complexities of global supply chains (Erlandsson et al., 2023; Ryberg et al., 2020).
- The lack of a common framework for industry use also represents an issue. While approaches have been suggested for national, regional and local planning regarding planetary boundaries, the absence of a common framework poses a significant challenge. Without a unified framework, industry stakeholders may struggle to adopt consistent metrics when addressing planetary boundaries at various scales (University of Cambridge Institute for Sustainability Leadership (CISL), 2019).
- The interconnectedness of different systems complicates the implementation of planetary boundaries at different scales. For example, exceeding local-scale boundaries in multiple locations can lead to cumulative global impacts, emphasising the interconnected nature of planetary boundaries in global supply chains (Clift et al., 2017; Ryberg et al., 2018).
- Another significant challenge is establishing a fair basis for allocating "safe operating space" or a "fair share" within each planetary boundary to individual activities and enterprises. This allocation necessitates a consensual ethical and political framework, requiring extensive deliberation and inclusive approaches similar to those seen in climate change initiatives (Lucas et al., 2020; Ryberg et al., 2018; University of Cambridge Institute for Sustainability Leadership (CISL), 2019).
- Implementing measures to respect planetary boundaries requires coordination at a global scale, which currently lacks comprehensive bodies. While efforts exist on climate change and biodiversity, achieving sustainable development within planetary boundaries demands enhanced global coordination beyond national and regional levels (University of Cambridge Institute for Sustainability Leadership (CISL), 2019; Whiteman et al., 2013).

5 Integrating Respect for Planetary Boundaries Through Life Cycle Assessment at Unilever

Despite the challenges of operationalising respect for planetary boundaries in global supply chains, a few tools have been proven suitable for integrating the logic of planetary boundaries. One of them is the life cycle assessment (LCA), and the company developing this tool further is Unilever.

Unilever operates in the fast-moving consumer goods (FMCG) industry, has over 400 brands, 128,000 employees worldwide, sales in 190 countries and 3.4 billion people using their products worldwide, based on data from 2024. The company's portfolio includes products in foods, beverages, ice cream, home care and personal care (Unilever, 2024a). Unilever's corporate purpose of "making sustainable living commonplace" explicitly addresses sustainability. Unilever's strategy for sustainable growth is driven by a vision to leverage its global reach to enhance health and well-being, reduce environmental impacts and improve livelihoods (Unilever, 2024b). Throughout its history, Unilever has pioneered sustainability efforts, consistently grounding its strategy and initiatives in scientific research.

Unilever's sustainability strategy is underpinned by its commitment to scientific rigour and innovation. The company has long employed LCA as a tool to systematically analyse the potential environmental impacts of its products and services throughout their entire life cycle—from raw material extraction to production, distribution, consumer use and disposal (Sim et al., 2016; Unilever, 2024c). LCA enables Unilever to consider all relevant environmental inputs and emissions, classify and quantify these for potential impacts in various categories and make informed decisions that minimise environmental footprints.

However, while essential, Unilever recognises that traditional LCA does not address all environmental impacts with equal robustness and requires continuous improvements in data availability, spatial and temporal resolution, and impact assessment. To address these gaps, Unilever has spearheaded the development and application of advanced LCA methodologies over the past three decades (Unilever, 2024c).

Unilever has introduced several significant advancements in LCA, notably the Land-Use Change Improved (LUCI)-LCA and Planetary Boundaries (PB)-LCA methodologies. LUCI-LCA integrates spatially explicit modelling of land change and ecosystem services into the life cycle inventory, replacing traditional elements with outputs from predictive land-change modelling and ecosystem services modelling using the InVEST software suite (The Natural Capital Project, 2024; Unilever, 2024c). This approach allows Unilever to assess the implications of large-scale growth in demand for bio-based raw materials, providing a more comprehensive understanding of the environmental impacts associated with agricultural practices.

PB-LCA represents another groundbreaking development. This approach is an absolute environmental sustainability assessment (AESA) method designed to compare the environmental impacts of products against an assigned share of the planet's environmental carrying capacity (Bjørn et al., 2020; Ryberg et al., 2018; Unilever,

2024c). Unilever has "developed and demonstrated advances in PB-LCA methods for specific impact categories, including freshwater use and eutrophication (nitrogen use)" (Unilever, 2024c). These methods enable a more holistic assessment of product-level environmental impacts, ensuring that Unilever's operations remain within the safe operating space defined by planetary boundaries.

Unilever's integration of the planetary boundaries framework into its sustainability strategy underscores its commitment to operating within the ecological limits of our planet. By aligning its sustainability efforts with these boundaries, Unilever aims to avoid irreversible environmental damage and ensure long-term sustainability.

Respecting planetary boundaries requires Unilever to adopt a comprehensive understanding of the interconnectedness of global supply chains and the environmental impacts at each stage of production and distribution. This approach involves recognising the inherent value of natural systems and the ethical duty to act in ways that promote sustainability, biodiversity and the longevity of ecosystems. Unilever's use of PB-LCA facilitates this process by providing a scientific basis for evaluating and mitigating the environmental impacts of its products, ensuring that its operations do not exceed the Earth's ecological limits.

6 Conclusion

This chapter focused on the principle of respect for the circular economy. After a brief description of the nature of respect, four different types of respect were presented, and subsequently, the respect for the rights of future generations and the respect for planetary and social boundaries were discussed. The challenges of integrating the principle of respect into the planetary and social boundaries within global supply chains have also been presented. Finally, a short description of how Unilever performs the integration through a special planetary boundaries life cycle assessment has been provided.

References

Bjørn, A., Chandrakumar, C., Boulay, A.-M., Doka, G., Fang, K., Gondran, N., Hauschild, M. Z., Kerkhof, A., King, H., Margni, M., McLaren, S., Mueller, C., Owsianiak, M., Peters, G., Roos, S., Sala, S., Sandin, G., Sim, S., Vargas-Gonzalez, M., & Ryberg, M. (2020). Review of life-cycle based methods for absolute environmental sustainability assessment and their applications. *Environmental Research Letters, 15*(8), Article 83001. https://doi.org/10.1088/1748-9326/ab89d7

Campos, A. S. (2018). Intergenerational Justice Today. *Philosophy Compass*, *13*(3), Article e12477. https://doi.org/10.1111/phc3.12477

Clift, R., Sim, S., King, H., Chenoweth, J., Christie, I., Clavreul, J., Mueller, C., Posthuma, L., Boulay, A.-M., Chaplin-Kramer, R., Chatterton, J., DeClerck, F., Druckman, A., France, C.,

Franco, A., Gerten, D., Goedkoop, M., Hauschild, M. Z., Huijbregts, M. A. J., Koellner, T., Lambin, E. F., Lee, J., Mair, S., Marshall, S., McLachlan, M. S., i Canals, L. M., Mitchell, C., Price, E., Rockström, J., Suckling, J., & Murphy, R. (2017). The challenges of applying planetary boundaries as a basis for strategic decision-making in companies with global supply chains. *Sustainability, 9*(2), Article 279. https://doi.org/10.3390/su9020279

Dillon, R. S. (2007). Respect: A philosophical perspective. *Gruppe. Interaktion. Organisation. Zeitschrift Für Angewandte Organisationspsychologie (GIO), 38*(2), 201–212. https://doi.org/10.1007/s11612-007-0016-5

Ege, C. C. (2019). Respect and business ethics. In D. C. Poff & A. C. Michalos (Eds.), *Encyclopedia of business and professional ethics* (pp. 1–3). Springer.

Ehrenstein, M., Galán-Martín, Á., Tulus, V., & Guillén-Gosálbez, G. (2020). Optimising fuel supply chains within planetary boundaries: A case study of hydrogen for road transport in the UK. *Applied Energy, 276*, Article 115486. https://doi.org/10.1016/j.apenergy.2020.115486

Erlandsson, J., Bergmark, P., & Höjer, M. (2023). Establishing the planetary boundaries framework in the sustainability reporting of ICT companies - A proposal for proxy indicators. *Journal of Environmental Management, 329*, Article 117032. https://doi.org/10.1016/j.jenvman.2022.117032

Ezeah, C., Fazakerley, J. A., & Roberts, C. L. (2013). Emerging trends in informal sector recycling in developing and transition countries. *Waste Management, 33*(11), 2509–2519. https://doi.org/10.1016/j.wasman.2013.06.020

Geissdoerfer, M., Savaget, P., Bocken, N. M., & Hultink, E. J. (2017). The circular economy – A new sustainability paradigm? *Journal of Cleaner Production, 143*, 757–768. https://doi.org/10.1016/j.jclepro.2016.12.048

Hausdorf, M., & Timm, J.-M. (2023). Business research for sustainable development: How does sustainable business model research reflect doughnut economics? *Business Strategy and the Environment, 32*(6), 3398–3416. https://doi.org/10.1002/bse.3307

Heikkurinen, P., Clegg, S., Pinnington, A. H., Nicolopoulou, K., & Alcaraz, J. M. (2021). Managing the anthropocene: Relational agency and power to respect planetary boundaries. *Organization & Environment, 34*(2), 267–286. https://doi.org/10.1177/1086026619881145

Hendlin, Y. H. (2014). The threshold problem in intergenerational justice. *Ethics and the Environment, 19*(2), 1–38. https://doi.org/10.2979/ethicsenviro.19.2.1

Hudson, S. D. (1980). The nature of respect. *Social Theory and Practice, 6*(1), 69–90. http://www.jstor.org/stable/23557720

Kirchherr, J., Yang, N.-H. N., Schulze-Spüntrup, F., Heerink, M. J., & Hartley, K. (2023). Conceptualising the circular economy (revisited): An analysis of 221 definitions. *Resources, Conservation and Recycling, 194*, Article 107001. https://doi.org/10.1016/j.resconrec.2023.107001

Kopnina, H., Washington, H., Taylor, B., & Piccolo, J. J. (2018). Anthropocentrism: More than just a misunderstood problem. *Journal of Agricultural and Environmental Ethics, 31*(1), 109–127. https://doi.org/10.1007/s10806-018-9711-1

Leach, M., Raworth, K., & Rockström, J. (2013). Between social and planetary boundaries for humanity: Navigating pathways in the safe and just space. In UNESCO & ISSC (Ed.), *Word Social Science Report 2013: Changing global environments* (pp. 84–89). OECD Publishing and UNESCO Publishing. https://doi.org/10.1787/9789264203419-10-en

Li, M., Wiedmann, T., Fang, K., & Hadjikakou, M. (2021). The role of planetary boundaries in assessing absolute environmental sustainability across scales. *Environment International, 152*, Article 106475. https://doi.org/10.1016/j.envint.2021.106475

Linzner, R., & Lange, U. (2013). Role and size of informal sector in waste management – A review. *Proceedings of the Institution of Civil Engineers - Waste and Resource Management, 166*(2), 69–83. https://doi.org/10.1680/warm.12.00012

Lucas, P. L., Wilting, H. C., Hof, A. F., & van Vuuren, D. P. (2020). Allocating planetary boundaries to large economies: Distributional consequences of alternative perspectives on distributive fairness. *Global Environmental Change, 60*, Article 102017. https://doi.org/10.1016/j.gloenvcha.2019.102017

Morseletto, P. (2020). Targets for a circular economy. *Resources, Conservation and Recycling, 153*, Article 104553. https://doi.org/10.1016/j.resconrec.2019.104553

Muñoz-Torres, M. J., Fernández-Izquierdo, M. Á., Rivera-Lirio, J. M., Ferrero-Ferrero, I., Escrig-Olmedo, E., Gisbert-Navarro, J. V., & Marullo, M. C. (2018). An assessment tool to integrate sustainability principles into the global supply chain. *Sustainability, 10*(2), Article 535. https://doi.org/10.3390/su10020535

Mwangi, G. M., Despoudi, S., Espindola, O. R., Spanaki, K., & Papadopoulos, T. (2022). A planetary boundaries perspective on the sustainability: Resilience relationship in the Kenyan tea supply chain. *Annals of Operations Research, 319*(1), 661–695. https://doi.org/10.1007/s10479-021-04096-y

Nilsen, H. R. (2024). Code Red for humanity: The role of business ethics as we transgress planetary thresholds. *Journal of Business Ethics, 189*(1), 1–7. https://doi.org/10.1007/s10551-023-05402-3

O'Neill, D. W., Fanning, A. L., Lamb, W. F., & Steinberger, J. K. (2018). A good life for all within planetary boundaries. *Nature Sustainability, 1*(2), 88–95. https://doi.org/10.1038/s41893-018-0021-4

Potting, J., Hekkert, M., Worrell, E., & Hanemaaijer, A. (2017). *Circular economy: Measuring innovation in the product chain*. PBL Netherlands Environmental Assessment Agency. https://www.pbl.nl/uploads/default/downloads/pbl-2016-circular-economy-measuring-innovation-in-product-chains-2544.pdf

Raworth, K. (2012). *A safe and just space for humanity: Can we live within the doughnut?* Oxfam Discussion Papers

Raworth, K. (2017). *Doughnut economics: Seven ways to think like a 21st century economist*. Chelsea Green Publishing.

Rockström, J., & Gaffney, O. (2021). *Breaking boundaries: The science of our planet*. Dorling Kindesley.

Rockström, J., Gupta, J., Qin, D., Lade, S. J., Abrams, J. F., Andersen, L. S., Armstrong McKay, D. I., Bai, X., Bala, G., Bunn, S. E., Ciobanu, D., DeClerck, F., Ebi, K., Gifford, L., Gordon, C., Hasan, S., Kanie, N., Lenton, T. M., Loriani, S., Liverman, D. M., Mohamed, A., Nakicenovic, N., Obura, D., Ospina, D., Prodani, K., Rammelt, C., Sakschewski, B., Scholtens, J., Stewart-Koster, B., Tharammal, T., van Vuuren, D., Verburg, P. H., Winkelmann, R., Zimm, C., Bennett, E. M., Bringezu, S., Broadgate, W., Green, P. A., Huang, L., Jacobsen, L., Ndehedehe, C., Pedde, S., Rocha, J., Scheffer, M., Schulte-Uebbing, L., de Vries, W., Xiao, C., Xu, C., Xu, X., Zafra-Calvo, N., & Zhang, X. (2023). Safe and just Earth system boundaries. *Nature, 619*(7968), 102–111. https://doi.org/10.1038/s41586-023-06083-8

Rockström, J., Steffen, W., Noone, K., Persson, Å., Chapin, F. S., III, Lambin, E., Lenton, T. M., Scheffer, M., Folke, C., Schellnhuber, H., Nykvist, B., De Wit, C. A., Hughes, T., van der Leeuw, S., Rodhe, H., Sörlin, S., Snyder, P. K., Costanza, R., Svedin, U., Falkenmark, M., Karlberg, L., Corell, R. W., Fabry, V. J., Hansen, J., Walker, B., Liverman, D., Richardson, K., Crutzen, P., & Foley, J. (2009). Planetary boundaries: Exploring the safe operating space for humanity. Ecology and Society, 14(2), 32. https://www.ecologyandsociety.org/vol14/iss2/art32/

Ryberg, M. W., Andersen, M. M., Owsianiak, M., & Hauschild, M. Z. (2020). Downscaling the planetary boundaries in absolute environmental sustainability assessments – A review. *Journal of Cleaner Production, 276*, Article 123287. https://doi.org/10.1016/j.jclepro.2020.123287

Ryberg, M. W., Owsianiak, M., Clavreul, J., Mueller, C., Sim, S., King, H., & Hauschild, M. Z. (2018). How to bring absolute sustainability into decision-making: An industry case study using a planetary boundary-based methodology. *The Science of the Total Environment, 634*, 1406–1416. https://doi.org/10.1016/j.scitotenv.2018.04.075

Sahan, E., Sanz Ruiz, C., Raworth, K., van Winden, W., & van den Buuse, D. (2022). *What doughnut economics means for business: Creating enterprises that are regenerative and dis-*

tributive by design. Doughnut Economics Action Lab. https://research.hva.nl/en/publications/what-doughnut-economics-means-for-business-creating-enterprises-t

Schaltegger, S., Beckmann, M., & Hockerts, K. (2018). Sustainable entrepreneurship: Creating environmental solutions in light of planetary boundaries. *International Journal of Entrepreneurial Venturing, 10*(1), 1–16. https://doi.org/10.1504/IJEV.2018.090990

Schlesier, H., Schäfer, M., & Desing, H. (2024). Measuring the doughnut: A good life for all is possible within planetary boundaries. *Journal of Cleaner Production, 448*, Article 141447. https://doi.org/10.1016/j.jclepro.2024.141447

Schneider, A., & Murray, J. (2024). Escaping the loop of unsustainability: Why and how business ethics matters for earth system justice. *Journal of Business Ethics*, 1–9. https://doi.org/10.1007/s10551-024-05728-6

Schöggl, J.-P., Stumpf, L., & Baumgartner, R. J. (2020). The narrative of sustainability and circular economy - A longitudinal review of two decades of research. *Resources, Conservation and Recycling, 163*, Article 105073. https://doi.org/10.1016/j.resconrec.2020.105073

Sembiring, E., & Nitivattananon, V. (2010). Sustainable solid waste management toward an inclusive society: Integration of the informal sector. *Resources, Conservation and Recycling, 54*(11), 802–809. https://doi.org/10.1016/j.resconrec.2009.12.010

Sim, S., King, H., & Price, E. (2016). The role of science in shaping sustainable business: Unilever case study. In R. Clift & A. Druckman (Eds.), *Taking stock of industrial ecology* (pp. 291–302). Springer.

Sjåfjell, B. (2020). Sustainable value creation within planetary boundaries—Reforming corporate purpose and duties of the corporate board. *Sustainability, 12*(15), Article 6245. https://doi.org/10.3390/su12156245

Steffen, W., Richardson, K., Rockström, J., Cornell, S. E., Fetzer, I., Bennett, E. M., Biggs, R., Carpenter, S. R., de Vries, W., de Wit, C. A., Folke, C., Gerten, D., Heinke, J., Mace, G. M., Persson, L. M., Ramanathan, V., Reyers, B., & Sörlin, S. (2015). Planetary boundaries: Guiding human development on a changing planet. *Science, 347*(6223), Article 1259855. https://doi.org/10.1126/science.1259855

The Natural Capital Project. (2024). *InVEST.* Natural Capital Project - Stanford University. https://naturalcapitalproject.stanford.edu/software/invest

Unilever. (2024a). *Our company*. https://www.unilever.com/our-company/

Unilever. (2024b). *Our strategy and Growth Action Plan*. https://www.unilever.com/our-company/strategy/

Unilever. (2024c). *Tools and Approaches*. https://seac.unilever.com/our-science/environmental-sustainability-sciences/tools-and-approaches/

University of Cambridge Institute for Sustainability Leadership (CISL). (2019). *Linking planetary boundaries to business: The first White Paper in Kering's series on Planetary Boundaries for Business*. Cambridge Institute for Sustainability Leadership. https://www.cisl.cam.ac.uk/system/files/documents/linking-planetary-boundaries.pdf

Wheeler, J., Galán-Martín, Á., Mele, F. D., & Guillén-Gosálbez, G. (2021). Designing biomass supply chains within planetary boundaries. *AIChE Journal, 67*(4), Article e17131. https://doi.org/10.1002/aic.17131

Whiteman, G., Walker, B., & Perego, P. (2013). Planetary boundaries: Ecological foundations for corporate sustainability. *Journal of Management Studies, 50*(2), 307–336. https://doi.org/10.1111/j.1467-6486.2012.01073.x

Wilson, D. C., Velis, C., & Cheeseman, C. (2006). Role of informal sector recycling in waste management in developing countries. *Habitat International, 30*(4), 797–808. https://doi.org/10.1016/j.habitatint.2005.09.005

Zavos, S., Lehtokunnas, T., & Pyyhtinen, O. (2024). The (missing) social aspect of the circular economy: A review of social scientific articles. *Sustainable Earth Reviews, 7*(1), 1–17. https://doi.org/10.1186/s42055-024-00083-w

Cristian R. Loza Adaui is a Professor of Management and Corporate Social Responsibility at the THI Business School of the Technische Hochschule Ingolstadt in Germany. He previously worked for the CBS International Business School in Cologne, Germany, where he also served as Deputy Director of the Centre for Advanced Sustainability Management (CASM). He has held teaching and research positions at the Institute of Economics of the Friedrich-Alexander University Erlangen-Nürnberg in Germany and the Universidad Católica San Pablo in Peru. Prof. Loza Adaui received his Doctoral Degree in Management from the Ingolstadt School of Management at the Catholic University of Eichstätt-Ingolstadt in Germany. His research and consulting activities focus on how businesses become more sustainable, responsible, human-centred and purpose-driven through strategic management, corporate social responsibility, sustainability management, non-financial reporting, circular economy and corporate purpose. His research has appeared in outlets such as *Business Strategy and the Environment*, *Journal of Business Research*, *Long Range Planning*, *Voluntas: International Journal of Voluntary and Nonprofit Organizations* and *Journal of Management Development*, among others. He serves on the editorial boards of journals such as *Technological Sustainability* and *Frontiers in Sustainability*.

The Principle of Responsibility in Circular Economy: Between Freedom and Duty

Cristian R. Loza Adaui

1 Introduction

In the current landscape of global supply chains, sustainability has emerged as a paramount concern amidst escalating environmental challenges. The traditional linear economy model, characterised by "take-make-dispose" practices, has led to resource depletion and extensive waste generation, necessitating a paradigm shift towards circular economy principles (Kirchherr et al., 2023). Central to this transition is the principle of responsibility, which mandates that businesses and stakeholders account for the entire life-cycle impacts of their products and operations (Savini & Giezen, 2020). However, integrating responsibility into the framework of a circular economy remains a complex endeavour with implications spanning ethical, legal, and operational domains.

As a guiding principle, responsibility offers a moral imperative for businesses to adopt sustainable practices that preserve resources and minimise environmental harm (Velenturf & Purnell, 2021). By embracing responsibility within the context of a circular economy, companies can not only enhance their environmental stewardship but also bolster their resilience against regulatory shifts and consumer demands for eco-friendly products. This proactive stance aligns with ethical norms and positions businesses strategically in a competitive market increasingly shaped by sustainability concerns.

Despite the critical importance of responsibility in driving sustainable business practices, scholarly research specifically addressing its integration into circular economy strategies remains limited (Centobelli et al., 2020). Existing literature predominantly focuses on broader aspects of corporate social responsibility and

C. R. Loza Adaui (✉)
THI Business School, Technische Hochschule Ingolstadt, Ingolstadt, Germany
e-mail: Cristian.LozaAdaui@thi.de

E. Fröhlich et al. (eds.), *Circular Economy in Sustainable Supply Chains*, CSR, Sustainability, Ethics & Governance,
https://doi.org/10.1007/978-3-031-70749-0_3

sustainability, often overlooking the nuanced intersections with circular economy principles. This gap underscores the need for comprehensive exploration and empirical studies that elucidate the practical implications, challenges, and benefits of embedding responsibility within circular business models.

This chapter proposes the following research questions to address these gaps: How can the responsibility principle be integrated into global supply chains transitioning towards a circular economy? What are the key challenges and opportunities faced by businesses in operationalising responsibility within circular economy frameworks? How do successful case studies and voluntary initiatives exemplify best practices in extended producer responsibility (EPR) and circular business responsibility?

This chapter unfolds by first delving into the philosophical underpinnings of responsibility, drawing insights from seminal works by Max Weber and Hans Jonas. It then examines the evolution of responsibility within corporate governance, highlighting pivotal debates and theoretical advancements that have shaped contemporary perspectives on business ethics and sustainability. Subsequently, the chapter explores the practical application of responsibility in the context of circular economy, proposing the concept of circular business responsibility showcasing the case study of Apple and empirical evidence that illustrates successful implementations of extended producer responsibility (EPR) and circular business practices. Finally, it synthesises key findings to propose actionable recommendations for businesses aiming to enhance their sustainability practices through responsible circular economy strategies.

2 The Principle of Responsibility

Responsibility is a multidimensional concept that includes normative and practical dimensions, deeply rooted in ethical, legal, and social frameworks. The concept of responsibility can be clarified through a constellation of three elements. First, it involves identifying the subject of responsibility, which answers the question: Who is responsible? The answer to this question identifies the individual or entity that holds the responsibility, often referred to as the responsible agent. The next element addresses the object of responsibility, answering the question of "for what?" by specifying the actions, duties, or outcomes for which the subject is responsible. Finally, it considers the instance of responsibility, which defines towards what or whom the responsibility is directed. It outlines the relationship or authority to which the subject is answerable, such as society, specific individuals, legal bodies, or ethical principles (Haase, 2015).

From the ethical perspective, responsibility can be studied within normative ethics, which examines the standards and principles that govern what individuals and organisations should do (Baumgärtner et al., 2018). Several philosophers have addressed the topic of responsibility over time; according to Dahl Rendtorff (2014), in Oriental and Eastern philosophical thinking, the concept of responsibility has already been discussed around 500 years before Christ, for example, by Confucius, whereas in the Western philosophical tradition, the discussion of responsibility was established in a more explicit way just during the modernity (Baumgärtner et al., 2018; Haase, 2015).

It is, for example, Max Weber who explicitly proposes an ethics of responsibility (*Verantwortungsethik*) as opposed to an ethics of convictions (*Gesinnungsethik*):

> We must be clear about the fact that all ethically oriented conduct may be guided by one of two fundamentally differing and irreconcilably opposed maxims: conduct can be oriented to an 'ethic of ultimate ends' [*Gesinnungsethik*] or to an 'ethic of responsibility' [*Verantwortungsethik*]. This is not to say that an ethic of ultimate ends is identical with irresponsibility or that an ethic of responsibility is identical with unprincipled opportunism. Naturally nobody says that. However, there is an abysmal contrast between conduct that follows the maxim of an ethic of ultimate ends—that is, in religious terms, 'The Christian does rightly and leaves the results with the Lord'—and conduct that follows the maxim of an ethic of responsibility, in which case one has to give an account of the foreseeable results of one's action. (Weber, 1946, p. 120)

This distinction generated a fruitful discussion in socio-political sciences and philosophical circles. The contrast between the ethics of convictions and the ethics of responsibility has prompted deeper exploration and debate on the fundamental differences between them and their respective implications for moral decision-making as well as human and organisational conduct assessment. However, the discussion on Weber's analysis took place initially and mainly in continental Europe because a popular translation and the extended reception of Weber's text occurred just after the Second World War in English-speaking countries (Borchert, 2007).

A more systematic analysis of responsibility was done by Hans Jonas (1984). In his book titled "The Principle of Responsibility" (*Das Prinzip Verantwortung* in German), which in English was translated as "The Imperative of Responsibility", he proposes a "new type of agency" re-elaborating the Kantian categorical imperative in four different ways:

> Act so that the effects of your action are compatible with the permanence of genuine human life. (Jonas, 1984, p. 11)

> Act so that the effects of your action are not destructive of the future possibility of such life. (Jonas, 1984, p. 11)

> Do not compromise the conditions for an indefinite continuation of humanity on earth. (Jonas, 1984, p. 11)

> In your present choices, include the future wholeness of Man among the objects of your will. (Jonas, 1984, p. 11)

Hans Jonas's concept of responsibility is deeply rooted in the preservation of life, emphasising an ethical framework that prioritises the well-being of future generations. Jonas argues that in the nuclear age, human freedom can pose significant threats to life, necessitating a robust ethical theory of responsibility that transcends religious faith and is grounded in an ontological basis. For Hans Jonas, the highest value is life itself, which must be preserved across all forms and species (Ianulardo et al., 2022). Accordingly, responsibility is inherently relational and normative, involving interactions among actors within a system of reciprocal expectations and identifications. This relational aspect is evident in environmental governance, where actors' responsibilities are defined by legal and political norms and their actions are strategically configured to achieve collective goals (Savini & Giezen, 2020).

This understanding of responsibility has been incorporated into evaluating business responsibility and agency when speaking about corporate sustainability or corporate social responsibility. In the corporate world, the growing challenges related to environmental care, climate change, and globalisation have transformed environmental responsibility into a crucial component of corporate social responsibility (CSR). Public expectations have shifted, demanding greater environmental sustainability from businesses, which are now pressured by various stakeholders to adopt more eco-friendly practices and even embrace a circular economy (Ivanova & Slavova, 2019). In this sense, Jonas's ethics of responsibility also involves a collectivist dimension, where responsibilities are attributed based on normative claims about what actors should do in specific situations, such as reducing $CO2$ emissions or investing in sustainable infrastructure. This process of responsibilisation often leads to political struggles over who should bear the risks and costs associated with long-term investments for future generations (Savini & Giezen, 2020). Ultimately, Jonas's framework underscores the duty to ensure the existence and quality of life for future humankind, highlighting the importance of responsible actions today to safeguard the well-being of those yet to be born (Ianulardo et al., 2022).

3 Towards a Circular Business Responsibility

One of the questions that emerge from the discussions on the principle of responsibility applied to management relates to considering business as the subject of responsibility. The discussion on responsibility applied to businesses can be traced, for example, in legal scholarship related to the function and purpose of corporations, notably in the debate between Adolph A. Berle and E. Merrick Dodd (Windsor, 2008). On the one hand, Berle (1931) argued that "all powers granted to a corporation or to the management of a corporation, or to any group within the corporation, whether derived from statute or charter or both, are necessarily and at all times exercisable only for the ratable benefit of all shareholders" (Berle, 1931, p. 1049), thereby A. Berle emphasises the fiduciary duty between managers and shareholders. In contrast, Dodd (1932) posited that "the corporation is an economic institution which has a social service as well as a profit-making function" (Dodd, 1932, p. 1148). According to Freeman et al. (2020), Dodd's viewpoint prevailed over time, influencing subsequent generations of managers in the United States to recognise businesses' broader social responsibilities (Freeman et al., 2020).

Besides legal scholarship, the principle of responsibility holds a prominent place in the academic literature on management. Howard Bowen's seminal work, "The Social Responsibilities of the Businessman", is widely recognised for transcending the legal discussion and initiating a discourse on corporate obligations to society from a more managerial perspective, asserting that there are "obligations of businessmen to pursue those policies, to make those decisions, or to follow those lines of action which are desirable in terms of the objectives and values to our society" (Bowen, 1953, p. 6). This foundational text underscored the need for businesses to integrate societal values into their decision-making processes, thus arguing in favour of the social responsibility of businesses.

However, later in the sixties and seventies, this social responsibility narrative was strongly criticised and questioned by Milton Friedman, who received the Nobel Prize in Economic Sciences in 1976. In his words:

> In a free-enterprise, private-property system, a corporate executive is an employee of the owners of the business. He has direct responsibility to his employers. That responsibility is to conduct the business in accordance with their desires, which generally will be to make as much money as possible while conforming to the basic rules of the society, both those embodied in law and those embodied in ethical custom. (Friedman, 1970, p. 17)

The debate generated about the responsibility of businesses, the role of profit, and the primacy of shareholders' interests as the primary guidance for managerial decision-making has been very active since then. A further development of the recognition of business responsibility has been provided by Edward Freeman, who developed the stakeholder management approach, a renewed understanding of capitalism and a whole new research field. For Freeman et al. (2010), the stakeholder-shareholder dichotomy is not the best way to address business responsibilities. On the one hand, they consider Friedman to be a predecessor of stakeholder management because of its emphasis on the interests of one stakeholder: the shareholders. Nevertheless, at the same time, they ensure that paradoxically, the exclusive search for profit maximisation is not the best way to create value and reach profitability; in their own words:

> Friedman may actually believe that if you try to maximise profits you will do so. We believe that trying to maximise profits is counterproductive, because it takes attention away from the fundamental drivers of value – stakeholder relationships. There has been considerable research that shows that profitable firms have a purpose and values beyond profit maximisation. Profit maximisation may be better thought of as a result or outcome. (Freeman et al., 2010, pp. 11–12)

On the other hand, Freeman et al. (2010) transcend the narrow view of management that provides primacy to shareholders' interest, arguing in favour of a stakeholder capitalism that embraces a renewed understanding of the responsibilities of businesses. They affirm:

> The idea of company stakeholder responsibility looks at business and society as intertwined, and it looks not just at corporations, but at many different forms of organisations, and promotes a pragmatic focus on managing the relations with all the organisation's stakeholders as a primary task for success. This required a detailed understanding of to whom exactly a firm is responsible and the nature of those responsibilities. (Freeman et al., 2010, p. 264)

The state-of-the-art discussions about the responsibilities of businesses incorporate the multidimensional nature of the responsibilities that companies bear, including at least economic, social, and environmental responsibilities. The immense body of knowledge and empirical evidence generated during the last 50 years about the way in which companies assume their responsibilities seems to confirm that the open question is no more whether companies should be considered as bearers of responsibility but how companies can better assume their responsibilities towards stakeholders.

Based on the Jonasian definition of responsibility and considering the multidimensional nature of business responsibilities, it appears plausible to argue in favour of circular business responsibility, that is, a corporate responsibility to contribute to the transition from a linear economy towards a circular economy. Integrating circular economy and business responsibility can be done in two different and interrelated ways.

First, the circular economy can be considered the *telos* of business responsibility, or, in other words, one of the critical *objects* of business responsibility. This perspective would partially answer the question of what the company is responsible for. In this regard, many companies explicitly commit to the circular economy as part of their core business. For example, the Chilean company TriCiclos describe itself as follows:

> Guided by the intention to work in the environmental and social spheres with the same relevance as the economic one, we design and implement solutions to eliminate the concept of waste. We do this by engineering for a circular economy in order to reduce the waste problem before it is generated or by ensuring that it has the most circular destination possible (through reuse, return and recycling). (TriCiclos, 2024)

Or as part of their business responsibility, for example, in the case of the furniture company IKEA:

> At IKEA, our main priorities are to drastically reducing greenhouse gas emissions across the IKEA value chain and move toward the use of only renewable or recycled materials by 2030. We also want to inspire and enable the many people to live a better everyday life within the boundaries of the planet. This is a huge change that impacts every aspect of what we do, from how we meet customers to which products and services we develop. It will affect our complete IKEA value chain and the sourcing of energy and materials. (IKEA, 2024)

Second, the circular economy also provides concrete guidance for implementing social and environmental responsibility; in other words, the circular economy can also contribute to answering the questions about how to be a more responsible company or what to do to be responsible. There are many ways of integrating circular economy into decision-making; in this regard, the r-principles play a critical role because they allow for the implementation of circularity in different stages of corporate activities. For example, businesses can be responsible for designing products with recyclable or biodegradable materials, implementing take-back programs to recover used products, or investing in renewable energy sources to power their operations. Additionally, companies can be responsible for promoting circular business models, such as product-service systems or sharing platforms, that keep materials and products in circulation for as long as possible.

4 Implementing the Principle of Responsibility in Global Supply Chains Through Extended Producer Responsibility

Extended Producer Responsibility (EPR) is a policy instrument that can significantly contribute to the transition towards a circular economy. By shifting the responsibility for end-of-life product management from consumers and governments to producers, EPR incentivises manufacturers to design and produce more sustainable products (Hickle, 2017; Massarutto, 2014). This approach holds producers accountable for the entire life cycle of their products, from design and manufacturing to disposal and recycling, encouraging them to minimise environmental impacts and enhance product sustainability (Lifset et al., 2013).

EPR has been implemented in various forms across different countries, addressing the environmental impacts of waste, mainly plastic and electronic waste. For example, in Italy, EPR schemes focused on plastic packaging waste have demonstrated environmental benefits when collection rates are maximised and impurities minimised, although economic challenges remain for recycling certain plastics (Andreasi Bassi et al., 2020). In China, stronger government regulations and penalties have significantly improved compliance among manufacturers within EPR frameworks for electronic waste (Peng et al., 2019). Indonesia's experience highlights the need for better integrating responsibilities among stakeholders to prevent waste from ending up in landfills despite existing recycling regulations (Septianingrum et al., 2023).

A systematic review of EPR policies suggests innovative measures for policy and operational improvements, emphasising the importance of EPR in enhancing sustainability practices across different sectors (Cai & Choi, 2021). In Australia, EPR policies for construction and demolition waste face barriers such as costs and stakeholder diversity, yet there is substantial support for expanding these regulations to improve waste management and promote a circular economy (Shooshtarian et al., 2021).

While mandatory EPR programmes have been established in many jurisdictions, voluntary adoption of EPR frameworks within global supply chains can also drive circular economy practices. Successful voluntary initiatives showcase how producers can proactively take responsibility for their products throughout their life cycle, ultimately reducing waste and enabling the recovery and reuse of materials (Choi, 2017).

One prominent example is the voluntary e-waste take-back programmes implemented by leading electronics manufacturers. These programmes allow consumers to return their used devices to the producer, who then ensures proper recycling and recovery of materials (Garlapati, 2016). This model not only diverts e-waste from landfills but also incentivises design for the disassembly and reuse of components (Sahajwalla & Hossain, 2023). Another successful case is the voluntary battery stewardship programme adopted by the battery industry. This initiative facilitates the collection and recycling of used batteries, recovering valuable materials like lithium, cobalt, and nickel that can be reintroduced into the manufacturing process (Sahajwalla & Hossain, 2023).

The voluntary adoption of EPR is crucial in assessing circular business responsibility in global supply chains. From a regulatory perspective, it is challenging and nearly impossible to create regulations with a global reach (Gupt & Sahay, 2015). This is particularly problematic given that the issues addressed by the circular economy, such as resource depletion and waste management, are inherently global challenges. Without a global governing body to enforce uniform regulations across all nations, reliance on voluntary EPR adoption by multinational corporations becomes essential.

Voluntary EPR allows companies with global reach and extensive supply chains to take proactive steps in assuming responsibility for their environmental impacts. These companies can set industry standards and influence suppliers, partners, and even competitors to adopt similar practices. By voluntarily implementing EPR, companies can address the shortcomings of fragmented national regulations and ensure consistent sustainability practices across their global operations.

Moreover, voluntary EPR can drive innovation and foster a culture of continuous improvement within industries. Companies that lead in voluntary EPR adoption can develop more efficient recycling processes, design products for easier disassembly, and create closed-loop systems that reduce reliance on virgin materials. This benefits the environment, enhances the companies' reputations, and meets the growing consumer demand for sustainable products.

Therefore, it is plausible to conclude that the voluntary adoption of EPR is a vital component of circular business responsibility, especially for companies operating on a global scale. It allows them to address environmental challenges that transcend national borders and contribute meaningfully to the global circular economy. Through voluntary EPR, businesses can demonstrate leadership in sustainability, drive systemic change, and pave the way for a more sustainable future.

5 Voluntary Assumption of Extended Producer Responsibility at Apple

Apple has been at the forefront of implementing voluntary Extended Producer Responsibility (EPR) as part of its broader commitment to sustainability and environmental stewardship. Apple's approach to EPR focuses on taking full responsibility for the life cycle of its products, from design and manufacturing to end-of-life management, ensuring that products are recycled and disposed of in an environmentally friendly manner.

Apple's voluntary EPR initiatives include the Apple Renew program, which allows customers to return any Apple device to Apple stores or through mail for recycling, free of charge. This program aims to recover valuable materials and ensure that electronic waste is processed responsibly. Apple's recycling process is notable for its use of advanced robotics, such as Daisy, a disassembly robot that can take apart 1.2 million devices annually, recovering materials like cobalt, aluminium, and tin for reuse in new products (Apple, 2021).

In addition to recycling, Apple claims to design its products with longevity and repairability in mind. The company provides extensive support for product repairs and offers refurbished products to extend the life cycle of its devices. Apple's focus on using recycled materials is also a key component of its EPR strategy. For instance, the latest iPhones use 100% recycled rare earth elements in all magnets, 100% recycled tungsten in the Taptic Engine, and 100% recycled tin in the solder of the main logic board (Apple, 2021).

Apple's voluntary EPR efforts are part of its broader environmental goals, which include becoming carbon neutral across its entire business, manufacturing supply chain, and product life cycle by 2030. This comprehensive approach involves transitioning to recycled and renewable materials, reducing reliance on mining, and investing in renewable energy projects (Apple, 2021).

The effectiveness of Apple's EPR initiatives is underscored by its transparent reporting and continuous improvement efforts. Apple publishes an annual Environmental Progress Report detailing its sustainability initiatives, achievements, and areas for improvement. This commitment to transparency helps build trust with consumers and stakeholders, demonstrating Apple's dedication to environmental responsibility (Apple, 2021).

However, despite these positive steps, Apple faces several criticisms regarding its EPR practices. Critics argue that while Apple's recycling programs are advanced, they are not as widespread or accessible as they could be. For instance, many consumers are unaware of the Apple Renew program, limiting its impact. Additionally, some argue that Apple's focus on product design and innovation often results in devices that are difficult to repair or upgrade, which can counteract its sustainability goals (iFixit, 2023).

Another criticism is related to the transparency and actual effectiveness of Apple's recycling efforts. While Apple reports high material recovery rates, there is limited independent verification of these claims. Critics suggest that third-party audits and more detailed reporting could enhance the credibility of Apple's sustainability efforts (Greenpeace, 2017).

Lastly, some environmental advocates highlight that Apple's use of proprietary components and software restrictions limits the ability of third-party repair shops to fix Apple devices, leading to higher rates of device turnover and waste. This practice, known as planned obsolescence, has been criticised for contributing to electronic waste despite Apple's recycling programs (European Environmental Bureau, 2020).

To sum up, Apple's voluntary assumption of Extended Producer Responsibility sets a benchmark for the electronics industry, highlighting how proactive corporate strategies can drive significant environmental benefits. By focusing on recycling, product longevity, and using recycled materials, Apple minimises its environmental impact and supports the transition towards a more sustainable and circular economy. However, addressing the criticisms regarding accessibility, transparency, and repairability will be crucial for enhancing the effectiveness and credibility of its EPR initiatives.

6 Conclusion

The responsibility principle is crucial for guiding conduct within individual and organisational contexts. Historically, philosophical contributions from intellectuals like Max Weber and Hans Jonas have profoundly shaped our understanding of responsibility, emphasising accountability and the ethical imperative to preserve life for future generations.

In the corporate sphere, the evolution of corporate social responsibility (CSR) reflects a growing recognition of businesses' multiple responsibilities, encompassing economic, social, and environmental dimensions. This shift from a shareholder-centric to a stakeholder-inclusive approach has been crucial in addressing modern global challenges and promoting sustainability.

Incorporating the principle of responsibility into the circular economy involves rethinking corporate roles and practices to support sustainable resource use and waste elimination. Companies like TriCiclos and IKEA exemplify how businesses can successfully integrate circular principles into their operations, demonstrating a commitment to sustainability and reducing their environmental footprint.

Extended Producer Responsibility (EPR) is an important policy instrument in this transition, encouraging sustainable product design and life-cycle management. Both mandatory and voluntary EPR initiatives highlight the potential for producers to proactively manage their products' end-of-life stages, reducing waste and promoting material recovery and reuse.

This chapter underscores businesses' possibilities and needs to comprehensively understand their responsibility in a way that aligns with circular economy principles. By doing so, companies can fulfil their ethical obligations, drive positive environmental and social outcomes, and contribute significantly to the well-being of current and future generations.

References

Andreasi Bassi, S., Boldrin, A., Faraca, G., & Astrup, T. F. (2020). Extended producer responsibility: How to unlock the environmental and economic potential of plastic packaging waste? *Resources Conservation and Recycling, 162*, 105030. https://doi.org/10.1016/J.RESCONREC.2020.105030

Apple. (2021). Environmental Progress Report. https://www.apple.com/environment/pdf/Apple_Environmental_Progress_Report_2021.pdf

Baumgärtner, S., Petersen, T., & Schiller, J. (2018). The concept of responsibility: Norms, actions and their consequences. *SSRN Journal.* Advance online publication. https://doi.org/10.2139/ssrn.3157667

Berle, A. A. (1931). Corporate powers as powers in trust. *Harvard Law Review, 44*(7), 1049. https://doi.org/10.2307/1331341

Borchert, J. (2007). From Politik als Beruf to Politics as a Vocation. *Contributions to the History of Concepts, 3*(1), 42–70. https://doi.org/10.1163/180793207X209075

Bowen, H. R. (1953). *Social responsibilities of the businessman. University of Iowa faculty connections*. University of Iowa Press. http://www.jstor.org/stable/j.ctt20q1w8f. https://doi.org/10.2307/j.ctt20q1w8f

Cai, Y.-J., & Choi, T.-M. (2021). Extended producer responsibility: A systematic review and innovative proposals for improving sustainability. *IEEE Transactions on Engineering Management, 68*(1), 272–288. https://doi.org/10.1109/TEM.2019.2914341

Centobelli, P., Cerchione, R., Chiaroni, D., Del Vecchio, P., & Urbinati, A. (2020). Designing business models in circular economy: A systematic literature review and research agenda. *Business Strategy and the Environment, 29*(4), 1734–1749.

Choi, T. (2017). Environmental impact of voluntary extended producer responsibility: The case of carpet recycling. *Resources Conservation and Recycling, 127*, 76–84. https://doi.org/10.1016/j.resconrec.2017.08.020

Dahl Rendtorff, J. (2014). The importance of responsibility in times of crisis. *Nordicum-Mediterraneum. Icelandic E-Journal of Nordicum and Mediterranean Studies, 9*(3), 1–15. https://nome.unak.is/wordpress/09-3/c73-conference-paper/the-importance-of-responsibility-in-times-of-crisis/

Dodd, E. M. (1932). For whom are corporate managers trustees? *Harvard Law Review, 45*(7), 1145–1163.

European Environmental Bureau. (2020). Planned Obsolescence. https://eeb.org/tag/planned-obsolescence/

Freeman, R. E., Harrison, J. S., Wicks, A. C., Parmar, B. L., & de Colle, S. (2010). *Stakeholder theory: The state of the art*. Cambridge University Press.

Freeman, R. E., Martin, K. E., & Parmar, B. L. (2020). *The power of and: Responsible business without trade-offs*. Columbia University Press.

Friedman, M. (1970, September 13). The social responsibility of business is to increase its profits. *The New York Times Magazine*, p. 17.

Garlapati, V. K. (2016). E-waste in India and developed countries: Management, recycling, business and biotechnological initiatives. *Renewable and Sustainable Energy Reviews, 54*, 874–881. https://doi.org/10.1016/j.rser.2015.10.106

Greenpeace. (2017). Guide to Greener Electronics. https://www.greenpeace.de/publikationen/20171016-greenpeace-guide-greener-electronics-englisch.pdf

Gupt, Y., & Sahay, S. (2015). Review of extended producer responsibility: A case study approach. *Waste Management & Research: The Journal of the International Solid Wastes and Public Cleansing Association, ISWA, 33*(7), 595–611. https://doi.org/10.1177/0734242X15592275

Haase, M. (2015). Responsibility. In S. O. Idowu, N. Capaldi, M. S. Fifka, L. Zu, & R. Schmidpeter (Eds.), *CSR, sustainability, ethics & governance. Dictionary of corporate social responsibility: CSR, sustainability, ethics and governance* (p. 447). Springer.

Hickle, G. (2017). Extending the boundaries: An assessment of the integration of extended producer responsibility within corporate social responsibility. *Business Strategy and the Environment, 26*(1), 112–124. https://doi.org/10.1002/bse.1908

Ianulardo, G., Stella, A., & de Angelis, R. (2022). Uncovering the dialogical dimension of corporate responsibility: Towards a transcendental approach to economics, with an application to the circular economy. In G. Faldetta, E. Mollona, & M. M. Pellegrini (Eds.), *Philosophy and business ethics: Organizations, CSR and moral practice* (pp. 535–567). Palgrave Macmillan. https://doi.org/10.1007/978-3-030-97106-9_21

iFixit. (2023). iFixit Celebrates 20 Years of Fixing the World. https://www.ifixit.com/News/72076/ifixit-celebrates-20-years-of-fixing-the-world

IKEA. (2024). *Our circular agenda – IKEA Global*. https://www.ikea.com/global/en/our-business/people-planet/our-circular-agenda/

Ivanova, V., & Slavova, I. (2019). Corporate environmental responsibility within the circular economy context: Opportunities for development and sustainability. *Economic Alternatives* (2), 184–204. https://www.unwe.bg/uploads/alternatives/3_ivanova_slavova_alternatives_en_2_2019.pdf

Jonas, H. (1984). *The imperative of responsibility: In search of an ethics for the technological age: in search of an ethics for the technological age*. University of Chicago Press.

Kirchherr, J., Yang, N.-H. N., Schulze-Spüntrup, F., Heerink, M. J., & Hartley, K. (2023). Conceptualising the circular economy (revisited): An analysis of 221 definitions. *Resources, Conservation and Recycling, 194*, Article 107001. https://doi.org/10.1016/j.resconrec.2023.107001

Lifset, R., Atasu, A., & Tojo, N. (2013). Extended producer responsibility. *Journal of Industrial Ecology, 17*(2), 162–166. https://doi.org/10.1111/jiec.12022

Massarutto, A. (2014). The long and winding road to resource efficiency – An interdisciplinary perspective on extended producer responsibility. *Resources Conservation and Recycling, 85*, 11–21. https://doi.org/10.1016/j.resconrec.2013.11.005

Peng, B., Wang, Y., Elahi, E., & Wei, G. (2019). Behavioral game and simulation analysis of extended producer responsibility system's implementation under environmental regulations. *Environmental Science and Pollution Research International, 26*(17), 17644–17654. https://doi.org/10.1007/S11356-019-05215-W

Sahajwalla, V., & Hossain, R. (2023). Rethinking circular economy for electronics, energy storage, and solar photovoltaics with long product life cycles. *MRS Bulletin, 48*(4), 375–385. https://doi.org/10.1557/s43577-023-00519-2

Savini, F., & Giezen, M. (2020). Responsibility as a field: The circular economy of water, waste, and energy. *Environment and Planning C: Politics and Space, 38*(5), 866–884. https://doi.org/10.1177/2399654420907622

Septianingrum, D., Mizuno, K., & Herdiansyah, H. (2023). Extended producer responsibility for waste management policy. *Jurnal Penelitian Pendidikan IPA (JPPIPA), 9*(5), 2686–2692. https://doi.org/10.29303/jppipa.v9i5.3469

Shooshtarian, S., Maqsood, T., Wong, P. S. P., Khalfan, M., & Yang, R. J. (2021). Extended producer responsibility in the Australian Construction Industry. *Sustainability, 13*(2), 620. https://doi.org/10.3390/SU13020620

TriCiclos. (2024). *Who we are*. https://triciclos.net/en/triciclos/

Velenturf, A. P., & Purnell, P. (2021). Principles for a sustainable circular economy. *Sustainable Production and Consumption, 27*, 1437–1457.

Weber, M. (1946). Politics as a vocation. In H. H. Gerth & C. W. Mills (Eds.), *From Max Weber: Essays in sociology* (pp. 77–128). Oxford University Press.

Windsor, D. (2008). Berle-Dodd Debate. In R. W. Kolb (Ed.), *Encyclopedia of business ethics and society*. Sage Online.

Cristian R. Loza Adaui is a Professor of Management and Corporate Social Responsibility at the THI Business School of the Technische Hochschule Ingolstadt in Germany. He previously worked for the CBS International Business School in Cologne, Germany, where he also served as Deputy Director of the Centre for Advanced Sustainability Management (CASM). He has held teaching and research positions at the Institute of Economics of the Friedrich-Alexander University Erlangen-Nürnberg in Germany and the Universidad Católica San Pablo in Peru. Prof. Loza Adaui received his Doctoral Degree in Management from the Ingolstadt School of Management at the Catholic University of Eichstätt-Ingolstadt in Germany. His research and consulting activities focus on how businesses become more sustainable, responsible, human-centred, and purpose-driven through strategic management, corporate social responsibility, sustainability management, non-financial reporting, circular economy and corporate purpose. His research has appeared in outlets such as *Business Strategy and the Environment*, *Journal of Business Research*, *Long Range Planning*, *Voluntas: International Journal of Voluntary and Nonprofit Organizations* and *Journal* *of Management Development*, among others. He serves on the editorial boards of journals such as *Technological Sustainability* and *Frontiers in Sustainability*.

The Principle of Regeneration in Circular Economy: Revitalising for Resilience

Cristian R. Loza Adaui

1 Introduction

One of the main challenges of moving from a linear economy towards a circular economy is intricately tied to the principle of regeneration. This principle is usually put into context, arguing that the linear economy is heavily reliant on extraction activities, often leading to unintended degradation of nature, whereas the circular economy actively aims for the regeneration of nature.

The regeneration of nature addresses major sustainability challenges, such as the Earth's degraded capacity, which, according to recurrent scientific evidence (Rockström et al., 2009; Steffen et al., 2015), is mainly driven by human activities. For example, in 2022, the human ecological footprint indicates that human beings are employing nature 1.75 times faster than ecosystems can regenerate (World Wildlife Fund, 2022).

According to the Global Footprint Network (2019), around 80% of the world's countries consumed more resources and generated more waste in 2019 than their ecosystems can manage sustainably. Moreover, if everyone lived the lifestyle of an average European in 2022, we would need the resources of more than three Earths (3.08 exactly) to endure global consumption standards and almost five Earths (4.98 exactly) in the case of an average US American lifestyle (Global Footprint Network & York University and Footprint Data Foundation, 2024).

Evidence of degraded nature is also provided by the Intergovernmental Science-Policy Platform on Biodiversity and Ecosystem Services (IPBES, 2018), which reported that nearly 33% of the Earth's soils are significantly degraded due to human

C. R. Loza Adaui (✉)
THI Business School, Technische Hochschule Ingolstadt, Ingolstadt, Germany
e-mail: Cristian.LozaAdaui@thi.de

E. Fröhlich et al. (eds.), *Circular Economy in Sustainable Supply Chains*, CSR, Sustainability, Ethics & Governance,
https://doi.org/10.1007/978-3-031-70749-0_4

activity. International organisations like the United Nations Convention for Combatting Desertification reveal that desertification is growing globally, with over 75% of the global land surfaces altered, and a pessimistic prognosis, according to which desertification may reach up to 90% by 2050 (Cherlet et al., 2018).

Beyond desertification, life on Earth has been affected over the last decades; the IPBES (2018) estimated that around one million animal and plant species are threatened with extinction, many within decades, and this represents the highest level in human history. Consequently, it is not surprising that a crucial goal of achieving a truly circular economy requires to shape economic activities and orient decision-making processes towards the regeneration of ecosystems.

The concept of regeneration goes beyond simply reducing waste and promoting recycling. It encompasses a holistic approach to restoring and replenishing natural resources, which have been heavily impacted by decades of unchecked industrialisation and consumption (World Wildlife Fund, 2022). In order to effectively implement this principle, there needs to be a shift towards sustainable and regenerative practices across all aspects of the global supply chain. This includes rethinking production processes to minimise resource extraction, implementing regenerative agricultural practices and prioritising the use of renewable and biodegradable materials. Furthermore, collaboration between governments, businesses and local communities is essential to implement strategies that promote ecosystem regeneration. This can involve initiatives such as reforestation projects, sustainable land management practices and the protection of biodiversity hotspots.

This chapter is designed with a dual purpose. Firstly, it aims to delve into various definitions and perspectives associated with the principle of regeneration within the context of the circular economy. Secondly, it seeks to elucidate the implications arising from the adoption of regenerative business practices in global supply chains. In essence, the chapter endeavours to address two fundamental research questions: what are the diverse interpretations of the principle of regeneration within the circular economy framework, and how does GreenWave operationalise the principle of regeneration across its strategy and business model? The first research question will be addressed in Sects. 2 and 3 of this microchapter, and the second research question will be addressed in Sects. 4 and 5. The paper ends by providing a short conclusion.

2 The Principle of Regeneration in the Circular Economy

According to a literature review by Kirchherr et al. (2017), one of the most common definitions of circular economy has been articulated by the Ellen MacArthur Foundation (2015, p. 2), characterising it as "an economy that is restorative or regenerative by intention and design". The principle of regeneration is therefore widely acknowledged as a central idea driving circular economy practices. Nevertheless, Morseletto (2020) highlights the paucity of efforts in precisely defining "restoration" and "regeneration" within the academic discourse on the circular economy, and fortunately, after a careful analysis of the literature, he proposes a definition for each one of these notions.

Morseletto (2020, p. 769) posits that regeneration, in the context of the circular economy, involves "the promotion of self-renewal capacity of natural systems, aiming to reactivate ecological processes damaged or over-exploited by human action". On the one hand, he contends that the principle of regeneration, with its connotation of "rebirth or restart", finds particular applicability in agricultural industries, where it has been developed as a field of research and where the experimentation of practical solutions has been more extensive. However, on the other hand, Morseletto (2020) suggests that the principle of regeneration may not be "easily" suitable for application to technical cycles, proposing that, for these cases, another principle would be more adequate: the principle of restoration. This exclusive application of the principle of regeneration to natural cycles is in line with the developments of the Ellen MacArthur Foundation, which places regenerative activities within the natural cycles on its butterfly diagram of circular economy.

Despite providing a comprehensive analysis and definition of regeneration, Morseletto (2020) rejects its designation as a core circular economy principle. One argument he posits is the potential overlap with other concepts employed to address sustainability challenges in agriculture, such as organic farming, biodynamic agriculture, low-input agriculture, resource-conserving farming, permaculture and agroecology.

Nevertheless, delving into the subtle nuances between the concepts of restoration and regeneration reveals a pivotal distinction with significant implications for sustainable environmental practices. Restoration, according to Morseletto (2020, p. 769), is "the return to a previous or original state"; this definition has several applications in technical cycles but is less applicable to natural cycles. When applied to ecological contexts, the restoration of damaged natural cycles may prove unattainable due to irreversible temporal effects and indelible anthropogenic imprints on the Earth's systems, as exemplified by the loss of floral and faunal species. Furthermore, the risks of socio-political manipulation of restorative goals, fostering calls for a radical return to an idealised historical pristine state of nature, pose new challenges. The socio-ecological idealisation of restorative aims may overlook ecosystems' complex and dynamic nature, which defies simplistic attempts at reversion. Moreover, the achievement of restoration necessitates the establishment of a global consensus regarding the precise definition of "restoration goals". This, in turn, relies on the political willingness of the parties involved in the discussions and decision-making processes related to goal setting.

In contrast, regeneration places explicit emphasis on human agency in sustaining and invigorating natural cycles dynamically. Rather than pursuing an unattainable restoration of pristine states, regeneration seeks to engage with natural processes, allowing for the amelioration or even reduction of negative impacts induced by human activities. The salient feature lies in the human commitment to supporting and protecting natural cycles, letting nature thrive. The regenerative paradigm envisions a role for humanity as stewards, navigating the intricate interplay between ecological systems and anthropogenic influence. Consequently, the discourse surrounding regeneration manifests as a paradigmatic shift from a backwards-oriented, idealised view of restoration to a future-oriented, dynamic and collaborative strategy, epitomising the necessary recalibration of human-environmental relations. This sophisticated understanding acknowledges the inherent dynamism of ecosystems, necessitating adaptive strategies for sustainable coexistence.

For the present study, the principle of regeneration is defined as a dynamic that aims to enhance and improve the health of socio-ecological systems. It operates through an interdependent approach that provides life-supporting and life-enhancing conditions for the global human community and the ecosystem. This definition underscores the interconnectedness and mutual reliance between human societies and the natural environment, emphasising the imperative of nurturing symbiotic relationships that foster resilience, vitality and sustainability.

3 Implementing the Principle of Regeneration in Business Practices

Beyond the discourse surrounding the definition of the regeneration principle, scholars have presented compelling examples of how businesses can actively contribute to regeneration efforts. For instance, Hawken (2021) emphasises the role of businesses in regeneration, highlighting their potential to serve as catalysts for positive environmental and social change. Hawken's work underscores the importance of integrating regenerative practices into business strategies to address pressing sustainability challenges and create lasting impact (Hawken, 2021). In his book *Regeneration: Ending the Climate Crisis in One Generation*, he presents a comprehensive roadmap for addressing the climate crisis through regenerative practices, advocating for a paradigm shift towards regenerative capitalism.

In a similar vein, Hahn and Tampe (2021) delve into the realm of regenerative business strategies, offering insights into the practical approaches that companies can adopt to align with regenerative principles. Their research explores the intersection of business strategy and sustainability, providing valuable guidance for organisations seeking to transition towards regenerative business models. To that aim, Hahn and Tampe (2021) introduce a framework for operationalising the regeneration principle within business strategies. It is important to remark that they advocate for viewing regenerative business strategies "not as a binary concept, but rather as a continuum" represented by the "restore-preserve-enhance" scale. This approach underscores a shift from business logic to systems logic, emphasising the importance of enhancing the health of social-ecological systems (SES) through adaptive management practices. By conceptualising business sustainability in terms of regeneration, Hahn and Tampe (2021) offer practical strategies for organisations to contribute to the enhancement of life-supporting conditions within SES.

Additionally, Konietzko et al. (2023) contribute to the discourse by focusing on one element of business strategy: the business model. By examining various business models that qualify as regenerative, Konietzko et al. (2023) elucidate the characteristics and attributes of regenerative business models, shedding light on the diverse approaches companies can take to integrate regenerative principles into their operations.

Konietzko et al. (2023) have made significant contributions to the field of regenerative business models by providing a comprehensive framework that distinguishes these models from sustainable and circular business models. They have conducted a systematic literature and practice review to identify key themes and establish a rigorous foundation for conceptualising regenerative business models. Through international and multicultural focus groups, including indigenous representatives, they have enriched the empirical understanding of regenerative practices and integrated diverse cultural perspectives into their research. Moreover, Konietzko et al. (2023) have also engaged in a detailed comparison of regenerative, sustainable and circular business models, highlighting their differences and overlaps, thereby offering a nuanced perspective on value creation for future-fit organisations.

Furthermore, they have contributed to the methodological approach in this emerging field by employing a structured data analysis using coding techniques to derive definitions and categories for regenerative business models. They have also addressed the potential for greenwashing and the importance of integrating indigenous knowledge, which is crucial for the authenticity and effectiveness of regenerative practices. The authors have provided practical illustrations of how businesses can integrate regenerative principles into their operations by including examples of companies that have adopted regenerative practices, such as Guayaki and Burt's Bees.

Regenerative business models are characterised by their focus on planetary health and societal well-being, which is reflected in their value proposition, value creation and delivery, and value capture mechanisms. These models prioritise a holistic approach to business that includes the health of the biosphere and human societies, recognising the deep interdependence between the two. A regenerative business model's value proposition is centred on contributing positively to both nature and society, aiming to give more than they take and striving for a net positive impact (Konietzko et al., 2023).

In terms of value creation and delivery, regenerative business models engage in activities that promote regenerative leadership, co-creative partnerships with nature, as well as justice and fairness. This involves a systemic perspective that goes beyond the organisation itself to include multiple stakeholders, such as nature, societies, customers, suppliers and partners, shareholders, investors and employees. The value capture in regenerative business models is achieved through multi-capital accounting, which seeks to measure and account for the value generated across different forms of capital, including natural, social, human and financial (Konietzko et al., 2023).

Furthermore, according to Konietzko et al. (2023), regenerative business models are distinguished from sustainable and circular models by their systemic perspective, which is rooted in social-ecological systems rather than socio-technical or closed-loop economic systems. They also differ in their main goals, with regenerative models focusing on planetary health and societal well-being, as opposed to the triple bottom line or material productivity. The design foci of regenerative business models include the regenerative biosphere, which emphasises ecosystem regeneration and the use of biological, biodegradable and non-toxic materials.

Furthermore, Ryan et al. (2023) present a comprehensive exploration of regenerative business practices, culminating in the development of a Framework for Regenerative Business. Through qualitative case studies and literature synthesis, they identify six core principles central to regenerative business models: empowered stakeholder relationships, connection to context and place, circular flows, trust and transparency, innovation and adaptability to ecosystems and building inner and outer capacity for systems resilience (Ryan et al., 2023). This framework provides a structured approach for companies seeking to transition from conventional sustainability practices to regenerative models that prioritise planetary well-being.

Howard et al. (2019) contribute to the discourse by examining and developing a framework for circular economy indicators, particularly in the context of large multinational organisations transitioning to regenerative and circular systems. Their research offers insights into the fragmented nature of existing literature on circular economy indicators and proposes a coherent framework linking indicators to the core goals and principles of a circular economy. Through analysis of company reports, case descriptions and interviews, they highlight the challenges and opportunities associated with circular economy implementation, providing valuable recommendations for indicator development aligned with circular economy goals.

4 Methodological Framework

In the previous sections of this microchapter, the question of the definition of the regeneration principle within the context of circular economy was addressed. To answer the second research question—how does the organisation GreenWave operationalise the principle of regeneration across its strategy and business model?—the case study research method has been adopted.

The case study method is chosen in this chapter because it offers a comprehensive understanding of complex contemporary phenomena within their real-world context (Yin, 2018). Given the novelty of the regenerative principle in a circular economy, an exploratory research design is necessary, and the case study research method is particularly suitable for addressing such research questions. The case selection follows a purposive sampling strategy. GreenWave was chosen based on its exemplary implementation of regenerative activities, a notable commitment to sustainability, innovative practices and demonstrable contribution to climate change mitigation. This sampling process is non-probabilistic, and therefore, some case selection bias should be taken into consideration (Campbell et al., 2020).

The analysis has been done based on secondary data, including the company website, documents, reports and relevant literature written by the founders of the organisation and by others covering the activities of GreenWave. The analysis of the data was thematic. Key themes and patterns related to the implementation of the regeneration principle, as described in Sects. 2 and 3 of this microchapter, have been contrasted with the information available about GreenWave's approach to business and circularity. In concrete, the theoretical contributions of Hawken (2021), Hahn and Tampe (2021), and Konietzko et al. (2023) served as operationalisation benchmarks for the regeneration principle in the circular economy.

The methodology adopted in this chapter, however, entails some limitations. Factors such as time constraints, limited access to information and GreenWave's unique context may have influenced the generation of the findings and their generalisability to other organisations. Consequently, the outcomes of the analysis serve the mere aim of showcasing a company striving to implement the regeneration principle of the circular economy.

5 The Regeneration Principle at GreenWave

GreenWave is an organisation at the forefront of sustainable aquafarming, embodying a commitment to innovation and environmental stewardship. In their own words, "GreenWave replicates and scales regenerative ocean farms to create jobs and protect the planet" (GreenWave, 2024a).

GreenWave has adopted a unique approach to farming, known as polyculture, which cultivates a variety of seaweeds and shellfish without the need for any additional resources. This method is considered one of the most eco-friendly ways to produce food on a global scale. The farms are designed to be submerged under the water, producing a high yield in a compact area. According to GreenWave (2024b), establishing a new farm requires just 20 acres of water, a boat and a budget between 20 to 50 thousand US dollars.

Bren Smith, founder of GreenWave, describes his farm in the following way:

> Imagine a vertical underwater garden with hurricane-proof anchors on the edges connected by floating horizontal ropes across the surface. From these lines, kelp, Gracilaria, and other kinds of seaweeds grow vertically downward next to scallops in hanging nets that look like Japanese lanterns and mussels held in suspension in mesh socks. Staked below the vertical garden are oysters in cages and then clams buried in the sea floor. (Smith, 2015)

This innovative aquaculture system generates food, has regenerative effects and contributes to climate change mitigation. Moreover, seaweed aquaculture has received scholarly attention for several sustainability reasons. Among the benefits of seaweed farming is its contribution to carbon sequestration (Chung et al., 2017; Hurd et al., 2022); it also helps prevent ocean deoxygenation and acidification, benefiting marine life and ecosystem resilience (Yong et al., 2022). Additionally, seaweed aquaculture serves as an environmental sentinel, reducing aquatic nutrient loadings and promoting socio-economic development (Subasinghe et al., 2009). Last but not less importantly, seaweed farming contributes to sustainable food, feed, bioenergy and material production, supporting economic growth and livelihoods in coastal communities (Costello et al., 2020; Greene et al., 2022; Nunes et al., 2003).

GreenWave's business model is versatile. It is used for both reforestation and commercial farming. The seaweed and shellfish harvested are utilised in a wide range of industries, including food, fertiliser, animal feed and bioplastics. In evaluating GreenWave's operational model with Hawken's (2020) vision of integrating regenerative practices into business strategy to tackle sustainability challenges, the company emerges as a catalysator for the regeneration of the environment below water and as a promoter of social change in the coastal communities. This practice aligns closely with Hawken's call for a paradigm shift towards regenerative capitalism.

GreenWave's innovative and sustainable farming model aligns closely with regenerative business strategies, as outlined by Hahn and Tampe (2021). The organisation's commitment to replicating regenerative ocean farms and creating jobs while protecting the planet reflects a clear focus on societal well-being and the health of the biosphere. In terms of the "restore-preserve-enhance" scale created by Hahn and Tampe (2021), GreenWave appears as a case of an organisation implementing an enhancing strategy for two main reasons. First, on a systems-based level, GreenWave aspires to reach a net positive impact. It demonstrates symbiotic embeddedness with the aquatic ecosystem and promotes a rationale in which business and ecosystem co-evolve mutually. Second, in terms of the management approach, GreenWave is also a good case for nestedness of place within the social-ecological system, focusing on aquafarming, the well-being of the farmers and their impact on coastal regions. The organisation has a long-term orientation that follows seasonal environmental changes and promotes participative experimentation through a series of initiatives with farmers and partners to improve the business model over time.

Considering the regenerative business model framework proposed by Konietzko et al. (2023) as a benchmark for regenerative organisations, GreenWave embodies several elements of the framework. Firstly, regarding the value proposition, GreenWave demonstrates a commitment to holistic value creation. This is built around three types of activities: an ocean farming hub, which is a platform for regenerative ocean farming for individuals and organisations interested in launching their own farms; a training programme for seaweed farmers and an explicit commitment to economic justice. To showcase this last element, a description of GreenWave's purpose in their own words is presented:

> As the emerging regenerative ocean farming industry matures, GreenWave's goal is to ensure that under-resourced communities directly affected by climate change benefit from the industry's growth and have pathways to become leaders in the field. From paid hatchery mentorships to laying the groundwork for the first two Indigenous-owned seaweed hatcheries in the country, we focus our programming and resources on supporting these communities. (GreenWave, 2024c)

With regard to value creation and delivery, the characteristics of regenerative leadership are clearly depicted by the foundational storytelling of the organisation and especially of the founder, Bren Smith:

> I used to be a commercial fisherman, chasing your dinner on the high seas for a living, but now I farm twenty acres of saltwater, growing a mix of sea greens and shellfish. I've paid my debt to the sea. I dropped out of high school to fish and spent too many nights in jail. My body is beat to hell: I crawl out of bed like a lobster most mornings. I've lost vision in half my right eye from a chemical splash in Alaska. I'm an epileptic who can't swim, and I'm allergic to shellfish. But every shiver of pain has been worth it. It's a meaningful life. I'm proud to spend my days helping feed my community, and if all goes well, I will die on my boat one day. Maybe get a small obit in the town paper, letting friends know that I was taken by the ocean, that I died a proud farmer growing food underwater. That I wasn't a tree hugger but spent my days listening to and learning from waves and weather. That I believed in building a world where we can all make a living on a living planet. (Smith, 2019, pp. 3–4)

The element of co-creative partnerships with nature could also be considered embodied by GreenWave, especially in its contribution to climate resilience and the connection to local nature. In terms of justice and fairness, GreenWave promotes nature-based solutions, diversity and inclusion, with activities like the ocean farming community they design and run, as well as online training programmes given for free and designed to support ocean farmers during the whole lifecycle of ocean farming and hatchery management.

Concerning value capture, the business model of GreenWave appears to emphasise natural capital, cultural capital and social capital focus. However, a more in-depth analysis and additional information would be necessary to comment on financial and human capital. Table 1 summarises the elements of the regenerative business model framework by Konietzko et al. (2023) and evaluates the extent of their presence in GreenWave's business model and operations.

Table 1 Analysis of GreenWave according to the regenerative business model framework (based on Konietzko et al., 2023)

		Stakeholder Levels				
		Nature	Societies	Customers, Partners and Suppliers	Shareholders and Investors	Employees
Value Proposition	Planetary Health and Societal Wellbeing	Planetary health	Human health and wellbeing	Healthy products and services	Organisational purpose	Meaning and values
Value Creation and Delivery	Regenerative Leadership	In service to nature	In service to society	Trust and reputation	Demonstrated purpose	Inner work
	Co-Creative Partnerships with Nature	Nature-inclusive design	Climate resilience	Nature-inspired design	Natural capital investment	Connection to local nature
	Justice and Fairness	Nature-based solutions Nature rights and animal rights advocacy	Indigenous knowledge Justice advocacy	Responsible sourcing	Community investment	Diversity and inclusion
Value Capture	Multi-Capital Accounting and Net Positive Impact	Natural capital	Cultural capital	Social capital	Financial capital	Human capital

Legend: ■ Clearly found in GreenWave's organisational model; □ Not found or not enough information to complete evaluation

In summary, GreenWave exhibits a strong resemblance to the core features of a regenerative business model, as outlined by Konietzko et al. (2023). It demonstrates a high degree of accordance with the characteristics of value proposition, value creation and value capture at diverse stakeholder levels. GreenWave's methods for aquafarming and commitment to environmental stewardship serve as a compelling example of a company striving to integrate the principle of regeneration into its purpose, strategy, business model and operations.

6 Conclusion

This chapter aimed to elucidate the principle of regeneration within the circular economy, emphasising its role in addressing global sustainability challenges by fostering the restoration and replenishment of natural resources. It underscored the urgency of integrating regenerative practices into global supply chains and business models. As an exemplar case study, GreenWave's innovative approach to aquafarming, which not only supports ecological stewardship but also promotes social well-being through sustainable food production systems, was presented. The analysis revealed that regenerative business models, characterised by their systemic perspective and focus on planetary health, offer a distinct and necessary shift towards a net positive impact on both nature and society. However, the challenges in operationalising the regeneration principle, including the need for clear definitions and the difficulty of applying it uniformly across different industries and contexts, are duly recognised.

Furthermore, some limitations of this chapter's empirical part must be acknowledged. Firstly, the analysis relies on secondary data, the unique context of GreenWave and the supply chain that its business model generates. Accordingly, while the findings provide valuable insights into the application of the regeneration principle, they may not be universally generalisable. To conclude, this chapter contributes to the academic discourse by clarifying the conceptual underpinnings of the regeneration principle within the circular economy and by illustrating its practical application through the case study of GreenWave. However, as Das and Bocken (2024) demonstrate, there are many different strategies that companies are using to implement the regeneration principle, and the dynamic of business model experimentation with renerative aims is growing over time.

References

Campbell, S., Greenwood, M., Prior, S., Shearer, T., Walkem, K., Young, S., Bywaters, D., & Walker, K. (2020). Purposive sampling: Complex or simple? Research case examples. *Journal of Research in Nursing, 25*(8), 652–661. https://doi.org/10.1177/1744987120927206

Cherlet, M., Hutchinson, C., Reynolds, J., Hill, J., Sommer, S., & von Maltitz, G. (2018). *World Atlas of Desertification* (JRC111155). Publication Office of the European Union. https://publications.jrc.ec.europa.eu/repository/handle/JRC111155

Chung, I. K., Sondak, C. F. A., & Beardall, J. (2017). The future of seaweed aquaculture in a rapidly changing world. *European Journal of Phycology, 52*(4), 495–505. https://doi.org/10.1080/09670262.2017.1359678

Costello, C., Cao, L., Gelcich, S., Cisneros-Mata, M., Free, C. M., Froehlich, H. E., Golden, C. D., Ishimura, G., Maier, J., Macadam-Somer, I., Mangin, T., Melnychuk, M. C., Miyahara, M., de Moor, C. L., Naylor, R., Nøstbakken, L., Ojea, E., O'Reilly, E., Parma, A. M., Plantinga, A. J., Thilsted, S. H., & Lubchenco, J. (2020). The future of food from the sea. *Nature, 588*(7836), 95–100. https://doi.org/10.1038/s41586-020-2616-y

Das A, Bocken N (2024) Regenerative business strategies: A database and typology to inspire business experimentation towards sustainability. *Sustainable Production and Consumption*, 49, 529–544. https://doi.org/10.1016/j.spc.2024.06.024

Ellen MacArthur Foundation. (2015). *Towards a circular economy: Business rationale for an accelerated transition*. https://www.ellenmacarthurfoundation.org/towards-a-circular-economy-business-rationale-for-an-accelerated-transition

Global Footprint Network. (2019). *National Footprint and Biocapacity Account 2019*. https://us9.campaign-archive.com/?u=6b189d93d725df997bc7ec31d&id=d71750d29c

Global Footprint Network, York University and Footprint Data Foundation. (2024). *Compare countries*. https://api.footprintnetwork.org/v1/data/231,2002/2022/earth

Greene, C. H., Scott-Buechler, C. M., Hausner, A. L., Johnson, Z. I., Gen Lei, X., & Huntley, M. E. (2022). Transforming the future of marine aquaculture: A circular economy approach. *Oceanography, 35*(2), 26–34. https://doi.org/10.5670/oceanog.2022.213

GreenWave. (2024a). *Our model*. https://www.greenwave.org/our-model

GreenWave. (2024b). *Breathing life back into our planet*. https://www.greenwave.org/

GreenWave. (2024c). *Programs*. https://www.greenwave.org/work

Hahn, T., & Tampe, M. (2021). Strategies for regenerative business. *Strategic Organization, 19*(3), 456–477. https://doi.org/10.1177/1476127020979228

Hawken, P. (2021). *Regeneration: Ending the climate crisis in one generation*. Penguin.

Howard, M., Hopkinson, P., & Miemczyk, J. (2019). The regenerative supply chain: A framework for developing circular economy indicators. *International Journal of Production Research, 57*(23), 7300–7318. https://doi.org/10.1080/00207543.2018.1524166

Hurd, C. L., Law, C. S., Bach, L. T., Britton, D., Hovenden, M., Paine, E. R., Raven, J. A., Tamsitt, V., & Boyd, P. W. (2022). Forensic carbon accounting: Assessing the role of seaweeds for carbon sequestration. *Journal of Phycology, 58*(3), 347–363. https://doi.org/10.1111/jpy.13249

IPBES. (2018). The IPBES Assessment Report on Land and Restoration. In L. Montanarella, R. Scholes, & A. Brainich (Eds.), *Secretariat of the intergovernmental science-policy platform on biodiversity and ecosystem services*. https://www.ipbes.net/assessment-reports/ldr

Kirchherr, J., Reike, D., & Hekkert, M. (2017). Conceptualizing the circular economy: An analysis of 114 definitions. *Resources, Conservation and Recycling, 127*, 221–232. https://doi.org/10.1016/J.RESCONREC.2017.09.005

Konietzko, J., Das, A., & Bocken, N. (2023). Towards regenerative business models: A necessary shift? *Sustainable Production and Consumption, 38*, 372–388. https://doi.org/10.1016/j.spc.2023.04.014

Morseletto, P. (2020). Restorative and regenerative: Exploring the concepts in the circular economy. *Journal of Industrial Ecology, 24*(4), 763–773. https://doi.org/10.1111/jiec.12987

Nunes, J. P., Ferreira, J. G., Gazeau, F., Lencart-Silva, J., Zhang, X. L., Zhu, M. Y., & Fang, J. G. (2003). A model for sustainable management of shellfish polyculture in coastal bays. *Aquaculture, 219*(1–4), 257–277. https://doi.org/10.1016/S0044-8486(02)00398-8

Rockström, J., Steffen, W., Noone, K., Persson, Å., Chapin, F. S., Lambin, E. F., Lenton, T. M., Scheffer, M., Folke, C., Schellnhuber, H. J., Nykvist, B., de Wit, C. A., Hughes, T., van der Leeuw, S., Rodhe, H., Sörlin, S., Snyder, P. K., Costanza, R., Svedin, U., Falkenmark, M., Karlberg, L., Corell, R. W., Fabry, V. J., Hansen, J., Walker, B., Liverman, D., Richardson, K., Crutzen, P., & Foley, J. A. (2009). A safe operating space for humanity. *Nature, 461*(7263), 472–475. https://doi.org/10.1038/461472a

Ryan, N., Beesemyer, L., Caulliez, S., Waiyaki, J., Nayak, M., Chakrabarty, R., Kim, S. K. A., & Vladimirova, D. (2023). Introducing a novel framework for regenerative business. In *New Business Models Conference Proceedings*. Maastricht University Press. https://doi.org/10.26481/mup.2302.40

Smith, B. (2015). Ecological redemption: Ocean farming in the era of climate change. In H. Hannum (Ed.), *Thirty-Fifth Annual E. F. Schuhmacher Lectures*. Schumacher Center for a New Economics. https://centerforneweconomics.org/publications/ecological-redemption-ocean-farming-in-the-era-of-climate-change/

Smith, B. (2019). *Eat like a fish: My adventures farming the ocean to fight climate change*. Vintage.

Steffen, W., Richardson, K., Rockström, J., Cornell, S. E., Fetzer, I., Bennett, E. M., Biggs, R., Carpenter, S. R., de Vries, W., de Wit, C. A., Folke, C., Gerten, D., Heinke, J., Mace, G. M., Persson, L. M., Ramanathan, V., Reyers, B., & Sörlin, S. (2015). Planetary boundaries: Guiding human development on a changing planet. *Science, 347*(6223). https://doi.org/10.1126/science.1259855

Subasinghe, R., Soto, D., & Jia, J. (2009). Global aquaculture and its role in sustainable development. *Reviews in Aquaculture, 1*(1), 2–9. https://doi.org/10.1111/j.1753-5131.2008.01002.x

World Wildlife Fund. (2022). In R. E. A. Almond, M. Grooten, D. Juffe Bignoli, & T. Petersen (Eds.), *Living Planet Report 2022: Building a nature-positive society*. WWF. https://wwfint.awsassets.panda.org/downloads/embargo_13_10_2022_lpr_2022_full_report_single_page_1.pdf

Yin, R. K. (2018). *Case study research and applications* (6th ed.). Sage Publications.

Yong WTL, Thien VY, Rupert R, Rodrigues KF (2022) Seaweed: A potential climate change solution. *Renewable and Sustainable Energy Reviews*, 159, 112222. https://doi.org/10.1016/j.rser.2022.112222

Cristian R. Loza Adaui is a Professor of Management and Corporate Social Responsibility at the THI Business School of the Technische Hochschule Ingolstadt in Germany. He previously worked for the CBS International Business School in Cologne, Germany, where he also served as Deputy Director of the Centre for Advanced Sustainability Management (CASM). He has held teaching and research positions at the Institute of Economics of the Friedrich-Alexander University Erlangen-Nürnberg in Germany and the Universidad Católica San Pablo in Peru. Prof. Loza Adaui received his Doctoral Degree in Management from the Ingolstadt School of Management at the Catholic University of Eichstätt-Ingolstadt in Germany. His research and consulting activities focus on how businesses become more sustainable, responsible, human-centred and purpose-driven through strategic management, corporate social responsibility, sustainability management, non-financial reporting, circular economy and corporate purpose. His research has appeared in outlets such as *Business Strategy and the Environment*, *Journal of Business Research*, *Long Range Planning*, *Voluntas: International Journal of Voluntary and Nonprofit Organizations* and *Journal of Management Development*, among others. He serves on the editorial boards of journals such as *Technological Sustainability* and *Frontiers in Sustainability*.

The Principle of Rethink in Circular Economy: Redefining Business Solutions

Maud H. Schmiedeknecht

1 Introduction

Rethinking traditional linear approaches to production and consumption ("take-make-waste") is one main characteristic of a circular economy. The rethink principle emphasises the importance of creating a continuous cycle and the necessity of circular economy-oriented innovations.

Some rethink definitions focus on the product level. Potting et al. (2017, p. 5), for instance, define rethink as "mak[ing] product use more intensive (e.g. through sharing products, or by putting multi-functional products on the market)". Other definitions have a broader approach, highlighting the correlations to other R-principles and the effects on the strategy. According to Morseletto (2020, p. 8), "Rethink—i.e. proposing new ideas and solutions to provide certain product functions—can embrace the reframing and regenerating of potentially every aspect of CE systems. [...] [D]esign can be traced back to this strategy, as can (potentially) Refuse and Reduce. To do so, the analysis of the strategies and targets starts with Rethink".

In order to integrate the rethink principle into strategy and operations, the broader perspective will be chosen in this article. In the following, light will be shed on how companies can rethink their products/services, business models and ecosystems through sustainability-oriented and circular economy-oriented business model innovation (BMI). Then a card-based tool as a method to explore the circularity potential will be introduced. Finally, sufficiency-based approaches will be outlined, as sufficient consumption patterns rely on business offerings that enable changes—to rethink business in an advanced way.

M. H. Schmiedeknecht (✉)
ESB Business School, Reutlingen University, Reutlingen, Germany
e-mail: maud.schmiedeknecht@reutlingen-university.de

E. Fröhlich et al. (eds.), *Circular Economy in Sustainable Supply Chains*, CSR, Sustainability, Ethics & Governance,
https://doi.org/10.1007/978-3-031-70749-0_5

2 Rethinking Products/Services, Business Models and Ecosystems for Sustainability and Circular Economy

For years, researchers have been analysing approaches for business models in the context of sustainability and circular economy (Bocken et al., 2014; Geissdoerfer et al., 2020; Pieroni et al., 2019; Schaltegger et al., 2016).

A business model is understood as the "design or architecture of the value creation, delivery, and capture mechanisms" of a business (Teece, 2010, p. 172). It describes what the company offers, to whom, how it creates and delivers the offering and how it captures value. The Business Model Canvas has been established as a common visualisation of a business model with nine building blocks for value generation, organised into four pillars: product/value proposition, customer interface, financial aspects and infrastructure management (Osterwalder et al., 2005). Similarly, Gassmann et al. (2015) describe business models with four dimensions: who (segment: who is your target customer?), what (value proposition: what do you offer to the customer?), how (value chain: how is the value proposition created?) and value (profit mechanism: why does the business model generate profit?). According to their research, a business model innovation (BMI) occurs as the conscious change of at least two of these dimensions.

Sustainability-oriented and circular economy-oriented BMI integrates sustainability and/or circular principles as guidelines for business model design in order to incorporate a positive contribution to society and the environment (Pieroni et al., 2019).

For instance, Lüdeke-Freund et al. (2018) identified 45 patterns to support sustainability-oriented BMI and clustered them into 11 categories: (1) *pricing & revenue patterns* defining how the offerings of sustainable business model can be priced and revenues generated, (2) *financing patterns* suggesting different ways of acquiring equity, debt and operating capital to finance sustainable business models, (3) *ecodesign patterns* defining how activities, processes and offerings can be designed to improve the ecological performance of business models, (4) *closing-the-loop patterns* integrating the idea of circular material and energy flows into business model design, (5) *supply chain patterns* defining how resource inputs are sourced and target markets are reached, (6) *giving patterns* enabling donation of products or services, (7) *access provision patterns* supporting to create markets and providing offerings for otherwise neglected social target groups, (8) *social mission patterns* showing how to empower social groups to become productive partners and to turn their needs into demand, (9) *service & performance patterns* supporting dematerialising a business model by shifting from physical products to functions, services and results, (10) *cooperative patterns* integrating stakeholders as co-owners and co-managers and (11) *community platform patterns* substituting individual resource or product ownership with community-based access. Mainly in the closing-the-loop pattern, circular economy activities are captured.

Regarding circular economy-oriented BMI, often *narrow*, *slow*, *close* and *regenerate* strategies of material and energy flow are associated with business activities

(Bocken et al., 2016; Konietzko et al., 2020): companies can strive to use less materials and energy (narrow), use products and components longer (slow), use materials again (close) and use renewable energy and non-toxic materials (regenerate). Often these circular activities rely on interaction and collaboration of organisations. Therefore, the ecosystem perspective is important in a circular economy, where higher degrees of collaboration, cross-cycle and cross-sector cooperation and a shift from supply chains to value networks are essential. Takacs et al. (2020), for instance, have identified 38 circular ecosystem patterns and developed a classificatory framework that categorises these patterns into four pillars: (1) *close the loop* with patterns for closing the material cycle, e.g. product-reuse and waste as input, (2) *improve the loop* with patterns to reduce or solve environmental problems and avoid unwanted effects, e.g. increased longevity and dematerialisation, (3) *monetise the loop* with patterns to capitalise the value from circular solutions, e.g. take-back and crowdfunding and (4) *excite the loop* with patterns motivating users for circular change and incentives, such as experience selling and sharing.

3 Card-Based Tools as a Method to Explore the Circularity Potential

How can companies rethink their products/services, business models or ecosystems? One structured method for companies to explore the circularity potential is using card deck-based tools (Konietzko et al., 2020; Pieroni et al., 2020).

Based on a literature review of circular-oriented innovation, Konietzko et al. (2020), for instance, have developed the Circularity Deck to facilitate the exploration of circular economy principles. Each card in the deck represents a specific circular-oriented innovation principle, offering practical examples of how these principles have been applied in real-world scenarios. In Table 1, the circularity principles are listed according to the innovation perspective that is needed to operationalise a principle on a product level, business model level or ecosystem level and how they contribute to the circular strategy. Besides the above-mentioned strategies—narrow, slow, close and regenerate—*inform* is added, describing the use of information technology to pursue circularity.

In workshops, participants have the opportunity to analyse and discuss if and how these circularity principles can be applied in their business context. Card deck-based tools, such as the Circularity Deck, can support generating new ideas to foster higher degrees of circular practices.

Table 1 Circularity principles for product, business model and ecosystem innovation (adapted from Konietzko et al., 2020, pp. 11–12)

Innovation perspective	Circular strategy	Circular principle
Product	Narrow	Design with low-impact inputs
		Design light-weight products
		Design for multiple functions
	Slow	Design for physical durability
		Design for emotional durability
		Design for ease of maintenance and repair
		Design for easy dis- and reassembly
		Design for upgradability
		Design for standardisation and compatibility
	Close	Design with recycled inputs
		Design components, where appropriate, with one material
		Design with materials suitable for primary recycling
		Design for easy disassembly at the end of the product lives
	Regenerate	Design with renewable materials
		Design self-charging products
		Design with living materials
		Design with non-toxic materials
	Inform	Use artificial intelligence to develop new materials with circular properties
		Virtualise
		Design connected products
Business Model	Narrow	Eliminate production waste
		Enable and incentivise users to consume less
		Organise light-weight urban transport
		Localise supply where appropriate
	Slow	Enable users to maintain and repair their products
		Remanufacture existing products and components
		Repurpose existing products and components
		Provide an unconditional lifetime warranty
		Encourage sufficiency
		Provide the product as a service
		Organise maintenance and repair services
		Upgrade and adapt existing products
	Close	Reuse and sell components and materials from discarded products
		Enable and incentivise product returns
		Recycle products in proper facilities
	Regenerate	Produce and process with renewable energy
		Power transportation with renewable energy
		Power the use of the product with renewable energy
	Inform	Use product-in-use data for circular design
		Track the resource intensity of the product-in-use
		Track the condition, location and/or availability of the product

Table 1 (continued)

Innovation perspective	Circular strategy	Circular principle
Ecosystem	Narrow	Maximise capacity use of products
	Slow	Turn disposables into a reusable service
	Close	Build local waste-to-product loops
		Engage in industrial symbiosis
	Regenerate	Embed renewable energy production in the existing infrastructure
		Recover nutrients from urban areas
		Regenerate polluted ecosystems
		Manage and sustain critical ecosystem services
	Inform	Market circular products, components and materials through online platforms
		Build material database ecosystems
		Co-create products, components, materials and information via online platforms
		Use artificial intelligence to optimise circular infrastructure
		Operate service ecosystems via online platforms

4 Rethinking Consumption Patterns: Organisational Strategies Supporting Sufficiency

Rethinking products/services, business models and ecosystems may require courage, especially if they are counterintuitive to the business-as-usual approach of growth. In this context, the concept of sufficiency based on sustainable consumption has been discussed (Bocken et al., 2014; Reisch & Thøgersen, 2015; Sandberg, 2021). One challenging question is how companies can rethink their business models by fostering sufficient consumption patterns. Typical patterns to rethink how to consume differently are offering product use rather than ownership through rental/leasing/sharing or offering the so-called "green alternatives", such as products with fewer resource inputs, long-life products and/or local and regional products (Bocken & Short, 2016). Integrating patterns in a business model, Bocken and Short (2016, p. 46) describe sufficiency-driven BMI as "curbing consumption as part of the business model by moderating demand through education and consumer engagement, making products that last and avoiding built-in obsolescence; extending product lives to slow disposal and replacement, focusing on satisfying 'needs' rather than promoting wants' and fast-fashion, and reducing overall resource consumption through conscious changes in sales and marketing techniques, new revenue models, and innovative technology solutions".

In a study, Bocken et al. (2022) analysed 150 companies' organisational strategies supporting sufficiency. The results have been published as a sufficiency business database on the Circular X website, a research project focused on experimentation with circular service business models (Circular X, 2024; Niessen & Bocken, 2022). Within the apparel sector, outdoor apparel and gear companies, such as Devold, Globetrotter, Patagonia and VAUDE, have been analysed. Furthermore, Gossen and

Kropfeld (2022) observed that outdoor companies like Patagonia, Globetrotter and VAUDE increasingly implement sustainability strategies based on sufficiency and use marketing to advocate sufficiency-oriented consumption. Patagonia's Worn Wear programme is described in the microchapter on the principle of repair.

In the following, VAUDE Sport GmbH & Co KG is briefly introduced to illustrate how one of these outdoor companies integrated sufficiency strategies into their business. VAUDE, founded in 1979, is an outdoor apparel and gear company with headquarters in Tettnang, Germany. In alignment with the sufficient consumption patterns, the company offers green alternatives through durable and reparable clothing. Furthermore, offering product use rather than ownership is a part of their business model. In their 2020 sustainability report, VAUDE (2021) outlined their degrowth approaches and highlighted the promotion of consumption based on sufficiency and responsibility. They also emphasised their support of self-sufficiency and "do-it-yourself" (DIY) amongst other topics, such as recycling management, climate neutrality and their commitment to transforming the economic system (VAUDE, 2021):

Selection of VAUDE Degrowth Approaches

We promote consumption based on sufficiency and responsibility

- Quality, durability, timelessness and reparability are the core elements of our product philosophy; they guide our product development and are the basis for a long product life […]
- Our sustainability report and the information found directly on our products provide a great deal of transparency so that you can become informed and make responsible purchasing decisions […]
- You can give your VAUDE product a second life by selling it at the VAUDE Second-Hand Shop on eBay or by donating to our partner FairWertung […]
- We would like to provide motivation for conscious consumption and for using our products as long as possible. Therefore we provide information under the motto "Eternally yours" about what you can do yourself and how we provide support […]

We support self-sufficiency and DYI [*sic*]

- We offer customers the ability to have their damaged VAUDE products repaired in our Repair Workshop […]
- For simple repairs that you can make yourself, we offer DYI [*sic*] repair instructions at iFixit. You can also order all spare parts or, if required, special tools and have them mailed directly to your home […] Need assistance with the repair? You can take advantage of our cooperation with the network of Repair Cafés or visit our own Repair Café in our factory outlet
- You can buy VAUDE remnant materials for handicrafts and upcycling for a donation to Save the Children at the eBay Upcycling Store powered by VAUDE […]
- As an alternative to buying new products and the option of shared use, we offer "iRentit by VAUDE" services […]. (VAUDE, 2021)

Meanwhile, VAUDE built a platform where customers can decide to rent or buy second-hand outdoor products (VAUDE, 2024a). Therefore, raising customer awareness is not only achieved through marketing campaigns. Additionally, VAUDE has established the VAUDE Academy to multiply sustainable business practices by sharing their experience and expertise from their transformation process with companies, initiatives, universities and schools (VAUDE, 2024b).

It is obvious that integrating the rethink principle into strategy and operations—especially if they support sufficiency—requires effort. Let us stay curious about how companies embrace circular economy by rethinking their products/services, business models and ecosystems.

References

Bocken, N. M. P., de Pauw, I., Bakker, C., & van der Grinten, B. (2016). Product design and business model strategies for a circular economy. *Journal of Industrial and Production Engineering, 33*(5), 308–320. https://doi.org/10.1080/21681015.2016.1172124

Bocken, N. M. P., Niessen L., & Short S. W. (2022). The sufficiency-based circular economy—An analysis of 150 companies. *Frontiers in Sustainability, 3*, Article 899289. https://doi.org/10.3389/frsus.2022.899289

Bocken, N. M. P., & Short, S. W. (2016). Towards a sufficiency-driven business model: Experiences and opportunities. *Environmental Innovation and Societal Transitions, 18*, 41–61. https://doi.org/10.1016/j.eist.2015.07.010

Bocken, N. M. P., Short, S. W., Rana, P., & Evans, S. (2014). A literature and practice review to develop sustainable business model archetypes. *Journal of Cleaner Production, 65*, 42–56. https://doi.org/10.1016/j.jclepro.2013.11.039

Circular X. (2024). *Business for sufficiency database*. https://www.circularx.eu/en/tool/26/business-for-sufficiency-database

Gassmann, O., Frankenberger, K., & Csik, M. (2015). *The business model navigator: 55 models that will revolutionise your business*. FT Publishing International.

Geissdoerfer, M., Pieroni, M. P. P., Pigosso, D. C. A., & Soufani, K. (2020). Circular business models: A review. *Journal of Cleaner Production, 277*, Article 123741. https://doi.org/10.1016/j.jclepro.2020.123741

Gossen, M., & Kropfeld, M. I. (2022). "Choose nature. Buy less." Exploring sufficiency-oriented marketing and consumption practices in the outdoor industry. *Sustainable Production and Consumption, 30*, 720–736. https://doi.org/10.1016/j.spc.2022.01.005

Konietzko, J., Bocken, N., & Hultink, E. J. (2020). A tool to analyze, ideate and develop circular innovation ecosystems. *Sustainability, 12*(1), Article 417. https://doi.org/10.3390/su12010417

Lüdeke-Freund, F., Carroux, S., Joyce, A., Massa, L., & Breuer, H. (2018). The sustainable business model pattern taxonomy – 45 patterns to support sustainability-oriented business model innovation. *Sustainable Production and Consumption, 15*, 145–162. https://doi.org/10.1016/j.spc.2018.06.004

Morseletto, P. (2020). Targets for a circular economy. *Resources, Conservation and Recycling, 153*, Article 104553. https://doi.org/10.1016/j.resconrec.2019.104553

Niessen, L., & Bocken, N. (2022). A sufficiency business database as a tool to drive sustainable business models. In L. Michelini, A. Minà, & P. Alaimo Di Loro (Eds.), *Conference Proceedings of the 7th International Conference on New Business Models: Sustainable Business Model Challenges: Economic Recovery and Digital Transformation* (pp. 640–644). LUMSA University.

Osterwalder, A., Pigneur, Y., & Tucci, C. (2005). Clarifying business models: Origins, present, and future of the concept. *Communications of the Association for Information Systems, 16*, 1–25. https://doi.org/10.17705/1CAIS.01601

Pieroni, M. P. P., McAloone, T. C., & Pigosso, D. C. A. (2019). Business model innovation for circular economy: Integrating literature and practice into a conceptual process model. *Proceedings of the 22nd International Conference on Engineering Design (ICED19), 1*(1), 2517–2526. https://doi.org/10.1017/dsi.2019.258

Pieroni, M. P. P., McAloone, T. C., & Pigosso, D. C. A. (2020). From theory to practice: systematising and testing business model archetypes for circular economy. *Resources, Conservation and Recycling, 162*, Article 105029. https://doi.org/10.1016/j.resconrec.2020.105029

Potting, J., Hekkert, M., Worrell, E., & Hanemaaijer, A. (2017). *Circular economy: Measuring innovation in the product chain. Policy Report*. PBL Netherlands Environmental Assessment Agency. https://www.pbl.nl/uploads/default/downloads/pbl-2016-circular-economy-measuring-innovation-in-product-chains-2544.pdf

Reisch, L. A., & Thøgersen, J. (2015). Research on sustainable consumption: Introduction and overview. In L. A. Reisch & J. Thøgersen (Eds.), *Handbook of research on sustainable consumption* (pp. 1–17). Edward Elgar Publishing.

Sandberg, M. (2021). Sufficiency transitions: a review of consumption changes for environmental sustainability. *Journal of Cleaner Production, 293*, Article 126097. https://doi.org/10.1016/j.jclepro.2021.126097

Schaltegger, S., Hansen, E. G., & Lüdeke-Freund, F. (2016). Business models for sustainability: Origins, present research, and future avenues. *Organization & Environment, 29*(1), 3–10. https://doi.org/10.1177/1086026615599806

Takacs, F., Stechow, R., & Frankenberger, K. (2020). *Circular ecosystems: Business model innovation for the circular economy white paper*. Institute of Management & Strategy, University of St. Gallen. https://ifb.unisg.ch/wp-content/uploads/2020/01/Circular-Ecosystems_Takacs-Stechow-Frankenberger-2020.pdf

Teece, D. J. (2010). Business models, business strategy and innovation. *Long Range Planning, 43*(2–3), 172–194. https://doi.org/10.1016/j.lrp.2009.07.003

VAUDE. (2021). *2020 Sustainability Report*. https://nachhaltigkeitsbericht.vaude.com/2020/gri-en/vaude/degrowth.php

VAUDE. (2024a). *Outdoor-Ausrüstung mieten oder Second Hand kaufen!* https://rent-secondhand.vaude.com/

VAUDE. (2024b). *VAUDE Academy für nachhaltiges Wirtschaften*. https://academy.vaude.com/

Maud H. Schmiedeknecht is a Professor of Corporate Governance and Corporate Social Responsibility at the ESB Business School of the Reutlingen University in Germany and a Business Design Coach. What drives her is supporting and empowering business leaders and business school students to develop skills and learn methods to (re)design sustainable, circular and regenerative business models. Maud previously worked as a consultant at the global management consulting firm Kearney in Germany and abroad. As a research assistant at the Konstanz Institute for Intercultural Management, Values and Communication of Prof. Dr. Josef Wieland, she undertook consulting and research projects in the fields of corporate social responsibility and compliance and participated in national and international committees, e.g. the German CSR-Forum and ISO 26000 Working Group on Social Responsibility. She was an associate of the Stiftung Neue Verantwortung, (now interface) a think tank for digital technologies, politics and society, on the topic of global economic ethic. Her dissertation on the governance of multistakeholder dialogues was awarded the Max Weber Prize for Business Ethics. Maud's current teaching, research and consulting focus is on corporate sustainability management, (social) entrepreneurship and sustainable business model innovation, business transformation and circular economy.

The Principle of Redesign in Circular Economy: At the Heart of Circular Economy

Elisabeth Fröhlich

1 Introduction

"Design sits prominently at the heart of the circular economy. It requires us to redesign everything: products, business models, cities, and the linear systems that have lasted for the past centuries" (Ellen McArthur Foundation, 2024). Nowadays, however, most products are still designed for the linear model. This means that almost everything has to be redesigned according to the principles of circular economy.

The main goal of the circular economy is to promote the redesign of entire economies, with cradle-to-cradle thinking serving as the primary driver for this new business model (McDonough & Braungart, 2002, p. 165). Redesign can, therefore, be defined as the reframing of products, processes or entire systems to follow the three circular economy principles: regenerate nature by avoiding pollution and waste and optimising the use of materials to promote sustainability throughout their life cycle. Systems thinking is the key to realising the circular R—redesign. Coming back to the different R-strategies (Malooly & Daphne, 2023) mentioned in the 9R Framework, redesign can only be achieved by considering all R-strategies in order to fulfil the key objectives of the circular economy (Devadula & Chakrabarti, 2015, p. 7).

The just mentioned three major principles of circular economy can be transformed into the three principles of circular design (Interaction Design Foundation, 2023):

1. Aim to *eliminate* waste and pollutants, including harmful substances that contaminate our soil and water, while also striving to reduce emissions. Considering

E. Fröhlich (✉)
ispira Think Tank for Sustainable Supply Chains, Hürth, Germany
e-mail: lisa.froehlich@ispira-thinktank.com

E. Fröhlich et al. (eds.), *Circular Economy in Sustainable Supply Chains*, CSR, Sustainability, Ethics & Governance,
https://doi.org/10.1007/978-3-031-70749-0_6

that almost every product contributes to waste, a comprehensive redesign is essential. The role of design is crucial in achieving this goal.
2. Ensure the *extended circulation of products and materials*, maintaining their highest value for as long as feasible. Recycling efforts must be energy efficient and should not compromise the quality of the recycled product. For instance, creating durable machines that can be easily disassembled allows for efficient repair and reuse of components in new items. This approach also supports recycling, reducing the need to extract fresh resources from the planet.
3. Promote the *regeneration* of natural ecosystems by returning natural resources to the environment. An example of this would be transforming food waste into compost to enrich soil, thereby revitalising agricultural land.

2 Defining Circular Design: Process and Strategies

Circular design encompasses all phases of a product's life cycle to ensure that a product follows circular principles from manufacture to disposal. In the following, some examples are listed to illustrate that redesign can only be implemented by taking all R-strategies into account (Interaction Design Foundation, 2023):

- **Raw Materials** Which raw materials are required, and how can they be obtained sustainably?
- **Product Specifications** Can the product be reused, repaired or recycled after use?
- **Manufacturing Process** Is the process as energy efficient as possible, and what opportunities does it offer for waste avoidance?
- **Distribution and Use** How much energy is needed to transport and use the product/service? Is energy from renewable sources used to fuel those processes?
- **Disposal** Can the product be easily disassembled into its individual components? Can the parts be reused in the manufacturing process without major effort? Can we return the raw materials to the Earth's cycle?

The *circular design journey* encompasses four key phases, drawing on methodologies like design thinking and human-centred design. In the first stage—understand—one has to get familiarised with both the users and the overarching system. On the next level—define—the design challenge and objectives of the designer must be clearly stated. In the third stage—make—designers engage in the creation, design and prototyping of numerous iterations and versions. The final stage—release—introduces the design to the world and weaves the story to foster customer loyalty and enhance stakeholder engagement through a compelling narrative (Ellen MacArthur Foundation, 2024).

In the circular design process, the launch of a product or service marks the beginning of repeating loops. With each loop, essential insights into resource reduction and optimisation are gained, as well as acceptance by stakeholders within the system. For a product to function within the circular economy, all participants in the respective ecosystem must continuously contribute to its maintenance. Unlike in the linear economy, where designers typically have no further involvement with a product after its market introduction, there is no final moment in the circular economy (Golta, 2022).

This cycle of iteration demands on-going evaluation and refinement based on user interaction with the design and its integration within the broader system. However, examining successful examples of circular design in action highlights some strategies that tend to be more consistently successful than others. The subsequent sections detail six such strategies (Ellen MacArthur Foundation, 2024):

- **Designing for the Inner Loops** It comes as no surprise that the most effective examples of circular design often focus on the inner loops of the circular economy systems diagram (Ellen MacArthur Foundation, 2022). Through designing products for easy repair or remanufacture or by developing new business models that simplify sharing, these companies are tapping into new value for both themselves and their customers.
- **Moving From Product to Services** At the heart of the circular economy is the transition from owning products to accessing them. This understanding has given rise to numerous new businesses that provide various products to their customers on a temporary basis—via rental, subscription, sharing or leasing models—instead of offering permanent ownership.
- **Product Life Extension** Extending the lifespan of a product ensures that it remains in circulation for as long as possible and thus fulfils one of the three basic principles of circular economy. This may entail designing products to be both physically robust and emotionally appealing, or it may demand creative solutions that enable the product to adapt to the changing needs of the user over time.
- **Safe and Circular Material Choices** Certain materials are not suitable for the circular economy because they include hazardous chemicals that pose a risk to human health or the environment. However, there are strategies to eliminate these hazardous substances. By selecting materials that are both safe and consistent with circular economy principles, consumer offerings can be improved.
- **Dematerialisation** This is a key element of eco-design that is frequently ignored and minimises the material quantity required for manufacturing goods and services (FasterCapital, 2024). This might involve transitioning offers to a digital format rather than a tangible one. Alternatively, it could entail designing products or services that only require a small quantity of physical materials for its creation.
- **Modularity** This serves as an effective strategy for simplifying the repair, remanufacture and upgrading of products. By facilitating the removal of just a few parts of a product, disassembly becomes more straightforward, reducing both the cost and effort needed to replace damaged components. Moreover, modularity allows for easier customisation. This prevents products from becoming outdated, ensuring their prolonged use.

3 The Power of Redesign: FREITAG's Circular Truck Tarps

In an interview with Rocío de la Cruz García, Circular Economy and Innovation Project Portfolio Manager at Covestro, a project initiated by FREITAG (https://www.freitag.ch/de) is showcased, focusing on the development of circular truck tarps (Covestro, 2022). This innovative project is briefly introduced, emphasising that redesign is the foundational element of a circular economy. The project embodies the principle of circular design by prioritising the redesign of materials to ensure their longevity and circularity from the outset. By rethinking the design of truck tarps to make them circular, FREITAG addresses the core principle of circular design, which is to innovate in a way that products can be reused, recycled or repurposed, thereby reducing waste and the need for new resources. The strategy of circular design that this project implements involves creating products that are not only durable and long-lasting but also designed with their eventual disassembly and recycling in mind. This approach minimises environmental impact and supports a more sustainable, circular economy.

What is the project about? When we want to get to an ever more circular economy, we have to rethink and redesign existing products and solutions as well as entire value chains. By doing so, we can jointly find ways to, for example, prolong the life of the products, reduce their carbon footprint or make them recyclable.

The circular truck tarp project is a good example of this. Using thermoplastic polyurethane (TPU) instead of the standard tarp material (PVC), the project aims to prolong the lifespan of truck tarpaulins, find a way to reuse them (e.g. as bags and accessories) and ensure the recyclability of the material at the end of its life. Through all of this, we can reduce the carbon footprint of such tarps and create a potentially endless cycle. This means not only the product but also the business model and value chain have to be redesigned to find a fully circular tarp material (Fig. 1).

Everything started with the Swiss company FREITAG, a manufacturer creating bags and accessories made from upcycled truck tarps. They wanted not only upcycled and reused but fully circular material and thus triggered the development of a

Fig. 1 Circular truck tarp (Schmid, 2024)

circular, PVC-free tarp. For that purpose, they looked for raw material providers that have circular economy high on their agenda—and quickly found and met us at Covestro. Among many other polymer materials, we supply Desmopan® TPU, which meets the high requirements of processing this type of product.

This thinking in itself represents a mindset shift: the discussions with materials manufacturers like us and product designers do not take place towards the end of such projects anymore but right at the start. Companies like us can thereby brainstorm with designers and make them aware of the possibilities of our materials—for example, to explore their potential in terms of performance and circularity.

Which principle of circular design is addressed by your project? For this to become a reality, Covestro is exploring different chemical recycling routes. Roughly speaking, this aims at breaking a material down into its initial raw materials again via chemical processing methods. Such raw materials can then be used like virgin raw materials. Until now, TPU has been mechanically recycled for decades, meaning, for example, that old products got shredded, melted and made into new products. But this also, more often than not, means downcycling, as the quality of such a mechanically recycled material is lower and not suited for all final applications.

In our example of truck tarps, this is simply because the tarps have lived through a lot during their product life already, leading to deterioration due to exposure to sun and wind. If you shred such material, the quality of the resulting product can only be lower. Therefore, the cycle is to be closed with chemical recycling, breaking down the polymers into monomers and thereby preserving the qualities of the polymers. This can be done with a material like TPU, which is why TPU was identified as an attractive material for the circular truck tarp project. Figure 2 shows such a circular truck tarp prototype in action.

Our example covers the second principle of circular design—facilitating the prolonged circulation of products and materials, preserving their maximum value for the longest possible duration, which impacts the strategy of product life extension. Circular design involves considering every stage of a product's life cycle to guarantee adherence to circular principles from production to end-of-life. Particularly, our circular truck tarp project focuses on the raw material phase.

Which circular design strategy could be implemented in your project? *Product life extension* is one of the main goals of the project. TPU is very robust and has a long lifespan. It allows truck tarps to be used and to be "on the road" for at least half a decade. What is more, the second life of the material comes when FREITAG manufactures bags out of the used tarps. These can again last many years. Last but not least, the project aims to fully recycle these bags and make new truck tarps from them, creating an endless cycle.

The additional circular design strategy of *safe and circular material choices* also plays an important role here. Proof of this is that Covestro's Desmopan® TPU used for the project has obtained a Material Assessment Statement according to the Cradle to Cradle Certified™ Product Standard. This evaluation has shown that the TPU material is neither human nor eco-toxic and offers the possibility of mechanical recycling and high-quality chemical recycling. Other circular strategies might be explored in the future, for example, *moving from products to services*.

Fig. 2 Circular truck tarp test drive (Bötticher, 2024)

This project has a huge potential and is living proof that design can be the key factor for a visionary idea to fail or succeed. It will take some years, and we will have to go through some obstacles, but I personally cannot wait to wear one of those bags with an everlasting life.

References

Bötticher, E. (2024). *Die grüne Planenalternative auf ihrer ersten (Test-)Fahrt ins Grüne* [Photograph]. Freitag. https://media.freitag.ch/de/media/circulartarp

Covestro. (2022, June 21). *FREITAG initiiert kreislauffähige LKW-Plane* [Press release]. https://www.covestro.com/press/de/freitag-initiiert-kreislauffaehige-lkw-plane/

Devadula, S., & Chakrabarti, A. (2015, July 27–30). *Design strategies for circular economy* [Conference paper]. International Conference on Engineering Design (ICED15), Politecnico di Milano, Italy.

Ellen MacArthur Foundation. (2022). *The technical cycle of the butterfly diagram.* https://www.ellenmacarthurfoundation.org/articles/the-technical-cycle-of-the-butterfly-diagram

Ellen MacArthur Foundation. (2024). *Design and the circular economy – deep dive.* https://www.ellenmacarthurfoundation.org/design-and-the-circular-economy-deep-dive

FasterCapital. (2024). *Eco design: Enhancing Eco Design with Dematerialization Techniques.* https://fastercapital.com/content/Eco-design%2D%2DEnhancing-Eco-Design-with-Dematerialization-Techniques.html

Golta, K. (2022). So funktioniert Circular Design. *PAGE online.* https://page-online.de/tools-technik/circular-design-prozess-projekte/

Interaction Design Foundation – IxDF. (2023). *What is circular design?* https://www.interaction-design.org/literature/topics/circular-design
Malooly, L., & Daphne, T. (2023). *R-Strategies for a circular economy*. Circularise. https://www.circularise.com/blogs/r-strategies-for-a-circular-economy
McDonough, W., & Braungart, M. (2002). *Cradle to Cradle: Remaking the way we make things*. North Point Press
Schmid, Y. (2024). *Auch wie sich eine kreislauffähige Plane beschriften lässt, muss getestet werden* [Photograph]. Freitag. https://media.freitag.ch/de/media/circulartarp

Elisabeth Fröhlich is the founder of ispira Think Tank for Sustainable Supply Chains. She invented an innovative business coach concept to prepare companies for the challenges of the different European sustainability laws. Her major aim is to bridge the gap between research and practice in this field to enable companies to implement the sustainable transformation of global supply chains successfully. Prof. Dr. habil. Elisabeth Fröhlich was president of the CBS International Business School in Germany until 2022. She holds a full professorship in Sustainable Procurement and Supply Chains and is an internationally recognised expert in sustainable supply chain management. She is also highly engaged in the field of responsible management education. She serves as a board member of PRME, chairs the PRME Nomination and Governance Committee and is the PRME DACH Chapter Chair. Additionally, she leads the AOM Management Education Division Ambassador Program as vice-chair, leads the Board of JARO Institute and offers online training in the field of sustainable procurement. Her research focuses on sustainable supply chain management and green procurement, qualification in purchasing, Procurement 4.0 and strategic supplier relationship management. Innovative teaching formats of responsible management education are further main areas of her research. She has published several books and articles on the above-mentioned topics and supports several journals as an external reviewer.

The Principle of Reduce in Circular Economy: Simply Using Less?

Elisabeth Fröhlich

1 Introduction

In Europe, the four major industrial materials—steel, plastics, aluminium and cement—are responsible for 75% of the CO_2 emissions. By applying circular business models by 2050 those four categories can bring down the European carbon footprint by 56% (Enkvist et al., 2022). The R-strategies are classified according to the length of the waste loop they represent. Reduce—to minimise or lessen materials—belongs to the short-loop category, which is the most sustainable one due to the focus on smarter product use and production. It seems to be by far the most important step to save our planet because this strategy uses fewer materials to represent a high level of circularity (Malooly & Daphne, 2023). However, the cradle-to-cradle approach (EPEA, 2024) assigns the reduce strategy to the fossil-based company type. Reducing their ecological footprint in the most efficient way is still the major target of those companies. They go for a "zero emissions" or "free from" strategy without putting the focus on transition to create new business models. Companies should not aim to be "less bad" but to set up an eco-effective business model. So, the reduce strategy needs to encompass more than simply bringing down the amount of material used. Circular business models can only benefit from reduce if outdated business practices and processes are redefined.

Reduce stands for bringing down the consumption of natural resources, which decreases waste production. The major aim is to make production and use processes more efficient (Malooly & Daphne, 2023).

E. Fröhlich (✉)
ispira Think Tank for Sustainable Supply Chains, Hürth, Germany
e-mail: lisa.froehlich@ispira-thinktank.com

E. Fröhlich et al. (eds.), *Circular Economy in Sustainable Supply Chains*, CSR, Sustainability, Ethics & Governance,
https://doi.org/10.1007/978-3-031-70749-0_7

2 Rethinking Buyer–Supplier Relationships to Bring Down the Carbon Footprint

One major contribution of implementing circular business models with a focus on reduce is to bring down the carbon footprint of entrepreneurial activities. Climate change is one of the global threats humanity has to face. This challenge calls for collective action from all companies and countries around the globe. The Paris Agreement outlines the necessity to halve greenhouse gas emissions by 2030 and achieve net zero by 2050 to fulfil the 1.5-degree target (United Nations, 2024). The latest IPCC report (2023) clarifies "the devastating consequences of rising greenhouse gas (GHG) emissions around the world—the destruction of homes, the loss of livelihoods and the fragmentation of communities, for example—as well as the increasingly dangerous and irreversible risks should we fail to change course" (Boehm & Schumer, 2023). Nevertheless, the authors of the report also offer hope by highlighting different strategies to reduce GHG emissions.

Based on the previous classification of reduce, the focus of this section will be on Scope 3 reduction—one of the promising solutions also highlighted in the above-cited IPCC report (2023). Figure 1 provides an overview of the three scopes and emissions across the value chain. Scope 1 covers all GHG (direct) emissions generated by the companies' own production. Indirect emissions are defined as Scope 2 and 3. Scope 2 comprises emissions from purchased electricity, heat and steam.

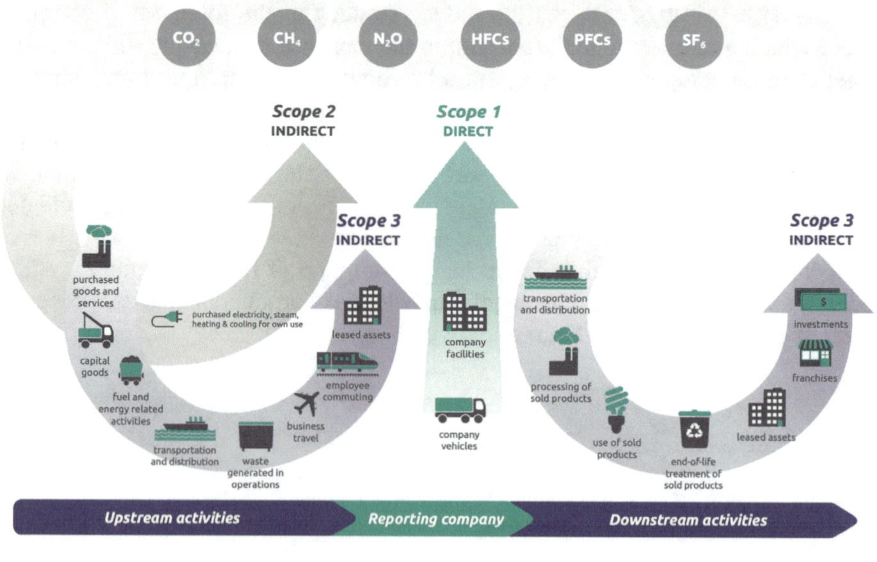

Fig. 1 Overview of scopes and emissions across the value chain (World Resources Institute & World Business Council for Sustainable Development, 2011, p. 5)

Scope 3 includes all GHG emissions generated in the upstream and downstream parts of the supply chain by dealing with suppliers and end consumers (Mazhar, 2023, p. 269).

Strategic supply management is essential for the success of a company because this function accounts for about 70% of a company's revenue. And even more importantly, purchasing is the only corporate function with a direct impact on profit (Howarth & Winfield, 2023, p. 186). The Carbon Disclosure Project (2024) estimates that GHG emissions released in the upstream part of the supply chain are ten times higher than those from the companies' own production processes. Therefore, the upstream activities covering mainly production and transportation of goods are responsible for up to 80% of the GHG emissions in global value chains.

The Ellen McArthur Foundation (2024) identified several key areas where supplier management can significantly reduce GHG emissions. Reduce has been defined as part of the short-loop strategies representing the early stages of a global supply chain. Bringing those two findings together, a practical example clarifying the importance of supplier engagement in bringing down Scope 3 emissions will be the focus of this chapter. Additionally, keeping in mind the numerous new European laws—Corporate Sustainability Reporting Directive (CSRD; European Commission, 2024b), EU Deforestation Regulation (European Commission, 2024c) or Carbon Border Adjustment Mechanism (CBAM; European Commission, 2024a)—the need for rethinking the upstream part of the supply chain becomes more crucial than ever. Aside from the fact that a significant reduction in GHG emissions is essential to save our planet, improving the sustainability of global supply chains is the only way to restore a certain level of resilience to them (EY, 2016, p. 15).

This requires a lot of effort in rethinking buyer supplier relationships because purchasing is the missing link in sustainable global supply chains. Villena and Gioia (2020) discuss four approaches which can support making global supply chains more sustainable with the help of strategic suppliers:

- **Direct Approach** First-tier suppliers are involved in establishing a joint sustainability strategy based on commonly defined sustainability goals and standards.
- **Indirect Approach** Accountability for the sustainability management of lower-tier suppliers is transferred to the first-tier suppliers. To this end, supplier trainings are offered, and incentives are created for the successful implementation of sustainability practices along the value chain.
- **Global Approach** Working with international organisations or NGOs, which share the sustainability goals of purchasers, enables companies to realise their sustainability vision for global supply chains.
- **Collective Approach** The challenges can only be overcome if purchasers work with their competitors and key suppliers to develop industry-wide sustainability standards. Individual company approaches are powerless when it comes to implementing increasingly stringent environmental and social practices in global supply chains.

3 Ørsted: Bringing Suppliers Along the Climate Journey

The practical example chosen in this chapter is taken from the report "Supplier Engagement in Action", published by the UNGC Local Network Denmark (2023). Ørsted—a Danish energy company—launched the project "Bringing Suppliers Along the Climate Journey" after realising that they can only achieve their emission reduction targets when actively collaborating with their suppliers. Ørsted's strategic suppliers account for 60% of the total procurement spend in high-impact categories (UNGC Denmark, 2023).

> Ørsted has been on a climate journey for over a decade and was previously known as DONG Energy, one of the most emission-intensive utilities in Europe. In the 2010s, DONG transitioned from fossil fuels into renewable energy and rebranded as Ørsted, which is now one of the largest RE [renewable energy] companies in the world. Ørsted was the first energy company to receive validation for its Net-Zero Target by Science Based Target Initiative (SBTi). The company already has plans to phase out their neutral gas business; their next frontier is to reduce emissions throughout their supply chain. (UNGC Denmark, 2023, p. 26)

Ørsted identified steel and fuels as their major contributors to GHG emissions, two materials hard to abate (HTA). They also had to learn that their strategic suppliers providing those two HTA materials were lacking knowledge, especially in the field of carbon calculation. Ørsted identified life cycle assessment (LCA) as a key tool to meet its 2040 net-zero goal and invented an LCA-based carbon tracking model, which helps to gather more accurate data from manufacturing, installation and transportation processes of renewable energy assets. So, the missing knowledge and the use of very complex tools call for long-term, strategic and R&D-based partnerships with their most important first-tier suppliers. In addition, Ørsted established a stepwise engagement programme with suppliers by setting up collaborations supported by continuous dialogue. Because renewable energy products are very complex and markets are hard to navigate in terms of decarbonisation, Ørsted supports their suppliers by offering concrete guidelines for achieving net-zero targets. The most important outcome of this engagement programme, however, is that Ørsted's first-tier suppliers can establish decarbonisation projects with their own suppliers. Referring to the aforementioned study from Villena and Gioia (2020), this corresponds to the indirect approach to carry Ørsted's sustainability vision into the entire global supply chain. Ørsted even went one step further, following the collective approach (Villena & Gioia, 2020) by establishing partnerships with organisations such as SteelZero (https://www.theclimategroup.org/steelzero) and First Mover Coalition (https://initiatives.weforum.org/first-movers-coalition/home).

Sustainable solutions for HTA materials can only be identified if shared approaches, including policies and standards, are developed and sustainably embedded into global supply chains with the support of all stakeholders (UNGC Denmark, 2023). "There is no global one-size-fits-all approach for reaching climate targets: technologies, access to RE, and supplier maturity vary greatly across regions […], with the fastest progress observed amongst European suppliers" (UNGC Denmark,

2023, p. 27). The engagement programme has led to a lot of improvement. When it started, only 36% of Ørsted's strategic suppliers disclosed their data, and none of them included the science-based target approach in their carbon calculation. The situation in 2023 completely changed: almost 90% of all strategically relevant suppliers disclose their data, and almost 50% are committed to the science-based target logic (UNGC Denmark, 2023).

Such an engagement programme is key to pushing sustainability into the whole global supply chain. Even though a company still lacks transparency in its supply chain, with the help of first-tier suppliers, a stepwise approach can be successfully implemented. The reduce strategy enables companies, especially in the upstream part of the supply chain, to achieve their net-zero targets with the support of their strategic suppliers. But this requires, as stated at the beginning of the chapter, the redefinition of outdated business models and processes. Partnership is the new leadership!

References

Boehm, S., & Schumer, C. (2023). *10 Big Findings form the 2023 IPCC Report on Climate Change*. World Resources Institute. https://www.wri.org/insights/2023-ipcc-ar6-synthesis-report-climate-change-findings

Carbon Disclosure Project. (2024). *Companies*. https://www.cdp.net/en/companies

Ellen McArthur Foundation. (2024). *Circular economy introduction. Overview*. https://www.ellenmacarthurfoundation.org/topics/circular-economy-introduction/overview

Enkvist, P.-A., Klevnäs, P., Westerdahl, R., & Ahlen, A. (2022). *How a 'materials transition' can support the net-zero agenda*. McKinsey Sustainability. https://www.mckinsey.com/capabilities/sustainability/our-insights/how-a-materials-transition-can-support-the-net-zero-agenda#/

EPEA. (2024). *Cradle to Cradle Mindset*. https://epea.com/ueber-uns/cradle-to-cradle

European Commission. (2024a). *Carbon Border Adjustment Mechanism*. https://taxation-customs.ec.europa.eu/carbon-border-adjustment-mechanism_en

European Commission. (2024b). *Corporate sustainability reporting*. https://finance.ec.europa.eu/capital-markets-union-and-financial-markets/company-reporting-and-auditing/company-reporting/corporate-sustainability-reporting_en

European Commission. (2024c). *Regulation on Deforestation-free products*. https://environment.ec.europa.eu/topics/forests/deforestation/regulation-deforestation-free-products_en

EY. (2016). *The state of sustainable supply chains. Building responsible and resilient supply chains*. https://d306pr3pise04h.cloudfront.net/docs/issues_doc%2Fsupply_chain%2Fstate-of-sustainable-supply-chains.pdf

Howarth, R., & Winfield, F. (2023). Procurement and supply: Pushing the boundaries to remove carbon emissions. In P. Moltan-Hill, F. Winfield, R. Howarth, & M. Mazhar (Eds.), *The handbook of carbon management* (pp. 185–209). Routledge.

IPCC. (2023). *Climate Change 2023: Synthesis Report. Contribution of Working Groups I, II and III to the Sixth Assessment Report of the Intergovernmental Panel on Climate Change* [Core Writing Team, H. Lee and J. Romero (Eds.)]. https://www.ipcc.ch/report/ar6/syr/downloads/report/IPCC_AR6_SYR_LongerReport.pdf

Malooly, L., & Daphne, T. (2023). *R-Strategies for a circular economy*. Circularise. https://www.circularise.com/blogs/r-strategies-for-a-circular-economy

Mazhar, M. (2023). Understanding Greenhouse Gas Management: Scopes and boundaries for carbon footprinting. In P. Moltan-Hill, F. Winfield, R. Howarth, & M. Mazhar (Eds.), *The handbook of carbon management* (pp. 265–287). Routledge.

UNGC Denmark. (2023). *Supplier Engagement in Action. Examples from Denmark on Scope 3 Reductions*. https://info.globalcompact.dk/supplier-engagement-in-action-2023

United Nations. (2024). *The Paris Agreement*. https://www.un.org/en/climatechange/paris-agreement

Villena, V. H., & Gioia, D. A. (2020, March–April). A more sustainable supply chain. *Harvard Business Review*. https://hbr.org/2020/03/a-more-sustainable-supply-chain

World Resources Institute & World Business Council for Sustainable Development. (2011). *Corporate Value Chain (Scope 3) Accounting and Reporting Standard*. https://ghgprotocol.org/sites/default/files/standards/Corporate-Value-Chain-Accounting-Reporing-Standard_041613_2.pdf

Elisabeth Fröhlich is the founder of ispira Think Tank for Sustainable Supply Chains. She invented an innovative business coach concept to prepare companies for the challenges of the different European sustainability laws. Her major aim is to bridge the gap between research and practice in this field to enable companies to implement the sustainable transformation of global supply chains successfully. Prof. Dr. habil. Elisabeth Fröhlich was president of the CBS International Business School in Germany until 2022. She holds a full professorship in Sustainable Procurement and Supply Chains and is an internationally recognised expert in sustainable supply chain management. She is also highly engaged in the field of responsible management education. She serves as a board member of PRME, chairs the PRME Nomination and Governance Committee and is the PRME DACH Chapter Chair. Additionally, she leads the AOM Management Education Division Ambassador Program as vice-chair, leads the Board of JARO Institute and offers online training in the field of sustainable procurement. Her research focuses on sustainable supply chain management and green procurement, qualification in purchasing, Procurement 4.0 and strategic supplier relationship management. Innovative teaching formats of responsible management education are further main areas of her research. She has published several books and articles on the above-mentioned topics and supports several journals as an external reviewer.

The Principle of Reuse in Circular Economy: Trash Is Treasure

Elisabeth Fröhlich

1 Introduction

Reuse means using a product more than once in its original form (International Organization for Standardization, 2013). The European Circular Economy Action Plan (European Commission, 2024a) also emphasises the reuse strategy to extend the life cycle of products and materials. According to the concept of different R-strategies, reuse is part of the second stage—the medium loop—with a focus on the consumption phase of a global supply chain. The main aim of reuse is to optimise usage and to maintain and lengthen the lifespan of products and their parts (Malooly & Daphne, 2023). Reuse is the utilisation of discarded products that are still in good shape and, therefore, fulfil their original function by different end users. Reuse takes many forms, from the second-hand market to the standardisation of products such as plastic packaging or pallets so they can be used internationally or across industries (Malooly & Daphne, 2023).

"Plastics, in many ways, contribute positively to society. There is, however, a dark side: the way we produce, use and dispose of plastics is polluting ecosystems, creating risks for human and animal health and destabilizing the climate" (United Nations Environment Programme, 2023, p. xii). While in the 1950s just under 1.5 million tonnes of plastic were produced each year, this figure has risen to almost 400 million tonnes today (NABU, 2024): from Lego bricks to yoghurt pots, from garden chairs to fishing nets, from tyres to mobile phones and many more, the world is full of plastic. If we pack all this plastic onto lorries, it would amount to a chain that would stretch three times around the Earth (BMUV, 2024). These lesser-known facts about plastic provide further reasons why the focus of this section is on plastic

E. Fröhlich (✉)
ispira Think Tank for Sustainable Supply Chains, Hürth, Germany
e-mail: lisa.froehlich@ispira-thinktank.com

E. Fröhlich et al. (eds.), *Circular Economy in Sustainable Supply Chains*, CSR, Sustainability, Ethics & Governance,
https://doi.org/10.1007/978-3-031-70749-0_8

waste: "Only 9% of all plastic produced is recycled, a whopping 2 million plastic bags are used every minute worldwide, every minute of every day a truckload of plastic is dumped into the ocean, one million plastic bottles are bought every minute, there could be more plastic than fish in the ocean by 2050, the average person eats 70,000 microplastics each year" (Giving Compass, 2024).

2 Reuse: The Best Way to Get Rid of Plastic Waste

Plastic waste is one of the biggest environmental challenges our society is currently facing. We desperately need innovative business models to deal with this problem, as only a small part of plastic waste is recycled (see microchapter on the principle of recycling). Among the various strategies proposed to mitigate this problem, recent research has emphasised the reuse approach as having the greatest potential for effectively preventing plastic waste. The latest report from the United Nations Environment Programme (2023, p. 20)—"Turning Off the Tap"—assumes that around 70% of plastic pollution can be reduced by introducing reuse, refilling and new delivery models. This focus on reusing materials not only presents a sustainable alternative to the conventional linear economy of produce-use-dispose but also significantly reduces the need for new plastic production, thereby lessening the environmental footprint. A critical barrier to realising the full potential of the reuse strategy is communicating with end consumers. Effective communication strategies are essential for raising awareness, changing consumer buying behaviour and fostering a culture of sustainability. However, despite the recognition of reuse's potential and the importance of consumer communication, the scaling of innovative circular business models remains limited (The Pew Charitable Trusts & Systemiq, 2020). Although there are successful examples of reuse models, they often face challenges in scaling up. These challenges can be caused by a variety of factors, including infrastructure limitations, market acceptance, regulatory barriers and the need for significant upfront investment. These issues will be addressed throughout this chapter and illustrated with examples in order to better understand the importance of the reuse strategy in the context of circular business models to minimise the negative environmental impact of plastic pollution.

To successfully implement a reuse strategy to get rid of plastic waste, some facts related to the current global plastics economy are briefly listed. One of the *structural weaknesses* of a linear plastic packaging supply chain is that 95% of the total value of plastic packaging is lost to the economy after a single-use cycle. Another problem of waste collection lies in the *chronic underfunding* by local authorities. For this reason, a significant share of plastic waste collection is carried out by the so-called informal recycling companies, which is often linked to undignified working conditions and considerable health risks. New business models that can contribute to the reduction of plastic waste have proven to be effective but have not reached the required scale so far. Although innovative solutions could cover the overall demand with a lower environmental impact, the *lack of incentives* prevents the use of these new materials or circular business models (United Nations Environment Programme, 2023, p. 3). Global supply chains of consumer goods usually *do not take into account the local conditions of the waste management infrastructure*. Design and packaging decisions by consumer goods manufacturers do not recognise the different waste infrastructures, which can vary greatly from municipality to municipality. Finally, there is a lack of consistent definitions and standards for the reporting of plastics data. This *lack of data transparency* currently restricts the effective and safe management of plastics throughout their entire life cycle, as many parameters are subject to great uncertainty. There is also a shortage of field data to measure plastic stocks and flows across the whole value chain (The Pew Charitable Trusts & Systemiq, 2020).

Furthermore, it is important to gain awareness of problematic or unnecessary plastic use:

- It is not reusable, recyclable or compostable in practice and at scale [...].
- It contains hazardous chemicals that pose a significant risk to human health or the environment [...].
- It can be avoided (or replaced by a reuse model) while maintaining utility.
- It hinders or disrupts the recyclability or compostability of other items.
- It has a high likelihood of being littered or ending up in the natural environment. (United Nations Environment Programme, 2023, p. 19)

To end plastic pollution by minimising this unnecessary plastic use and solving the problems created by linear plastic supply chains, the reuse strategy plays a major role, as it drives the acceleration of the market for reusable products. To transform the throwaway economy into a reuse society, business models for reuse must become more profitable compared to the market value of single-use plastics. Studies confirm that reuse systems offer the greatest potential for tackling plastic pollution. If the necessary conditions are created, a 30% reduction in plastic waste can be achieved by 2040 (The Pew Charitable Trusts & Systemiq, 2020). Figure 1 shows what such a transformation of plastic supply chains could look like.

Fig. 1 The systems change towards a new circular plastics economy (United Nations Environment Programme, 2023, p. xiv)

Reuse models are defined as innovative delivery schemes that allow individual products (e.g. packaging) to be used multiple times for their original purpose (International Organization for Standardization, 2013). Possible examples include the use of reusable water bottles, food containers and bags or the use of refill systems from dispensers and bulk packaging in the retail sector. Take-back systems with reverse vending machines, deposit refund systems or low-packaging subscription services are further examples of new delivery models (Ellen MacArthur Foundation, 2019, p. 9).

These reuse systems ensure that the resources used in the economy retain a higher value. This prevents the economic value of manufactured goods from being lost after they have been used once. The following aspects need to be considered to ensure that reuse systems can be implemented effectively (United Nations Environment Programme, 2021):

- Collection points must be set up.
- High return rates can only be ensured by granting return incentives.
- Reverse logistics (including washing and cleaning) need to be implemented.
- End consumers must be made aware of these innovative supply models through appropriate labelling and communication.

3 Designing Successful Reuse Business Models

As with all changes towards innovative circular business models, it is crucial to assess and address the potential negative impacts of this shift to the reuse strategy. The following section provides some ideas on how to deal with the challenges outlined previously. First, the four reuse models are briefly discussed (Ellen MacArthur Foundation, 2019, p. 13). When *refilling at home*, consumers replenish their reusable containers at home, for instance, by using refills that are provided via a subscription service. Consumers top up their reusable containers outside of their homes when *refilling on the go*, such as through a dispensing system located within a store. A collection service, such as a logistics company, retrieves packaging directly from homes, which is called *return from home*. *Return on the go* means that consumers bring back the packaging to a designated store or drop-off location, like through a deposit return machine or a designated mailbox.

The Ellen MacArthur Foundation (2019) report "Reuse–Rethinking Packaging" offers a lot of practical examples reflecting the four reuse approaches. Algramõ (https://algramo.com/en/) is an intelligent dispensing system on wheels—to give one concrete example of a "refill on the go" model. Here is how the process unfolds: Algramõ, in collaboration with Unilever and Nestlé, is revolutionising their original in-store vending machine concept, Algramõ 1.0, by integrating it into environmentally-friendly, smart-powered and mobile electric tricycles. These tricycles are designed to make doorstep deliveries directly to consumers. To participate, customers purchase reusable containers and set up an online account. This account facilitates the management of refill credits and accumulates rewards for repeated packaging use, which can be redeemed at the dispensing machines. Appointments for a tricycle visit can be easily scheduled through the Algramõ application, which is accessible on both Android and iOS platforms at no cost. The inaugural pilot project will introduce refill services for selected Unilever and Nestlé products, including Unilever's Omo and Quix homecare items, as well as Nestlé Purina's pet food. Algramõ is enthusiastic about expanding its offerings by incorporating additional brands into its system and tailoring the technology to fit various retail environments, including warehouses, mini-markets and supermarkets (Ellen MacArthur Foundation, 2019, pp. 46–47).

These novel reuse business models can only be scaled up if they are underpinned by local and international legislation. As an example, the European Parliament and Council's Directive from 2019 regarding single-use plastic products and fishing gear sets forth a detailed hierarchy of actions. These actions are specifically designed to take into account the availability of non-plastic alternatives for certain applications, alongside other strategies considered most effective for reducing the pollution from single-use plastic items. Key measures introduced include:

- a significant and on-going reduction goal for the use of single-use plastic versions of beverage cups and food containers;
- a prohibition on a range of widely used single-use plastic items, such as cutlery, plates, straws, stirrers and containers for food and beverages made from expanded polystyrene, as well as items made from oxo-degradable plastics and
- specific product standards for beverage containers that mandate the inclusion of recycled material in plastic bottles and labelling requirements for beverage cups.

The target for the collection of single-use plastic bottles is a collection rate of 77% by 2025 and 90% by 2029 (United Nations Environment Programme, 2021, p. 30).

The European Commission (2024b) has identified communication with consumers as one of the key triggers for the uptake of reuse schemes to reduce single-use plastic waste. To tackle this challenge, the European Commission initiated an awareness campaign to highlight the detrimental effects of single-use plastic products (SUPP) and the significant impact that changes in individual behaviours could have. Crafting a communication strategy was the foundational step. Insights from desk research indicated that while EU citizens, especially the younger demographic, are quite informed about the environmental toll of single-use plastics, this awareness seldom influences their everyday decisions. Given this context, launching another campaign on the same theme was deemed redundant. Instead, a new perspective was adopted, emphasising that single-use plastic products are a common part of our daily lives, yet opting for readily available alternatives to items such as plastic forks, plates, bottles and straws can not only simplify our lives but also help protect our oceans from plastic pollution. The campaign aimed to use humour to challenge the prevalent view of SUPP as convenient, pushing people to reconsider and alter their interaction with these items. An external review validated the campaign's success, highlighting its efficiency, effectiveness, relevance, coherence and value addition. Winning the EU Ombudsman Award, the campaign is celebrated as one of the European Commission's most effective communication efforts (United Nations Environment Programme, 2021, pp. 30–31).

Figure 2 highlights once again that only by implementing innovative circular business models—such as the reuse strategy—in our day-to-day business can we overcome today's ambitious challenges (2024). This chapter attempts to show how

Fig. 2 Reuse can … (Ellen MacArthur Foundation, 2019, pp. 10–11)

plastic waste, one of the biggest ecological problems of our time, can be tackled through the use of an innovative R-strategy, what challenges companies have to deal with and what role politics plays in the successful implementation of circular business models.

References

BMUV. (2024). *Plastikmüll – ein Problem, das uns alle angeht*. https://www.bmuv.de/kids/details/plastikmuell-ein-problem-das-uns-alle-angeht

Ellen MacArthur Foundation. (2019). *Reuse – rethinking packaging*. https://www.ellenmacarthurfoundation.org/reuse-rethinking-packaging

European Commission. (2024a). *Circular Economy Action Plan*. https://environment.ec.europa.eu/strategy/circular-economy-action-plan_en

European Commission. (2024b). *Single-use plastics*. https://environment.ec.europa.eu/topics/plastics/single-use-plastics_en

Giving Compass. (2024). *12 facts about plastic pollution you need to know*. https://givingcompass.org/article/10-facts-about-plastic-pollution-you-absolutely-need-to-know

International Organization for Standardization. (2013). *ISO 18603:2013. Packaging and the environment – Reuse*. https://www.iso.org/standard/55871.html

Malooly, L., & Daphne, T. (2023). *R-Strategies for a circular economy*. Circularise. https://www.circularise.com/blogs/r-strategies-for-a-circular-economy

NABU. (2024). *Plastikmüll und seine Folgen*. https://www.nabu.de/natur-und-landschaft/meere/muellkippe-meer/muellkippemeer.html

The Pew Charitable Trusts & Systemiq. (2020). *Breaking the plastic wave: A comprehensive assessment of pathways towards stopping plastic pollution*. https://www.systemiq.earth/breakingtheplasticwave/

United Nations Environment Programme. (2021). *Addressing Single-use Plastic Products Pollution using a Life Cycle Approach*. https://www.unep.org/resources/publication/addressing-single-use-plastic-products-pollution-using-life-cycle-approach

United Nations Environment Programme. (2023). *Turning off the Tap: How the world can end plastic pollution and create a circular economy*. https://www.unep.org/resources/turning-off-tap-end-plastic-pollution-create-circular-economy

Elisabeth Fröhlich is the founder of ispira Think Tank for Sustainable Supply Chains. She invented an innovative business coach concept to prepare companies for the challenges of the different European sustainability laws. Her major aim is to bridge the gap between research and practice in this field to enable companies to implement the sustainable transformation of global supply chains successfully. Prof. Dr. habil. Elisabeth Fröhlich was president of the CBS International Business School in Germany until 2022. She holds a full professorship in Sustainable Procurement and Supply Chains and is an internationally recognised expert in sustainable supply chain management. She is also highly engaged in the field of responsible management education. She serves as a board member of PRME, chairs the PRME Nomination and Governance Committee and is the PRME DACH Chapter Chair. Additionally, she leads the AOM Management Education Division Ambassador Program as vice-chair, leads the Board of JARO Institute and offers online training in the field of sustainable procurement. Her research focuses on sustainable supply chain management and green procurement, qualification in purchasing, Procurement 4.0 and strategic supplier relationship management. Innovative teaching formats of responsible management education are further main areas of her research. She has published several books and articles on the above-mentioned topics and supports several journals as an external reviewer.

The Principle of Repair in Circular Economy: Extending the Lifespan of Products

Kristina Steinbiß

1 Introduction

Repair is crucial to the circularity of material flows since it increases the longevity of products. In an economy designed to be regenerative, repair is better understood as a sociometric practice that reconfigures the interaction between people and things. As repair is one main focus of the transition to a regenerative circular system, it helps limit the damage previously done to the environment. What has already been produced, such as materials or products, is prolonged and sustained because of its capacity to repair, reorganise and purify the environment (van der Velden et al., 2023).

The principle of repair aims to make a defective product usable again in its original function. This is vital, especially with declining product durability and lifetime before reaching obsolescence. Simply put, obsolescence is the state of no longer being useful or in use. In this context, it is important to elucidate what the "optimal lifetime" of a product is. A closer look at the Eionet (2020, pp. 10–14) report reveals three different types of product lifetime: the useful/actual lifetime, the designed lifetime and the optimal lifetime.

- **Useful/Actual Lifetime** The period from the moment a product is sold until it is replaced or disposed of is referred to as its usable or actual lifetime. In addition to the quality of the product, the useful lifetime is greatly impacted by the user—how they use, maintain, repair and reuse the product—as well as societal norms and values. The useful/actual lifetime is covered with the term "product longevity".

K. Steinbiß (✉)
ESB Business School, Reutlingen University, Reutlingen, Germany
e-mail: kristina.steinbiss@reutlingen-university.de

E. Fröhlich et al. (eds.), *Circular Economy in Sustainable Supply Chains*, CSR, Sustainability, Ethics & Governance,
https://doi.org/10.1007/978-3-031-70749-0_9

	Past-oriented (purpose similar)	Future-oriented (purpose different)
Sustaining (materials the same)	**Reconstruction / Restoration**	**Reconciliation**
Remediation (material different)	**Remediation**	**Reconfiguration**

Fig. 1 A categorisation of forms of repair (McLaren et al., 2020, p. 4)

- **Designed Lifetime** The maximum functional life that a manufacturer plans for their product is known as the "designed lifetime", and it is mostly based on the design of the product, materials used, upgradability, after-sale support, etc. It is linked to the term "durability".
- **Optional Lifetime** This refers "to the lifetime when a product's lifecycle has achieved the optimal environmental, social and economic impact, taking into consideration the influencing factors on the product, consumer and socio-economic levels" (Eionet, 2020, p. 12).

In practical application, a product's actual lifetime rarely reaches its designed lifetime nor the upper threshold of the optimal lifetime, as a lot of factors, such as emotional, socio-economic or business conditions, influence the product's real lifetime. Once a product approaches the end of its lifespan, it becomes obsolete. By repairing a product its actual lifespan is increased in order to minimise its sustainable impact and to bring it as close as possible to its optimum lifetime.

In a recent literature review of repair research integrating literature that ran from 2010 to 2019, McLaren et al. (2020, p. 4) identified four categories of repair that are illustrated in Fig. 1.

On the one hand, repair as *reconstruction/restoration* relates to the process of restoring a subject to its former state of functioning and purpose while maintaining an authentic appearance through the use of traditional materials. While repair as *remediation* lowers material and cosmetic requirements, it also keeps the original goal in mind and may even improve functioning. On the other hand, repair as *reconciliation* uses the original materials or components for a new use, with an emphasis on mending the affected social relationships. It has an eye towards the future. Nevertheless, the greatest possibility for transformation and forward-thinking lies in repair as *reconfiguration*, which may incorporate new materials and objectives.

2 Advocating Repairability

Based on the previous classification of repair, this section will focus on advocating the repairability of a product, which is one of the important elements of a sustainable economy.

An initiative that delivers on the European Commission's priority of the green transition, specifically the European Green Deal and its objective of sustainable consumption, is the rules promoting the repair of goods. They aim to transform the EU into a circular economy with its main building blocks—the New Circular Economy Action Plan and the New Consumer Agenda. Both announced that the Commission will promote sustainable consumption by repair and work towards a new "right to repair" (European Commission, 2023).

Three initiatives covering the full life cycle of goods provide the right to repair. First, the proposal for a *Ecodesign for Sustainable Products Regulation* promotes the reparability of products starting in the production phase. Second, the proposal for a *Directive on empowering consumers for the green transition* enables consumers to be informed about the reparability and durability of goods when deciding to purchase them. Third, the proposal for a *Directive on common rules promoting the repair of goods* adds a new dimension: promoting repair in the after-sales context when the goods are used by consumers. Hence, the three initiatives jointly deliver on the right to repair. They will make repairs simpler, more accessible and more affordable. They also intend to send a valuable signal to companies that sustainable business models and investments in repairs pay off.

Aiming to support the Commission's proposal for an EU ecodesign to promote better reparability, the EU is planning a repair index (Spiliotopoulos et al., 2022). Customers may quickly assess a product's reparability with the help of the so-called Product Reparability Scoring System, which assigns a score based on factors like the product's simply navigable design, the equipment needed for repairs, the availability of replacement components and their cost. By 2030, more legislative actions should be taken to transform the EU Repair Index into a sustainability index by improving material knowledge and traceability.

To meet the requirements concerning consumer empowerment, several national and subnational repair funding programmes have been developed that provide comprehensive support for customers to repair their broken products (Repair, 2024). This financial assistance intends to make repairing products easier: simple, low-cost, accessible and more feasible. Repair funding programmes through which, for example, repair vouchers could be implemented are gaining popularity worldwide. By now more than 165,000 repairs were carried out under the programme in France and even over 840,000 in Austria (Repair, 2024), showing that more people have their devices repaired within the framework of these funding programmes. Improving some issues could expand the use of these programmes even further: an easy application process, an appropriate funding amount for the users, and low entry barriers and predictability for repair businesses.

For a deeper insight into different behavioural drivers that extend product life through repair, Parajuly et al. (2024) surveyed repair-minded individuals and identified some of the underlying factors and key parameters driving this behaviour using the example of e-product repair. Only a few selected aspects have been briefly addressed here.

3 Repair Is a Radical Act: The Repair Principle on the Example of Patagonia

The practical example chosen in this chapter is taken from the Patagonia website, where the headline of a 2015 appeal for a planet-friendly New Year's resolution is "Repair is a Radical Act", followed by the call to become a radical environmentalist (Marcario, 2015)!

Patagonia is known as an outdoor clothing company with its mission "We're in business to save our home planet". This starts by looking at their own production and the longevity of the gear Patagonia makes. All products are designed to last, and Patagonia customers are highly encouraged to care for and repair to enhance the longevity of their products. To achieve this, Patagonia has developed a variety of activities aimed at different interest groups, including a programme called "Worn Wear", which was launched in 2013. Some of its impressions are shown in Fig. 2. The programme intends to further reduce the environmental footprint through slowing resource loops with a focus on empowering the customers by making quality products that can be repaired. As repair has been an integral part of Patagonia's business philosophy since the foundation of the company in the 1970s, it has now become one key element of the Worn Wear programme besides reuse (which is described in the microchapter on the principle of reuse):

- Patagonia encourages their customers to repair their items when they break over the purchase of new ones. They organise Repair Events as part of the Worn Wear tour, where customers pick a used garment and repair it themselves or bring something they already own to be fixed for them. By now, they operate the largest garment repair facility in North America, and their goal is to increase the number of repairs fourfold to 100,000 per year over the next five years only in Europe (Bergstolz, 2023).

Fig. 2 Worn Wear programme (Berg-Freunde, 2017; Marcario, 2015)

- Additionally, Patagonia aims to empower their clients to perform self-repair. To do this, they collaborated with iFixit to produce repair guides that include detailed instructions.

The Worn Wear programme as an integral part of the company's philosophy perfectly fits the Patagonia mission of sustainable responsibility. If the active use of clothes can be extended for just nine additional months, according to the Waste & Resource Action Program (WRAP), CO_2 emissions, waste and water footprints are reduced by 20 to 30% each compared to buying a new item (Langenheim, n.d.).

Patagonia is still growing the Worn Wear programme by following recent trends and attracting younger customers to not just repair their old clothes but also to resell their used clothes. Over the last few years, Patagonia's line of fleece, especially vintage and rare styles, has become one of the most sought-after by Gen Z (Barkho, 2020). With all these initiatives, Patagonia is calling on its consumers to buy fewer products, take better care of them and be responsible for the product's lifespan. These activities did not harm the company at all. On the contrary, it has drawn loyal brand advocates.

Patagonia is by no means the only outdoor outfitter to offer a repair service. It is encouraging that there are now various brands that offer repair services.

References

Barkho, G. (2020). *Patagonia is investing in its Worn Wear resale platform*. ModernRetail. https://www.modernretail.co/retailers/patagonia-is-investing-in-its-worn-wear-resale-platform/

Berg-Freunde. (2017). *Reparieren statt wegschmeissen – Das Patagonia Worn Wear Programm*. www.berg-freunde.ch/blog/patagonia-nachhaltigkeit/

Bergstolz. (2023). *Patagonia: 100.000 Reparaturen pro Jahr 2028*. https://bergstolz.de/news/patagonia-100-000-reparaturen-pro-jahr-2028

Eionet. (2020). *Electronic Products and Obsolescence in a Circular Economy* (ETC/WMGE Report 3/2020). https://www.eionet.europa.eu/etcs/etc-wmge/products/etc-wmge-reports/electronics-and-obsolescence-in-a-circular-economy

European Commission. (2023). *Rules promoting the repair of goods. Promoting repair improves the sustainable consumption of goods by consumers*. https://commission.europa.eu/law/law-topic/consumer-protection-law/consumer-contract-law/rules-promoting-repair-goods_en

Langenheim, J. (n.d.). Laundry: Lightening the load. *National Geographic*. https://www.nationalgeographic.com/environment/article/partner-content-laundry-lightening-the-load

Marcario, R. (2015). *Repair is a Radical Act*. Patagonia. https://eu.patagonia.com/de/de/stories/repair-is-a-radical-act/story-17637.html

McLaren, D., Niskanen, J., & Anshelm, J. (2020). Reconfiguring repair: Contested politics and values of repair challenge instrumental discourses found in circular economies literature. *Resources, Conservation & Recycling: X, 8*, Article 100046. https://doi.org/10.1016/j.rcrx.2020.100046

Parajuly, K., Green, J., Richter, J., Johnson, M., Rückschloss, J., Peeters, J., Kuehr, R., & Fitzpatrick, C. (2024). Product repair in a circular economy: Exploring public repair behavior from a systems perspective. *Journal of Industrial Ecology, 28*(1), 74–86. https://doi.org/10.1111/jiec.13451

Repair. (2024). *A comprehensive overview of the current repair incentive systems: repair funds and vouchers.* https://repair.eu/news/a-comprehensive-overview-of-the-current-repair-incentive-systems-repair-funds-and-vouchers/

Spiliotopoulos, C., Alfieri, F., La Placa, M. G., Bracquene, E., Laps, E., Van Moeseke, T., Duflou, J., Dangal, S., Bolanos Arriola, J., Flipsen, B., Faludi, J., & Balkenende, R. (2022). *Product Reparability Scoring System: Specific application to Smartphones and Slate Tablets* (EUR 31057 EN). Joint Research Centre (European Commission). https://op.europa.eu/en/publication-detail/-/publication/0d25c914-d266-11ec-a95f-01aa75ed71a1/language-en

van der Velden, M., Maitre-Ekern, E., & Wanja, D. K. (2023). The role of independent repair in a circular and regenerative economy. *Circular Economy and Sustainability*. https://doi.org/10.1007/s43615-023-00304-y

Kristina Steinbiß holds a professorship in Management with a focus on Marketing at the ESB Business School of the Reutlingen University in Germany, where she teaches primarily in the Industrial Engineering programmes. Her fundamental teaching goal is to enable her students to think and act sustainably while adopting an integrative perspective combining business and technology. Kristina's current research focuses on implementing new business models to accomplish a company's sustainability goals, as well as on influencing and nudging consumers towards sustainable behaviour. She is the author of numerous publications in the above-mentioned fields, including the textbook *Marketing*, which is already available in its second edition. Next to her research activities, she is also enthusiastic about the Theory of Inventive Problem Solving, the TIPS innovation methodology. As a TIPS trainer and coach, she likes to combine this methodology with her marketing approach and adapt it to implement sustainable solutions in day-to-day business.

The Principle of Refurbish in Circular Economy: Products Better Than New

Kristina Steinbiß

1 Introduction

There is currently no standard definition of refurbishment in the literature. Refurbishment is often used in combination or as a synonym for reuse or remanufacture (Bridgestone et al., 2024). In the circular economy, refurbishment refers to the professional general overhaul of products for reuse to extract their maximum value. In the 9R Framework, it is located between repair and remanufacture (Potting et al., 2017, p. 5). While repair is only concerned with restoring broken components of a product, refurbishment is about more.

Products that are returned because they are defective or damaged, nearing the end of their useful life, or purchased wrongly are often sent back to a seller, disassembled into their various parts and cleaned. Wearing parts are replaced, and at the end of the process, a "new" product can be released to the market. Therefore, it appears that the term "refurbish" is most suited for situations in which a large, multi-component product's general structure is preserved, but numerous components are changed or fixed, leading to a total "upgrade" of the product (Reike et al., 2018, p. 247). In addition, a refurbished product usually includes functional testing to guarantee that it is back to near-new functioning quality, as well as correction of cosmetic flaws. Even though components are not reconditioned, refurbishing involves less raw materials, less energy, less pollution and lower prices for customers. In comparison with brand-new items, the production costs and the carbon footprint of refurbished goods are significantly lower. For example, refurbishing smartphones instead of producing new ones can reduce their carbon footprint by 87% (Wallner et al., 2024, p. 617).

K. Steinbiß (✉)
ESB Business School, Reutlingen University, Reutlingen, Germany
e-mail: kristina.steinbiss@reutlingen-university.de

E. Fröhlich et al. (eds.), *Circular Economy in Sustainable Supply Chains*, CSR, Sustainability, Ethics & Governance,
https://doi.org/10.1007/978-3-031-70749-0_10

A study by Fraunhofer Austria (2023) for the fastest-growing marketplace for refurbished products in the DACH region—refurbed—comes to similar conclusions. In a comparison of new and refurbished products, they found that *CO_2 savings* vary across different product categories, ranging from 69% for the refurbishment of a Lenovo Thinkpad to 78% in the case of an Apple iPhone 11. The study also refers to the average *water consumption* for a new Apple iPhone as 12,075 litres, whereas only 1695 litres are used when refurbishing the same device, equivalent to a saving of 86%. The result is even more significant for the Apple MacBook Air, as almost 57,000 litres of water are used for new product production compared to just 5385 litres for refurbishment—a saving of 91%. The Fraunhofer study considers *e-waste* to be the third relevant pillar of potential savings in refurbishment, which is constantly increasing, mainly due to increased consumption of electronic devices, short life cycles and few repair options. E-waste has turned into one of the fastest-growing waste streams in European households. Here too, refurbishment offers enormous savings potential. While the Lenovo Thinkpad can save at least 65% e-waste compared to a new purchase, the Apple MacBook Air offers the greatest savings potential at 80%. Figure 1 briefly illustrates the three savings potentials.

To gain an overview of the refurbishing process, six main steps using the example of refurbed are explained. Looking at the procedure in more detail, these stages can be broken down even further. As a basic overview, however, the following six-step process is sufficient (refurbed, 2024):

1. As soon as the returned product arrives at the seller, old data is deleted, the memory space is overwritten and the device is reset to the factory settings.
2. Technical experts test the device using approved and certified test software to ensure 100% functionality.
3. Components are replaced if necessary. Repairs are carried out by experts; individual parts are replaced with equivalent ones.
4. External signs of use (depending on the optical condition) are removed by polishing and cleaning. If necessary, upgrades to the operating system are carried out.

Savings in		Apple iPhone 11	Apple Macbook Air 2017 13,3"	Lenovo Thinkpad T460 i%
	CO2			
	in %	78%	83%	69%
	in kg	56,3 kg	281,3 kg	318,5 kg
	Water			
	in %	86%	91%	87%
	in Liter	10,380 liter	51,468 liter	67,043 liter
	E-Waste			
	in %	71%	80%	64%
	in g	169 g	1,286 g	1,080 g

Fig. 1 Savings refurbished vs new products (own illustration based on Fraunhofer Austria, 2023)

5. Technical experts or laser machines compare the exterior of the device with the guidelines and standards about optical conditions ("excellent", "very good", "good").
6. Finally, the device is sold and comes with a 30-day trial period and a minimum 12-month warranty.

2 Trends in the Refurbishment Market

In their studies on the factors affecting consumer behaviour when purchasing refurbished products, Alyahya et al. (2023, pp. 1–2) referred to the following: by 2025, the refurbished products market is estimated to generate sales close to $150 billion, as in 2020, the value of returned goods was around $428 billion and continues to rise significantly. This trend is being further stimulated by the steady price increases and the supply bottlenecks for new components and products, which have persisted for months and continue to worsen (metoda, 2022, p. 5).

Similar data is provided by the European Refurbishment Association (EUREFAS, 2024). They revealed that the refurbishment market in Europe is currently worth around €5.2 billion and is expected to surge from $27 billion globally (in 2019) to $67 billion in 2023. One of the most traded products on the refurbishment market is second-hand smartphones. They represent up to 10% of total smartphone sales in countries such as France or Germany, with an expected annual growth rate of around 10% in the upcoming years. The trend to buy refurbished products is particularly noticeable among young consumers seeking increasingly responsible behaviour (EUREFAS, 2024).

A trend that has long been taken for granted by many consumers is now making its way into the business sector: instead of buying new smartphones, tablets or PCs, more and more companies are turning to refurbished devices. This is the key finding of a study by the digital association Bitkom. According to Bitkom (2024), one in seven companies in Germany already uses refurbished IT (15%). In 2002, by contrast, only 4% of companies reported using refurbished laptops, monitors or smartphones. Today, an additional 15% (2022: 13%) of companies are at least considering the use of refurbished IT.

3 Philips Sustainability Campaign: Refurb Is the Future

On 5 June 2023—World Environment Day—Royal Philips, a global leader in health technology, launched its "Better Than New" campaign to create awareness for Philips' refurbished products (Fig. 2). To meet the demands of an increasingly sustainable customer base, enhance the company's environmental commitment and promote a more circular economy, Philips aims to provide goods that can be refurbished. To be precise, this campaign intends to challenge the assumption that

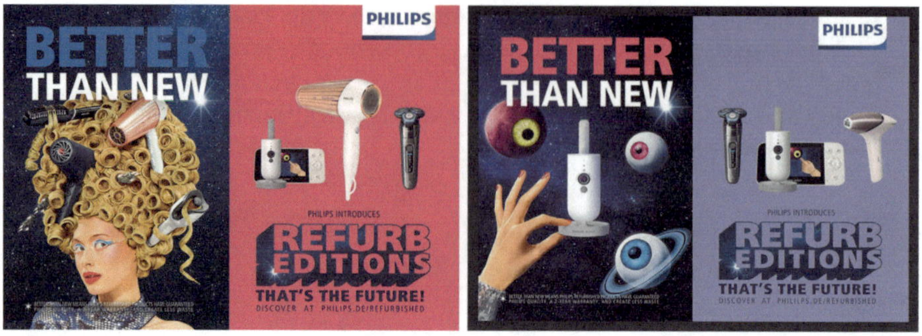

Fig. 2 "Better Than New" campaign: Philips Sustainability—Refurb is the future (Philips, 2023)

"new = better" by educating consumers about refurbished products as a sustainable and cost-effective alternative to new products (Philips, 2023).

Germany has been the first market to launch the brand's campaign. On 20 June 2023, a distinctive pop-up store opened in Hamburg's bustling Rindermarkthalle St Pauli, beckoning visitors to "come in and discover the future". The store offered refurbished products known as Philips Refurb Editions, which boast the same certified high-quality and two-year worldwide Philips warranty as every new Philips product. In comparison with the new products, however, Philips Refurb Editions are saving customers money and lessening their environmental impact (Philips, 2024a). The prices of refurbished Philips products are at least 10% below the recommended retail price (Philips, 2024b).

With the refurb campaign, Philips is sensitising consumers to sustainable alternatives when buying products. Damla Saricoban, Head of Marketing Personal Health at Philips DACH, gave the following statement in a press release (Philips, 2023):

> With our 'Better Than New' campaign, we want to change the mindset of consumers in the long term and show them new purchasing options. Philips stands for the development of innovations that improve people's lives. We are convinced that such innovations are only truly good if they aim to improve life and protect the environment in equal measure. Especially in times like these, when many people are concerned about climate change and at the same time about their money, a new impetus is needed. With the idea behind 'Better Than New', we want to show that refurbished products have many advantages over new products. With our initiative, we are making sustainable purchasing decisions accessible to everyone and are working to achieve this goal. This event is one of many steps to create long-term awareness and mindfulness for the purchase of refurb products.

The Philips refurbishment process is transparent: after having the returned products carefully inspected for possible damage, they are either dismantled into their reusable individual parts or given a general overhaul. Those products selected for refurbishment are thoroughly cleaned and subjected to several tests to ensure that they are fully functional and meet the quality standards. If necessary, repairs are carried out and all components are replaced with original Philips parts. Finally, the refurbished product is thoroughly cleaned again and put back into the original packaging where possible. If this is damaged, recycled packaging is used.

According to the company's webpage, refurbishing has so far helped to ensure that many of its products could continue to be used and did not have to end up in the recycling pile. In 2022 alone, Philips prevented 139 tons of electronic waste from being generated through refurbishing—that's more than 39,000 refurbished Philips products finding a second life.

References

Alyahya, M., Agag, G., Aliedan, M., & Abdelmoety, Z. H. (2023). Understanding the factors affecting consumers' behaviour when purchasing refurbished products: A chaordic perspective. *Journal of Retailing and Consumer Services, 75*, Article 103492. https://doi.org/10.1016/j.jretconser.2023.103492

Bitkom. (2024). *Mehr Unternehmen nutzen Refurbished-IT.* https://www.bitkom-research.de/news/mehr-unternehmen-nutzen-refurbished-it

Bridgestone, Dell, ecobee, General Electric, HP, Lexmark, Philips, & Renault. (2024). *Remanufactured & Refurbished Parts: Busting myths surrounding their impact on new product sales.* https://emf.thirdlight.com/link/y7a02gbqpz5d-w1unbp/@/preview/1?o

EUREFAS. (2024). *Refurbishment Industry. European Refurbishment Association.* https://www.eurefas.com/refurbishment_industry/

Fraunhofer Austria. (2023). *Sustainability Impact Measurement Refurbed GmbH.* https://www.fraunhofer.at/content/dam/austria/images/presse-news/pressemitteilungen/refurbed-05(2023/2023_Fraunhofer%20Austria%20Sustainability%20Impact%20Measurement%20Report%20DE.pdf

metoda. (2022). *Der Market für Refurbished Produkte in Deutschland. Eine Studie der metoda GmbH im Auftrag der eBay Marketplaces GmbH.* https://metoda.com/wp-content/uploads/2022/07/Refurbed-Studie-2022-eBay-4.pdf

Philips. (2023, June 5). *Besser als Neu: Philips launcht crossmediale Refurb Kampagne – weniger Abfall durch Generalüberholung der Produkte* [Press release]. https://www.philips.com/c-dam/corporate/newscenter/de/press-releases/2023/202306-philips-startet-crossmediale-refurb-Kampagne/Pressemitteilung-Philips-Besser-als-Neu-refurb-Kampagne.pdf

Philips. (2024a). *Campaign for refurbished products highlights Philips' increasing focus on circular solutions.* https://www.philips.com/a-w/about/news/archive/standard/news/articles/2023/20230706-campaign-for-refurbished-products-highlights-philips-increasing-focus-on-circular-solutions.html

Philips. (2024b). *Wie werden die Produkte von Philips generalüberholt?* https://www.philips.de/c-f/XC000019525/wie-werden-die-produkte-von-philips-general%C3%BCberholt

Potting, J., Hekkert, M., Worrell, E., & Hanemaaijer, A. (2017). *Circular economy: Measuring innovation in the product chain. Policy Report.* PBL Netherlands Environmental Assessment Agency. https://www.pbl.nl/uploads/default/downloads/pbl-2016-circular-economy-measuring-innovation-in-product-chains-2544.pdf

refurbed. (2024). *Umweltschutz bei refurbed.* https://nachhaltigkeit.refurbed.de/#lp-pom-block-323

Reike, D., Vermeulen, W. J. V., & Witjes, S. (2018). The circular economy: New or refurbished as CE 3.0? — Exploring controversies in the conceptualization of the circular economy through a focus on history and resource value retention options. *Resources, Conservation & Recycling, 135*, 246–264. https://doi.org/10.1016/j.resconrec.2017.08.027

Wallner, T. S., Haslbeck, J. M. B., Magnier, L., & Mugge, R. (2024). A network analysis of factors influencing the purchase intentions for refurbished electronics. *Sustainable Production and Consumption, 46*, 617–628. https://doi.org/10.1016/j.spc.2024.03.009

Kristina Steinbiß holds a professorship in Management with a focus on Marketing at the ESB Business School of the Reutlingen University in Germany, where she teaches primarily in the Industrial Engineering programmes. Her fundamental teaching goal is to enable her students to think and act sustainably while adopting an integrative perspective combining business and technology. Kristina's current research focuses on implementing new business models to accomplish a company's sustainability goals, as well as on influencing and nudging consumers towards sustainable behaviour. She is the author of numerous publications in the above-mentioned fields, including the textbook *Marketing*, which is already available in its second edition. Next to her research activities, she is also enthusiastic about the Theory of Inventive Problem Solving, the TIPS innovation methodology. As a TIPS trainer and coach, she likes to combine this methodology with her marketing approach and adapt it to implement sustainable solutions in day-to-day business.

The Principle of Remanufacture in Circular Economy: Same as New Product

Kristina Steinbiß

1 Introduction

Producers are becoming more and more important to the well-being of our planet in a society where economic growth and consumption are prioritised by many. One approach that businesses can use to address and lessen the environmental impact of their products is remanufacturing. In remanufacturing, producers take used products back and utilise an industrial process to return them to a condition that is like-new or even better than new (Lindkvist Haziri & Sundin, 2020, pp. 57–58).

According to King et al. (2006, p. 261), remanufacturing is the process where "used products are brought at least to original equipment manufacturer (OEM) performance specification from the customer's perspective and, at the same time, are given warranties that are equal to those of equivalent new products". Bringing old products to like-new condition while maintaining an equivalent guarantee requires disassembling, cleaning and inspecting the product completely, as well as replacing obsolete or damaged parts of the product (Nasr, 2024). Mainly, remanufacturing is useful for complex mechanical and electro-mechanical parts whose cores have the potential to add significant value when recovered, both in terms of their market worth and their initial cost (King et al., 2006, p. 261). However, in contrast to the sales market, remanufacturing strategies are not globalised to all fields of production or countries (Lahrour et al., 2019, p. 555). The high complexity of the reverse supply chain (APICS Foundation, 2014, p. 6) is one of the most important factors that block the implementation of remanufacturing by the OEM.

The remanufactured product's life cycle begins at the end of another product's life, as is shown in Fig. 1. Although there is no standard definition of the

K. Steinbiß (✉)
ESB Business School, Reutlingen University, Reutlingen, Germany
e-mail: kristina.steinbiss@reutlingen-university.de

E. Fröhlich et al. (eds.), *Circular Economy in Sustainable Supply Chains*, CSR, Sustainability, Ethics & Governance,
https://doi.org/10.1007/978-3-031-70749-0_11

Fig. 1 Remanufacturing process (own illustration based on Shanshan et al., 2023, p. 6)

remanufacturing process in the literature, it generally comprises six to seven stages, starting with the collection of parts and ending with the resale of the remanufactured product. The sequence of the process chain depends, among other things, on the parts to be reconditioned. For the book on hand, we chose a seven-step remanufacture process (Shanshan et al., 2023, p. 6) that starts with the customer no longer using a product and disposing of it in the sense of a circular economy:

1. The step of core collection includes the need for end-of-life products to be returned to a remanufacturing facility for the process to take place. Wholesalers very often collect their customers' old parts.
2. At the remanufacturing facility, the old product is completely dismantled into its components.
3. All components are carefully inspected to determine their condition for possible reuse. Non-reusable and non-refurbishable components are disposed of and recycled appropriately.
4. In the following cleaning stage, the dismantled parts that have been sorted out for the further process are degreased, de-oiled, de-rusted or stripped of paint. This is done by wet or dry cleaning systems (Lange, 2017, p. 19).
5. The clean parts are refurbished (see microchapter on the principle of refurbish) using typical processes such as drilling, milling, honing, grinding or welding. The method of mechanical reconditioning depends on the part in question and is determined during the process design stage.
6. All parts are thoroughly tested.
7. After the parts have been refurbed and tested, the component reassembly can be carried out on the production lines of the OEM into the original product. The refurbished parts and materials are used to create an entirely "same-as-new", remanufactured product. If needed, new parts are included.

Although remanufacturing has a lot of driving forces, such as saving energy and resources, there are also numerous barriers. Many companies are still reluctant to implement remanufacturing due to challenges and uncertainties they face, e.g. effective collection of used products, development of an efficient remanufacturing process including disassembly and reverse logistics, and the incentivisation of the OEMs (Matsumoto et al., 2016, pp. 130–131).

Furthermore, compared with conventional manufacturing, remanufacturing shows not just a high complexity but also a high uncertainty by determining the returned products as work blanks. Hence, the most significant differences between manufacturing and remanufacturing lie primarily in the varying availability and quality of parts and components, which are difficult to predict and time-consuming (Lahrour et al., 2019, p. 556). Components may be corroded, worn, deformed or damaged. In summary, this leads to the general assumption that operation planning, control and management in remanufacturing processes are more complex than in conventional manufacturing (Zhao et al., 2021, p. 2).

2 Design for Remanufacturing

The requirements for remanufacturing should be considered at an early stage of the development and production of products. To make the remanufacture process profitable, a lot of design guidelines exist. Shanshan (2023, pp. 5–10) has assigned requirements to each stage of the process that must be fulfilled to offer effective and efficient remanufacturing. However, the following paragraph only touches on a few aspects to illustrate some challenges of this principle.

Disassembly is not a simple reversal of assembly. Many permanent techniques that have been developed to realise and accelerate the assembly process, such as plugging, pressing, forming, sonic welding and adhesive, can cause problems when disassembling. To enable a flawless remanufacturing process, nonpermanent joints are generally preferred over adhesives since they are simple to loosen, and the components are expected to be separated without being damaged or causing damage to other parts of the product. In addition, the use of non-corrosive materials and clear instructions also simplifies the disassembly process (Shanshan et al., 2023, p. 8). In the process step, *inspection and sorting*, parts are divided into three categories: reusable without or with refurbishing and not reusable. To ease the sorting, parts performing the same function should have identical features. For instance, the components could be constructed using colour-coding or bar-coding methods to make them easier to distinguish from one another (Shanshan et al., 2023, p. 8). As *cleaning* is the most energy- and labour-intensive process in remanufacturing, it is important to consider it during design. With the use of easy-cleaning materials and structures that require fewer cleaning methods, the process can be simplified (Shanshan et al., 2023, p. 9).

In their research, Lindkvist Haziri and Sundin (2020, p. 60) have identified a total of nine categories that influence the most appropriate design of the remanufacturing process: ease of identification, ease of verification, ease of access, ease of handling, ease of separation, ease of securing, ease of alignment, ease of stacking and wear resistance. As the main goal in remanufacturing is part reuse, it is most effective to already incorporate these requirements in the design phase when developing new products.

3 The Remanufacture Principle Using the Example of mtu Engines from Rolls-Royce Power Systems

It is natural to think twice about throwing away a high-quality asset whose "service life" is coming to an end in order to replace it with an expensive new product. In the case of large, powerful diesel engines, it has been common practice for decades to dismantle, assess, repair, reassemble and test them to put them back into operation. But these are one-off repairs.

The Power Systems division of the British technology group Rolls-Royce has been pursuing the remanufacturing approach for its mtu brand combustion engines for a relatively long time, in addition to the usual repair services. This involves the standardised reconditioning of engines in predefined industrial processes. The company's Reman Technology Centre was officially opened in Magdeburg (Saxony-Anhalt, Germany) in 2013. Today, the facility is a prime example of the circular economy in the field of sophisticated off-highway drives and energy systems. A second remanufacturing plant is currently being built at the Aiken site (USA).

The products to be refurbished include high-speed diesel engines or complete power generators with an output of up to 3.5 megawatts. The engines, which weigh several tonnes and often cost large six-figure sums when new, are used to power, for example, commercial ships, luxury yachts and locomotives. They are also used to drive emergency power generators to safeguard safety-critical infrastructure, such as data centres, airports, water supplies and hospitals. Magdeburg also refurbishes the so-called railway power packs—integrated drive systems for multiple units consisting of an engine, gearbox, electrics, hydraulics, cooling and control system.

Remanufacturing is particularly suitable for products that are used in large quantities, making the development of standardised processes worthwhile. In addition, the products must not have exceeded the predefined operating time, so they must have potential for remanufacturing.

During disassembly, the components are assessed to determine whether they are suitable for remanufacturing or need to be replaced with new components. All components—from connecting rods to pistons, add-on parts, crankshafts and crankcases weighing several tonnes—go through this process. Once all components have been reconditioned or replaced and reassembled, the customer has a like-new product at a significantly lower price than a new equivalent would cost and with a full two-year warranty. In contrast to a regular repair, the customer does not necessarily get all their components back, but they do get completely equivalent components. For instance, a remanufactured piston can also come from a different engine. Figure 2 illustrates how an engine is disassembled at mtu and the before-and-after look of the remanufactured engine.

This all happens in a comparatively short time. Once the core and peripheral processes have been synchronised, a large diesel engine, for example, is ready for use again just a few weeks after delivery, depending on customer requirements and scope.

Remanufacturing not only saves time and money but is also an integral part of the corporate strategy for circular economy and on the way to becoming climate-neutral.

Saving CO_2 is practically part of the business model here. Around 600 engines and systems are reconditioned at the Magdeburg site every year. This saves 3500 tonnes of CO_2 per year. In addition to the remanufactured engine, the customer receives an emissions certificate. The majority of components can also be

Fig. 2 Disassembling an engine at mtu and engine before and after

remanufactured several times: depending on the engine, the degree of remanufacturing is between 30% and 70%.

References

APICS Foundation. (2014). *Examining remanufacturing in supply chain and operation mangement.* http://www.apics.org/docs/default-source/industry-content/apics_reverseman_report_full_version.pdf

King, A. M., Burgess, S. C., Ijomah, W., & McMahon, C. (2006). Reducing waste: Repair, recondition, remanufacture or recycle? *Sustainable Development, 14*(4), 257–267. https://doi.org/10.1002/sd.271

Lahrour, Y., Brissaud, D., & Zwolinski, P. (2019). The strategy for implementing remanufacturing process in a commercial enterprise, the case study of a French company. 26th CIRP Conference on Life Cycle Engineering (LCE) Purdue University, West Lafayette, IN, USA May 7-9, 2019 [Special issue]. *Procedia CIRP, 80*, 554–559. https://doi.org/10.1016/j.procir.2019.01.046

Lange, U. (2017). *Kurzanalyse Nr. 18: Ressourceneffizienz durch Remanufacturing – Industrielle Aufarbeitung von Altteilen.* VDI Zentrum Ressourceneffizienz. https://www.ressource-deutschland.de/fileadmin/user_upload/1_Themen/h_Publikationen/Kurzanalysen/VDI_ZRE_Kurzanalyse_18_Remanufacturing_bf.pdf

Lindkvist Haziri, L., & Sundin, E. (2020). Supporting design for remanufacturing - A framework for implementing information feedback from remanufacturing to product design. *Journal of Remanufacturing, 10*, 57–76. https://doi.org/10.1007/s13243-019-00074-7

Matsumoto, M., Yang, S., Martinsen, K., & Kainuma, Y. (2016). Trends and research challenges in remanufacturing. *International Journal of Precision Engineering and Manufacturing-Green Technology, 3*, 129–142. https://doi.org/10.1007/S40684-016-0016-4

Nasr, N. (2024). *Remanufacturing and the circular economy*. Ellen MacArthur Foundation. https://emf.thirdlight.com/link/mnxp9x4dat3w-h9yjw4/@/preview/1?o

Shanshan, Y., Ong, S. K., & Nee, A. Y. C. (2023). *Remanufacturing and remanufacturability assessment for the circular economy. A solutions guide*. CRC Press.

Zhao, J., Xue, Z., Li, T., Ping, J., & Peng, S. (2021). An energy and time prediction model for remanufacturing process using graphical evaluation and review technique (GERT) with multivariant uncertainties. *Environmental Science and Pollution Research*. Advance online publication. https://doi.org/10.1007/s11356-021-13438-z

Kristina Steinbiß holds a professorship in Management with a focus on Marketing at the ESB Business School of the Reutlingen University in Germany, where she teaches primarily in the Industrial Engineering programmes. Her fundamental teaching goal is to enable her students to think and act sustainably while adopting an integrative perspective combining business and technology. Kristina's current research focuses on implementing new business models to accomplish a company's sustainability goals, as well as on influencing and nudging consumers towards sustainable behaviour. She is the author of numerous publications in the above-mentioned fields, including the textbook *Marketing*, which is already available in its second edition. Next to her research activities, she is also enthusiastic about the Theory of Inventive Problem Solving, the TIPS innovation methodology. As a TIPS trainer and coach, she likes to combine this methodology with her marketing approach and adapt it to implement sustainable solutions in day-to-day business.

The Principle of Recycling in Circular Economy: Transforming Waste

Maud H. Schmiedeknecht

1 Introduction

Within the R-Framework, recycling is part of the long loops. In other words: "recycling begins at the end—the 'get rid' stage of a product's lifecycle" (Ellen MacArthur Foundation, 2021a).

Various recycling definitions exist: the European Union, for instance, defines recycling as "any recovery operation by which waste materials are reprocessed into products, materials or substances whether for the original or other purposes. It includes the reprocessing of organic material but does not include energy recovery and the reprocessing into materials that are to be used as fuels or for backfilling operations" (European Commission, 2008, p. 5). This definition emphasises the importance of recycling in the circular economy by highlighting the transformation of waste materials into valuable resources for continued use. Similarly, the Ellen MacArthur Foundation (2021a) describes recycling as "the action or process of converting waste into reusable material". The International Organization for Standardization (ISO) defines the recycling process in the ISO standard 18604 on packaging and the environment as a "physical or chemical process which converts collected and sorted used packaging, together in some instances with other material, into secondary (recycled) raw materials, products, or substances, excluding energy recovery and the use of the product as a fuel" (ISO, 2013a). This definition highlights the structured and organised nature of the recycling process, emphasising the need for control and documentation to ensure the efficient transformation of waste into valuable resources. All definitions emphasise the transformation of waste materials into reusable products or materials, while excluding energy recovery and the

M. H. Schmiedeknecht (✉)
ESB Business School, Reutlingen University, Reutlingen, Germany
e-mail: maud.schmiedeknecht@reutlingen-university.de

E. Fröhlich et al. (eds.), *Circular Economy in Sustainable Supply Chains*, CSR, Sustainability, Ethics & Governance,
https://doi.org/10.1007/978-3-031-70749-0_12

use of waste as a fuel. This common thread underlines the importance of converting waste into reusable materials and products to reduce environmental impact and promote a circular economy.

Targets are crucial in driving the transition towards a more sustainable and circular economy. According to the report from the World Business Council for Sustainable Development (WBCSD, 2022) titled "A 'Paris Agreement' for Recycling the Earth's Resources", worldwide currently 25–35% of the waste streams, based on both value and volume, are recycled from the industries concrete, metals, biowaste, wood, paper, plastics, e-waste, electric vehicle batteries and others (rubber, packaging glass, nylon). They estimated that, on average, an additional 55% of all the waste streams must be recycled by 2040 to keep from depleting the Earth's resources. Hence, a total average recycling rate of 80–90% for all materials is needed by 2040 (WBCSD, 2022). Government bodies, as one of the requests of the report, must create strong regulations and incentives to promote recycling. This should involve setting minimum required recycled content or recycling rates, as well as implementing extended producer responsibility (EPR) policies.

The European Union has specific targets for the reduction of waste generation and recycling rates in the Waste Framework Directive (WFD). The directive, for instance, aims for the preparation for re-use and the recycling of municipal waste to be increased to a minimum of 55% by 2025, 60% by 2023 and 65% by 2035 by weight (European Commission, 2018).

Furthermore, the concept of EPR is introduced in the WFD. In Article 8, the intention for establishing the EPR and some measures is described (European Commission, 2008, p. 10):

> In order to strengthen the re-use and the prevention, recycling and other recovery of waste, Member States may take legislative or non-legislative measures to ensure that any natural or legal person who professionally develops, manufactures, processes, treats, sells or imports products (producer of the product) has extended producer responsibility. Such measures may include an acceptance of returned products and of the waste that remains after those products have been used, as well as the subsequent management of the waste and financial responsibility for such activities. These measures may include the obligation to provide publicly available information as to the extent to which the product is re-usable and recyclable.

It holds producers accountable for the management of the products they bring to the market, encouraging them to design products with reusability and recyclability in mind. The directive also outlines requirements for waste collection, sorting and recycling. It addresses the management of specific waste streams, such as electronic waste and hazardous waste, ensuring proper treatment and disposal to reduce environmental and health risks.

In addition to municipal waste, the EU also established targets for the recycling of packaging waste. It sets a recycling target of 65% for packaging waste by 2025 and 70% by 2030, published in the EU Directive on packaging and packaging waste (European Commission, 2024). Moreover, it outlines specific targets for recycling various packaging materials, such as glass, paper and cardboard, metals, wood and plastics. These targets provide clear benchmarks for EU member states and businesses to work towards, guiding their efforts in waste reduction and recycling. Meeting these targets not only contributes to environmental protection but also supports the shift towards a more resource-efficient and circular economy, in line with the broader goals of the WFD.

The WFD of the EU outlines the waste hierarchy, which provides a prioritised approach to waste prevention and management legislation and policy: (a) prevention, (b) preparing for re-use, (c) recycling, (d) other recovery (e.g. energy recovery) and (e) disposal (European Commission, 2008). Recycling holds a position within the waste hierarchy due to its environmental and resource-conservation benefits. Through recycling, materials are diverted from landfills and reintroduced into the production cycle, contributing to the conservation of natural resources and reducing the environmental impact associated with raw material extraction and production.

Nevertheless, as Alexandre Lemille (2019), co-founder and general secretary of the African Circular Economy Network, stated on the World Economic Forum platform: "In a properly built circular economy, one should rather focus on avoiding the recycling stage at all costs. It may sound straightforward, but preventing waste from being created in the first place is the only realistic strategy".

2 Circular Economy Standards and Recycling Practices: Opportunities and Challenges

Recycling is a component of circular systems, as it helps to close the material loop and maximise resource efficiency. It provides opportunities for recovering valuable materials from waste, reducing the need for virgin resources and minimising the environmental impact of production and consumption processes.

But why is it so challenging to increase the recycling efforts? This is demonstrated by packaging and packaging waste practices. The complexity influencing the assessment of the recyclability of packaging is multifaceted and encompasses various factors (DIN, DKE and VDI, 2023), including:

- diverse packaging materials, such as plastics, paper, cardboard, natural materials, aluminium and glass;
- varied interpretation of terminology, such as design-4-recycling and recyclability;
- diverse perspectives and priorities through involvement of multiple stakeholders (e.g. plastic industry, waste management systems and non-governmental organisations) and
- various rules, laws, standards, assessment methods and labels in the packaging industry.

Navigating the landscape of rules, laws, standards, assessment methods and labels is essential and currently challenging for organisations committed to promoting sustainable packaging practices.

Standards play a critical role in guiding the organisations towards more sustainable practices. ISO is one globally recognised player in this standardisation field. In 2013, for instance, the ISO published ISO 18604:2013 in the field of packaging and recycling, which "specifies the requirements for packaging to be classified as recoverable in the form of material recycling while accommodating the continuing development of both packaging and recovery technologies and sets out procedures for assessment of meeting its requirements" (ISO, 2013a). Additionally, the ISO published in the same year ISO/TR 17098:2013, a non-exhaustive overview of substances and materials which may impede recycling and offers essential information for identifying and mitigating factors that hinder the recycling process (ISO, 2013b). However, it is important to note that these examples are limited by the fact that recycling operations can differ regionally, technology is constantly evolving and the intended use of the recycled material also determines whether the presence of certain substances and materials poses a problem (ISO, 2013b).

Currently, in response to the growing global interest in promoting circular economy practices, the ISO has initiated the development of the ISO 59000 series on circular economy (status: under development):

- ISO 59004 Circular Economy—Vocabulary, principles and guidance for implementation (ISO, 2024a)
- ISO 59010 Circular Economy—Guidance on the transition of business models and value networks (ISO, 2024b)
- ISO/DIS 59040 Circular Economy—Product circularity data sheet (ISO, 2024c)
- ISO 59020 Circular Economy—Measuring and assessing circularity performance (ISO, 2024d)
- ISO/DIS 59014 Environmental management and circular economy—Sustainability and traceability of secondary materials recovery—Principles, requirements and guidance (ISO, 2024e)
- ISO/CD TR 59031 Circular Economy—Performance-based approach—Analysis of cases studies (ISO, 2024f)
- ISO/TR 59032 Circular Economy—Review of existing value networks (ISO, 2024g)

This initiative aims to create a common language and conceptual framework of circular economy based on input from various stakeholders.

Collaboration among stakeholders is crucial in creating robust and interconnected recycling systems that maximise resource recovery and minimise waste. There are challenges in optimising recycling practices that need to be addressed, such as the need for effective waste management infrastructure, proper sorting and collection systems, and the development of new technologies to recycle complex materials. As outlined, standardisation organisations are one important actor. The WBCSD report stresses in their call to action that a wide range of actors will need to be involved to achieve the circularity aspirations of an overall recycling of 80–90% by 2040 (WBCSD, 2022). Table 1 summarises the report's required measures from actors along the value chain.

Table 1 Selection of required recycling measures from actors along the value chain (adapted from WBCSD, 2022, pp. 38–39)

Stakeholder	Measures
Governing bodies	• Enact regulatory measures and impose sanctions to create stronger incentives for circularity and higher recycling rates. • Invest in improved recycling facilities and infrastructure and establish extended producer responsibility (EPR) policies. • Incentivise both businesses and households through taxes and subsidies designed to encourage recycling. • Discourage, reduce and restrict the transboundary movements of hazardous waste. • Increase requirements for industries to extend the lifecycle of materials.
Producers	• Invest in R&D aimed at optimising, recycling and designing products with a long lifecycle. • Use recyclable materials that are easy to disassemble. • Create environment, health and safety (EHS) assurance schemes for secondary materials, with standardised material tracking and traceability platforms for all the materials they use by defining the business case for recycled content and partnering with recyclers to continue investing in circular innovation.
Industry organisations	• Develop harmonised definitions and reporting standards for waste. • Be influential as proponents for increased governance of takeback and collection globally and improved classification of waste at borders through trade facilitation programmes and capacity building.
Investors	• Incentivise circularity in industry ecosystems and encourage collaboration among industry players along the value chain. • Encourage their portfolio companies to adopt closed-loop recycling. • Invest in opportunities to improve collection, sorting and recycling facilities, and infrastructure. • Seek opportunities to invest in innovations that address challenges to increased circularity, e.g. cutting-edge collection mechanisms or sorting and recycling technologies.
Recyclers	• Support the formalisation of their industry and an increase in transparency. • Develop sorting, pre-processing and recycling operations at the regional and global level. • Increase transparency regarding the secondary material demand and supply. • Increase public–private cooperation in the development of effective EPR regulation.
Customers	• Follow at-source waste separation. • Participate in takeback and collection programmes.
Non-governmental organisations (NGOs)	• Build awareness of the need for circularity. • Promote knowledge and consistent application of circular procurement.

3 The Potential of New Technologies in Recycling

How can new technologies, such as artificial intelligence (AI), accelerate the transition to a circular economy? A study by the Ellen MacArthur Foundation (2019, pp. 4–5) highlights that AI can enhance and enable circular economy innovation across industries in the following three ways:

1. Design circular products, components and materialsL: AI can enhance and accelerate the development of new products, components and materials fit for a circular economy through iterative machine-learning-assisted design processes that allow for rapid prototyping and testing.
2. Operate circular business models: AI can magnify the competitive strength of circular economy business models, such as product-as-a-service and leasing. By combining real-time and historical data from products and users, AI can help increase product circulation and asset utilisation through pricing and demand prediction, predictive maintenance and smart inventory management.
3. Optimise circular infrastructure: AI can help build and improve the reverse logistics infrastructure required to 'close the loop' on products and materials by improving the processes to sort and disassemble products, remanufacture components and recycle materials.

Looking closely at the infrastructure, especially the supply chain, one may notice that AI can optimise logistics and supply chain operations, leading to more efficient transportation and processing of recyclable materials (Toorajipour et al., 2021). This can result in reduced costs, improved resource allocation and increased overall efficiency in recycling processes. The integration of AI in recycling also offers the potential for better monitoring and tracking of recyclable materials throughout the entire recycling process. By using AI algorithms, waste management systems can track and monitor the movement of materials from collection to processing, ensuring that they are properly recycled and preventing potential environmental harm, as the following examples show:

- Predictive analytics and data-driven waste sorting: AI can be used to analyse data and predict trends in waste generation, recycling rates and material recovery, enabling better planning and decision-making in the recycling industry. Furthermore, AI-powered sorting machines equipped with sensors and cameras can analyse and categorise various types of recyclable materials with high precision and speed. By leveraging machine learning algorithms, these systems can continuously improve their sorting accuracy, leading to better separation of materials and higher-quality recycled products (Gundupalli et al., 2017).
- Quality control and contamination detection: AI-powered systems can inspect recycled materials for impurities and contamination, ensuring that only high-quality recyclables are processed. By detecting and eliminating contaminants early in the recycling chain, these algorithms contribute to the production of cleaner recycled materials and mitigate potential negative impacts on downstream processes (Ni et al., 2021).
- Integration with robotic systems and automation: AI algorithms can be integrated with robotic systems to automate various tasks in recycling facilities, such as material handling, shredding and packaging. This automation not only improves operational efficiency but also enhances worker safety by minimising manual labour in potentially hazardous environments (Sarc et al., 2019).
- Optimisation of collection routes and schedules: AI algorithms can optimise waste collection routes and schedules based on historical data and real-time

factors, such as traffic conditions and demand fluctuation. This optimisation reduces fuel consumption, minimises carbon emissions and improves the overall logistics of waste collection, leading to cost savings and environmental benefits (Ghahramani et al., 2022).
- Consumer engagement and behaviour analysis: AI can be employed to analyse consumer behaviour and preferences related to recycling, allowing for targeted educational campaigns and personalised feedback. By understanding consumer habits, waste management systems can develop strategies to encourage proper recycling practices, reduce waste generation and increase recycling participation, for instance, through gamification (Hsu & Chen, 2021).

Companies such as ZenRobotics (https://www.terex.com/zenrobotics), Sadako Technologies (https://sadako.es) and AMP (https://ampsortation.com) developed systems that leverage machine learning to identify materials based on visual input, allowing for more effective separation of waste when combined with traditional sorting systems and robotics. In the following, AMP—a company based in Colorado, USA, and founded by Matanya Horowitz in 2014—will briefly be introduced, as AMP is published as a circular example on the Ellen MacArthur Foundation (2021b) platform and has been listed by Forbes (Ohnsman, 2020) in their "AI 50" list as a pioneer in AI and robotics for the recycling industry.

"At AMP, we're modernizing and scaling the world's recycling infrastructure. By applying AI and automation to the challenges of waste sortation, our solutions enable the economical recovery of commodities for the global supply chain" (AMP, 2024a). Through the innovative use of AI and automation, AMP has developed cutting-edge solutions that address the pressing challenges of waste sortation (Cai, 2020). In order to scan and identify different materials, AMP's AI platform uses cameras. Waste streams can be identified, categorised and continuously improved due to the deep learning capabilities.

Founder and CEO Matanya Horowitz (2024) describes his intention and explains why they rebranded from "AMP Robotics" to "AMP" in 2024:

> Since AMP was founded nearly a decade ago, we've grown from a small team attempting to prove the application of artificial intelligence (AI), machine learning, and robotics in recycling, to a company with a fleet of hundreds of automated sorting systems operating across the United States, Canada, Japan, and Europe. We've expanded our product portfolio to include solutions for material characterization, film removal and recovery, and compact spaces in materials recovery facilities, all enabled by AI. Most recently, we've launched a fully-automated, facility-scale sortation solution for waste management customers integrating our AI-powered sortation technology into custom or turnkey facilities that deliver maximum resource recovery with minimal labor and cost. In short, our business has evolved, as has our focus as a company. We're selling more than robots—we're providing solutions that address the waste industry's biggest challenges, and we're ready to introduce a new brand identity that reflects our delivery of AI-powered sortation at scale. Going forward, we're simply AMP.

The mentioned fully-automated, facility-scale solution is branded as "AMP Smart Sortation™ at Facility Scale", in short, "AMP ONE" (AMP, 2024b). Various technologies are being incorporated (AMP, 2024c):

- AMP Vision™ is a modular computer vision system.
- AMP Jet™ is a high-volume air-jet sortation solution.
- AMP Microjet™ is an air-jet sortation for sortation on lower-volume lines.
- AMP Delta™ is a robotic sortation system that processes material.
- AMP™ Vac is a vacuum sortation that removes thin film.

Artificially intelligent robotic sorters are the future of data-driven recycling. With their ability to efficiently and accurately sort through waste materials, they play a crucial role in enhancing recycling rates and minimising environmental impact. Let us continue to watch out for further examples of the transformative power of AI in the recycling industry, paving the way for a more efficient and environmentally conscious approach to waste management.

References

AMP. (2024a). *About*. https://ampsortation.com/about

AMP. (2024b). *AMP ONE™*. https://ampsortation.com/products/amp-one

AMP. (2024c). *Technologies*. https://ampsortation.com/technologies

Cai, K. (2020, November 12). Rise of the recycling robots. *Forbes*. https://www.forbes.com/sites/kenrickcai/2020/11/12/rise-of-the-recycling-robots/

DIN, DKE & VDI. (2023). *Standardization roadmap circular economy*. https://www.din.de/resource/blob/906910/0d691bed63405ae85f281336ed71162c/standardization-roadmap-circular-economy-data.pdf

Ellen MacArthur Foundation. (2019). *Artificial intelligence and the circular economy. AI as a tool to accelerate the transition*. https://www.ellenmacarthurfoundation.org/artificial-intelligence-and-the-circular-economy

Ellen MacArthur Foundation. (2021a). *Recycling and the circular economy: What's the difference?* https://www.ellenmacarthurfoundation.org/articles/recycling-and-the-circular-economy-whats-the-difference

Ellen MacArthur Foundation. (2021b). *Artificial intelligence for recycling: AMP Robotics*. https://www.ellenmacarthurfoundation.org/circular-examples/artificial-intelligence-for-recycling-amp-robotics

European Commission. (2008). *Directive 2008/98/EC of the European Parliament and of the Council of 19 November 2008 on waste and repealing certain Directives* (32008L0098). https://eur-lex.europa.eu/eli/dir/2008/98/oj/eng

European Commission. (2018). *Directive (EU) 2018/851 of the European Parliament and of the Council of 30 May 2018 amending Directive 2008/98/EC on waste* (32018L0851). http://data.europa.eu/eli/dir/2018/851/oj

European Commission. (2024). *Packaging waste. EU rules on packaging and packaging waste, including design and waste management*. https://environment.ec.europa.eu/topics/waste-and-recycling/packaging-waste_en#publications

Ghahramani, M., Zhou, M., Mölter, A., & Pilla, F. (2022). IoT-based route recommendation for an intelligent waste management system. *IEEE Internet of Things Journal, 9*(14), 11883–11892. https://doi.org/10.1109/jiot.2021.3132126

Gundupalli, S. P., Hait, S., & Thakur, A. (2017). A review on automated sorting of source-separated municipal solid waste for recycling. *Waste Management, 60*, 56–74. https://doi.org/10.1016/j.wasman.2016.09.015

Horowitz, M. (2024). *We are AMP: A message from our CEO*. AMP. https://ampsortation.com/articles/we-are-amp-a-message-from-our-ceo

Hsu, C.-L., & Chen, M.-C. (2021). Advocating recycling and encouraging environmentally friendly habits through gamification: An empirical investigation. *Technology in Society, 66*, Article 101621. https://doi.org/10.1016/j.techsoc.2021.101621

ISO. (2013a). *ISO 18604:2013(en) Packaging and the environment — Material recycling.* https://www.iso.org/obp/ui/#iso:std:iso:18604:ed-1:v1:en

ISO. (2013b). *ISO/TR 17098:2013. Packaging material recycling — Report on substances and materials which may impede recycling.* https://www.iso.org/standard/59137.html

ISO. (2024a). *ISO 59004. Circular Economy — Vocabulary, principles and guidance for implementation.* https://www.iso.org/standard/80648.html

ISO. (2024b). *ISO 59010. Circular Economy — Guidance on the transition of business models and value networks.* https://www.iso.org/standard/80649.html

ISO. (2024c). *ISO/DIS 59040. Circular Economy — Product circularity data.* https://www.iso.org/standard/82339.html

ISO. (2024d). *ISO 59020. Circular Economy — Measuring and assessing circularity performance.* https://www.iso.org/standard/80650.html

ISO. (2024e). *ISO/FDIS 59014. Environmental management and circular economy — Sustainability and traceability of secondary materials recovery — Principles, requirements and guidance.* https://www.iso.org/standard/80694.html

ISO. (2024f). *ISO/CD TR 59031. Circular Economy — Performance based approach — Analysis of cases studies.* https://www.iso.org/standard/81183.html

ISO. (2024g). *ISO/TR 59032. Circular Economy — Review of existing value networks.* https://www.iso.org/standard/83044.html

Lemille, A. (2019, November 15). For a true circular economy, we must redefine waste. *World Economic Forum.* https://www.weforum.org/agenda/2019/11/build-circular-economy-stop-recycling/

Ni, D., Xiao, Z., & Lim, M. K. (2021). Machine learning in recycling business: An investigation of its practicality, benefits and future trends. *Soft Computing, 25*, 7907–7927. https://doi.org/10.1007/s00500-021-05579-7

Ohnsman, A. (2020, July 3). AI 50 2020: America's most promising artificial intelligence companies. *Forbes.* https://www.forbes.com/sites/alanohnsman/2020/07/03/ai-50-americas-most-promising-artificial-intelligence-companies/

Sarc, R., Curtis, A., Kandlbauer, L., Khodier, K., Lorber, K. E., & Pomberger, R. (2019). Digitalisation and intelligent robotics in value chain of circular economy oriented waste management – A review. *Waste Management, 95*, 476–492. https://doi.org/10.1016/j.wasman.2019.06.035

Toorajipour, R., Sohrabpour, V., Nazarpour, A., Oghazi, P., & Fischl, M. (2021). Artificial intelligence in supply chain management: A systematic literature review. *Journal of Business Research, 122*, 502–517. https://doi.org/10.1016/j.jbusres.2020.09.009

WBCSD. (2022). *A "Paris Agreement" for recycling the Earth's resources.* https://www.wbcsd.org/Pathways/Products-and-Materials/Resources/A-Paris-Agreement-for-recycling-the-Earth-s-resources

Maud H. Schmiedeknecht is a Professor of Corporate Governance and Corporate Social Responsibility at the ESB Business School of the Reutlingen University in Germany and a Business Design Coach. What drives her is supporting and empowering business leaders and business school students to develop skills and learn methods to (re)design sustainable, circular and regenerative business models. Maud previously worked as a consultant at the global management consulting firm Kearney in Germany and abroad. As a research assistant at the Konstanz Institute for Intercultural Management, Values and Communication of Prof. Dr. Josef Wieland, she undertook consulting and research projects in the fields of corporate social responsibility and compliance and participated in national and international committees, e.g. the German CSR-Forum and ISO 26000 Working Group on Social Responsibility. She was an associate of the Stiftung Neue Verantwortung (now interface), a think tank for digital technologies, politics and society, on the topic of global economic ethic. Her dissertation on the governance of multistakeholder dialogues was awarded the Max Weber Prize for Business Ethics. Maud's current teaching, research and consulting focus is on corporate sustainability management, (social) entrepreneurship and sustainable business model innovation, business transformation and circular economy.

The Principle of Recovery in Circular Economy: Generating Energy

Maud H. Schmiedeknecht

1 Introduction

Recovery is the last R-strategy. This is also in line with the waste hierarchy of the EU, describing a priority order in waste prevention and management legislation and policy, where "other recovery, e.g. energy recovery" comes fourth behind "prevention", "preparing for re-use" and "recycling" (European Commission, 2008, 2018). The fifth place, "disposal"—sending waste to landfill—is the least favoured option. The hierarchy logic is embedded in Article 10 of the EU Waste Directive by stressing that "Member States shall take the necessary measures to ensure that waste undergoes preparing for re-use, recycling or other recovery operations" (European Commission, 2008, p. 17).

Why is the term "other recovery" used? To understand this, it is helpful to look at the definitions provided in the EU Waste Directive: recovery is the overarching term, meaning "any operation the principal result of which is waste serving a useful purpose by replacing other materials which would otherwise have been used to fulfil a particular function, or waste being prepared to fulfil that function, in the plant or in the wider economy" (European Commission, 2008, p. 5). Material recovery and energy recovery ("other recovery") are part of this recovery definition and can be distinguished as followed: material recovery "means any recovery operation, other than energy recovery and the reprocessing into materials that are to be used as fuels or other means to generate energy. It includes, inter alia, preparing for re-use, recycling and backfilling" (European Commission, 2008, p. 5).

Thus, in the following article, recovery refers to "other recovery" with waste as a source of energy, e.g. by incineration of material with energy recovery (Potting et al., 2017) and not waste that is re-used or recycled. Energy, such as electricity, heat,

M. H. Schmiedeknecht (✉)
ESB Business School, Reutlingen University, Reutlingen, Germany
e-mail: maud.schmiedeknecht@reutlingen-university.de

E. Fröhlich et al. (eds.), *Circular Economy in Sustainable Supply Chains*, CSR, Sustainability, Ethics & Governance,
https://doi.org/10.1007/978-3-031-70749-0_13

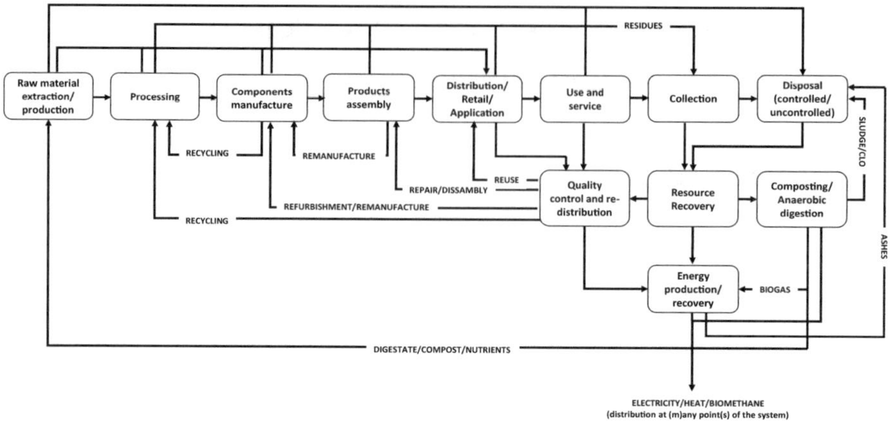

Fig. 1 Resource recovery from waste system (adapted from Iacovidou et al., 2017, p. 915)

steam or biomethane, can be generated through various methods (Iacovidou et al., 2017), as the simplified resource recovery from waste system in Fig. 1 illustrates.

Organic waste, such as agricultural or food waste, can recover energy through composting/anaerobic digestion by producing, for instance, biogas or by utilising the digestate/compost/nutrients as soil conditioners or fertilisers. Other waste, such as municipal and industrial waste, can recover energy through incineration with various existing waste-to-energy technologies (Boloy et al., 2021).

2 Waste Treatment Pathways and the Interconnectedness of the Rs

In alignment with the Paris Agreement, countries worldwide are committed to implementing policies and taking measures to reduce greenhouse gas emissions in all sectors (United Nations, 2016). Waste sector emissions are a key consideration in greenhouse gas inventories, as mentioned in the 2006 IPCC guidelines (IPCC, 2006): the main sources of waste sector emissions included in GHG inventories are solid waste disposal, biological treatment of solid waste, incineration and open burning of waste, and wastewater treatment and discharge. It is important to note that emissions from energy recovery of waste, processing recycled materials and transporting waste are not reported in the waste sector but in the energy, manufacturing and transport sectors.

A study from the European Environment Agency (2024) highlights those policies and measures in the waste sector that can help reduce emissions in other sectors. The following table summarises cross-sectoral impacts and the interconnectedness of waste treatment pathways with waste hierarchy stages and emissions reporting sectors. Recovery, as highlighted in Table 1 in italics, counts mainly to the energy recovery process.

Table 1 Waste treatment pathways and potential cross-sectoral impact (adapted from European Environment Agency, 2024)

Pathways of waste	Options to reduce emissions	Other sectors impacted
Waste Prevention	**Reduce**: measures to reduce waste generation, including extending the lifetime of products. This affects all other pathways.	Waste prevention reduces emissions across all sectors.
Waste Transport	**Optimisation** of waste transport and collection systems.	Measures related to waste treatment pathways may have a knock-on impact on transport of waste, goods and materials. Transport emissions are accounted for in the transport sector.
Processing and Recycling Residues from sorting processes are usually either landfilled or incinerated	**Recycling**: optimisation of sorting and processing to reduce impacts.	Emissions due to fuel/energy needed for the recycling process accounted for in the energy sector. Where recycled materials replace the need for virgin materials, the production of virgin materials and related emission from manufacturing and industry will be reduced (industrial processes and product use (IPPU)/energy).
Biological Treatment Composting and anaerobic digestion contribute methane and nitrous oxide emissions. Mechanical-biological treatment has to be reported under its component processes e.g. sorting and biological processes	***Recovery/recycling****: use biological processes to recover energy from waste. Biological treatment of bio-waste is a recycling process.*	Clean compost/digestate can substitute mineral fertilisers/soil improvers (IPPU/agriculture). However, the output of mechanical-biological treatment is too contaminated to be used as fertiliser or soil conditioner on land.
	Optimise processing and especially reduce methane and nitrous oxide leakage.	Methane from anaerobic digestion can be used as fuel for energy production, replacing alternative fuels.
Incineration CO2 emissions during treatment: • Without energy recovery • With energy recovery (emissions not reported under waste sector)	**Recycling**: optimise separate collection or extraction of recyclables before incineration.	Energy recovery through incineration leads to new/replacement activity data for energy generation.
	Recycling: extract some recyclables from the incineration slag.	Incineration slags can in some cases be treated and replace some virgin (mineral) materials (IPPU/energy).
	Recovery*: optimise energy recovery process.*	Metals extracted from slags can replace some virgin metals—metal sector (IPPU/energy).
	Reduce: prevent especially fossil-based waste materials.	Preventing fossil-based products from being produced can reduce upstream emissions in manufacturing (e.g. plastics).

(continued)

Table 1 (continued)

Pathways of waste	Options to reduce emissions	Other sectors impacted
Landfill Generation of methane during degradation of biodegradable waste	**Recycling**: improve collection, sorting and processing.	Increased recycling may have an impact on waste transfer routes (transport), energy consumption and replacement of virgin materials (IPPU/energy).
	Recovery: *capture methane.*	Methane captured from landfills can be used for energy recovery.
	Disposal: divert waste from landfill to other pathways, pre-treatment to reduce biological activity, closing and remediation of landfills.	Downstream impacts of diversion to other pathways.
	Reduce: especially prevention of biodegradable waste.	Prevention can reduce upstream impacts in manufacturing/agriculture due to decreased production emissions.

The allocation of emissions to different sectors underscores the interconnectedness of waste management with other economic activities. By understanding the specific sources of emissions and their allocation to different sectors, policymakers can better target interventions and initiatives to mitigate the environmental impact of waste.

A report of the Panel for the Future of Science and Technology (STOA) of the European Parliament examines the role of waste management in the context of a circular economy transition and observes the question of how circular economy policy frameworks can be integrated into existing management approaches of the EU member states (European Parliament, 2017). As Table 2 shows, various policy instruments ranging from legislation over economic incentives to voluntary tools are used within the EU on different levels—regional level, national level and EU-wide.

Within the policy instrument legislation, setting targets is one important measurement. In order to strive towards zero-waste approaches, it is necessary—besides recycling targets and landfill reduction targets—to define incineration targets at the lowest level of unavoidable waste. Unavoidable waste is currently not otherwise valuable in other circular economy strategies (non-reusable, non-repairable, non-recyclable, etc.). Morseletto (2020) notes that a realistic target range for waste-to-incineration is 0–10%, with a preference for lower values. Additionally, it is possible to establish further objectives to incentivise a reduction in incineration overcapacity and to penalise it, ensuring that incineration does not compete with other waste management strategies for resources. Under these conditions, incineration can play a positive role in the transition to a circular economy (Morseletto, 2020).

Table 2 Examples of policy instruments used to manage waste (adapted from European Parliament, 2017, p. 22)

Policy instrument	Examples used to manage waste
Legislation	Directives and regulations used to: • set targets for and reporting requirements for individual waste streams (e.g. recycling targets and landfill reduction targets) • establish extended producer responsibility schemes • establish economic instruments • encourage improved eco-design
Economic incentives	Investment in waste collection infrastructure supported through the Cohesion Fund, funding for R&D and innovation
Market-based instruments	Landfill tax and gate fees, incineration tax and fees, plastic bag taxes; Pay As You Throw (PAYT) schemes
Information requirements	Consumer recycling information on packaging, voluntary reporting of waste production and target setting by companies
Voluntary tools	Awareness raising campaigns for the public, voluntary industry commitments, product design and labelling (e.g. through the EU Ecolabel) provision of good practice information, business led initiatives

3 Waste Recovery from Industrial Processes: The Kalundborg Symbiosis

On a technological level, a large number of techniques within industrial processes exist to increase energy recovery, ranging from optimisation of overall energy efficiency and energy recovery over low-temperature flue-gas heat exchangers to special configurations of the water/steam cycle with external power plants (Neuwahl et al., 2019). On a business model level, one innovative waste recovery model within the circular economy field is the eco-industrial parks (IRENA, n.d.). These parks are examples of the industrial symbiosis pattern (Lüdeke-Freund et al., 2022).

Industrial symbiosis is defined by Chertow (2000, p. 314) as engaging "traditionally separate industries in a collective approach to competitive advantage involving physical exchange of materials, energy, water, and by-products. The keys to industrial symbiosis are collaboration and the synergistic possibilities offered by geographic proximity". Lüdeke-Freund et al. (2022, p. 111) highlight that companies in different industries "employ a shared or cascaded use of resources, by-products, and waste materials". Hence, waste or by-products from companies become the inputs for other companies. The goal of these ecosystems of companies is to optimise material flows, reduce waste and share the costs of investment.

In the following, one well-known industrial symbiosis example will be briefly introduced, as it is referred to as good practice of a circular approach to production by the European Circular Economy Stakeholder Platform (2019): the Kalundborg Symbiosis. In 1972, it started with an agreement between the refinery Statoil and Gyproc to supply Gyproc with excess gas from the former's production for drying plasterboard produced in their ovens (Kalundborg, n.d.). Subsequently, Dong Energy (then the Asnæs Plant) was connected to the Statoil water pipe. Over time, more public and private companies became part of this collaboration. Meanwhile,

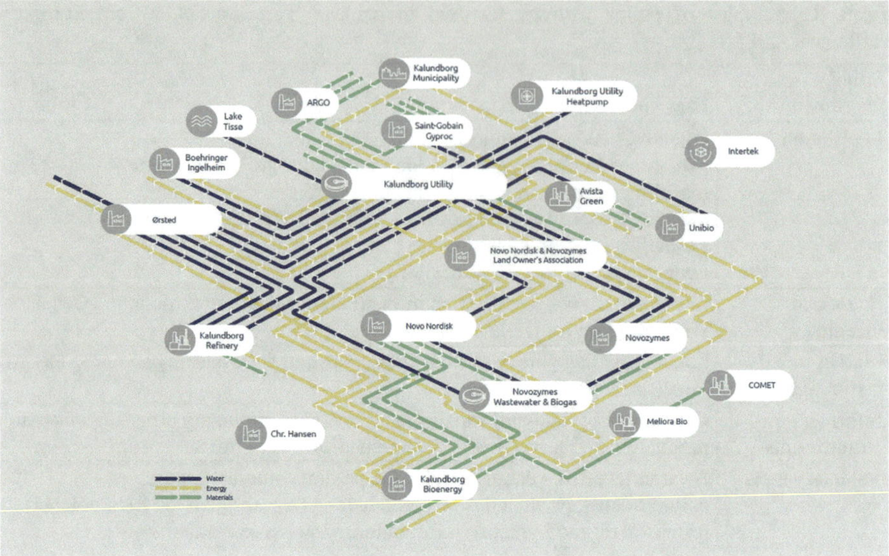

Fig. 2 Exchange of water, energy and materials among partners of Kalundborg Symbiosis (Kalundborg, 2024a)

seventeen companies exchange resources, including Novo Nordisk and Boehringer Ingelheim as pharmaceutical companies, Kalundborg Refinery as a refinery, Avista Green as a refinery for recycling waste oil, Ørsted-owned Asnæs Power Plant as a producer of steam, electricity and heat (converted from coal to biomass in 2019), Meliora Bio as a bioethanol plant, Gyproc as a producer of plasterboards and Kalundborg Municipality. Currently, the symbiosis has around 30 waste streams among these partners (Transition ApS, 2021). Figure 2 illustrates the exchange of water, energy and materials among the participants. For a detailed view of the exchange of resources, check the interactive graphic on Kalundborg's homepage (Kalundborg, 2024a).

In 1996, Kalundborg Symbiosis Centre was established under the auspices of Kalundborg Municipality as an organisation facilitating industrial symbiosis. It is responsible for external communication and with a secretariat managing the facilitation of the symbiosis, e.g. organising the board meetings and fundraisers, securing progress on projects and conducting administrative work. Later, the facilitating organisation was named Symbiosis Centre Denmark. In 2019, Kalundborg Symbiosis decided to restructure itself as a private association (Transition ApS, 2021).

The Kalundborg Symbiosis vision is ambitious as they strive to "be the world's leading industrial symbiosis with a circular approach to production. By symbiosis, we mean a local partnership to source, share and recycle resources for shared and increased value creation" (Kalundborg, 2024b). In this context, knowledge transfer is particularly relevant. Therefore, Kalundborg Symbiosis participates in projects capturing practical advice on how to create and facilitate symbiotic partnerships, such as the "Guide for Industrial Symbiosis Facilitators" (Transition ApS, 2021), and created a "screening tool" to check companies' technical development potential, motivation and readiness (Kalundborg, 2024c).

A systematic literature review on the energy-based industrial symbiosis approaches identified three categories of symbiotic synergies (Fraccascia et al., 2021): (1) energy cascades between two processes, e.g. waste energy produced by one is used by the other, such as in the Kalundborg case, (2) fuel replacement, e.g. by using waste materials instead of traditional fuels in existing fuel-based energy production processes and (3) bioenergy production, e.g. by using organic wastes to produce bioenergy, such as recently integrated in the Kalundborg Symbiosis. Drivers, barriers and enablers of industrial symbiosis have been identified for each of the three categories, as outlined in Table 3.

Drivers are mainly the economic advantages for the participating companies—besides reducing greenhouse gas emissions and preserving natural resources. As the study results show, barriers associated with industrial symbiosis are manifold (Fraccascia et al., 2021). As industries evolve and technologies advance, maintaining symbiotic relationships and adapting to changes in production processes requires on-going monitoring, flexibility and the ability to innovate collaboratively. Financial considerations, especially related to investments in infrastructure and technology

Table 3 Drivers, barriers and enablers of industrial symbiosis for each energy-based category (adapted from Fraccascia et al., 2021, p. 4803)

Category	Drivers	Barriers	Enabler
Energy cascade	• Additional revenues from selling energy (waste energy producer) • Reducing energy costs (waste energy user)	• Geographic distance among companies • Need to adjust business model for energy sources (waste energy user) • Building and managing infrastructures • Need to implement technical changes in production processes • Uncertainty in the amount of produced waste energy	• Economic incentives from the government • Regulations
Fuel replacement	• Reducing waste disposal costs (waste energy producer) • Reducing fuel purchase costs (waste energy user)	• Uncertainty in waste production • Need to implement technical changes in production processes • Regulations on waste-fuel mix	• Economic incentives from the government • Regulations • Technical support from the government
Bioenergy production	• Reducing waste disposal costs • Additional revenues from selling energy	• Bioenergy production chains made by different actors with different interests • Economic feasibility affected by several factors • Access to bioenergy production technologies	• Economic incentives from the government • Regulations

for waste heat recovery and resource exchange, present barriers to the adoption of symbiotic practices. Policy targets or strategies as well as legislative changes are seen as enablers to foster the implementation of industrial symbiosis (Transition ApS, 2021). Harmonising environmental regulations, waste management policies and economic incentives across industries and geographical regions represents an opportunity to create a conducive environment for industrial symbiosis.

In conclusion, while industrial symbiosis offers considerable economic and environmental benefits, its successful implementation relies on addressing multidimensional challenges. Overcoming these obstacles demands proactive collaboration, innovative solutions and supportive policies to foster the widespread adoption of industrial symbiosis within the circular economy framework as one waste recovery model.

References

Boloy, R. A. M., da Cunha Reis, A., Rios, E. M., de Araújo Santos Martins, J., Soares, L. O., de Sá Machado, V. A., & de Moraes, D. R. (2021). Waste-to-energy technologies towards circular economy: A systematic literature review and bibliometric analysis. *Water, Air, & Soil Pollution, 232*, Article 306. https://doi.org/10.1007/s11270-021-05224-x

Chertow, M. R. (2000). Industrial symbiosis: Literature and taxonomy. *Annual Review of Environment and Resources, 25*, 313–337. https://doi.org/10.1146/annurev.energy.25.1.313

European Circular Economy Stakeholder Platform. (2019). *Kalundborg Symbiosis: Six decades of a circular approach to production.* https://circulareconomy.europa.eu/platform/en/good-practices/kalundborg-symbiosis-six-decades-circular-approach-production

European Commission. (2008). Directive 2008/98/EC of the European Parliament and of the Council of 19 November 2008 on waste and repealing certain Directives (32008L0098). https://eur-lex.europa.eu/eli/dir/2008/98/oj/eng

European Commission. (2018). *Directive (EU) 2018/851 of the European Parliament and of the Council of 30 May 2018 amending Directive 2008/98/EC on waste* (32018L0851). http://data.europa.eu/eli/dir/2018/851/oj

European Environment Agency. (2024). *Capturing the climate change mitigation benefits of circular economy and waste sector policies and measures*. https://www.eea.europa.eu/publications/capturing-the-climate-change-mitigation

European Parliament. (2017). *Towards a circular economy – Waste management in the EU*. Panel for the Future of Science and Technology (STOA). https://www.europarl.europa.eu/stoa/en/document/EPRS_STU(2017)581913

Fraccascia, L., Yazdanpanah, V., van Capelleveen, G., & Yazan, D. M. (2021). Energy-based industrial symbiosis: A literature review for circular energy transition. *Environment, Development and Sustainability, 23*, 4791–4825. https://doi.org/10.1007/s10668-020-00840-9

Iacovidou, E., Velis, C. A., Purnell, P., Zwirner, O., Brown, A., Hahladakis, J., Millward-Hopkins, J., & Williams, P. T. (2017). Metrics for optimising the multi-dimensional value of resources recovered from waste in a circular economy: A critical review. *Journal of Cleaner Production, 166*, 910–938. https://doi.org/10.1016/j.jclepro.2017.07.100

IPCC. (2006). Volume 5 Waste. In H. S. Eggleston, L. Buendia, K. Miwa, T. Ngara, & K. Tanabe (Eds.), *2006 IPCC Guidelines for National Greenhouse Gas Inventories Prepared by the National Greenhouse Gas Inventories Programme*. IGES. https://www.ipcc-nggip.iges.or.jp/public/2006gl/vol5.html

IRENA. (n.d.). *Eco-industrial parks and waste heat recovery from industrial processes*. https://www.irena.org/Innovation-landscape-for-smart-electrification/Power-to-heat-and-cooling/32-Eco-industrial-parks-and-waste-heat-recovery-from-industrial-processes

Kalundborg. (2024a). *Partners*. https://www.symbiosis.dk/en/partnerne-bag/

Kalundborg. (2024b). *Our ambitions*. https://www.symbiosis.dk/en/strategi/

Kalundborg. (2024c). *Numbers are worth a thousand words*. https://www.symbiosis.dk/en/inspiration/

Kalundborg. (n.d.). *Surplus from circular production*. https://www.symbiosis.dk/en/

Lüdeke-Freund, F., Breuer, H., & Massa, L. (2022). *Sustainable Business Model Design - 45 Patterns*

Morseletto, P. (2020). Targets for a circular economy. *Resources, Conservation and Recycling, 153*, Article 104553. https://doi.org/10.1016/j.resconrec.2019.104553

Neuwahl, F., Cusano, G., Benavides, J. G., Holbrook, S., & Roudier, S. (2019). *Best Available Techniques (BAT) Reference Document for Waste Incineration* (EUR 29971 EN). Joint Research Centre (European Commission). https://op.europa.eu/en/publication-detail/-/publication/075477b7-329a-11ea-ba6e-01aa75ed71a1/language-en

Potting, J., Hekkert, M., Worrell, E., & Hanemaaijer, A. (2017). *Circular Economy: Measuring Innovation in the Product Chain. Policy Report*. PBL Netherlands Environmental Assessment Agency. https://www.pbl.nl/uploads/default/downloads/pbl-2016-circular-economy-measuring-innovation-in-product-chains-2544.pdf

Transition ApS. (2021). *Guide for industrial symbiosis facilitators*. https://symbiosecenter.dk/wp-content/uploads/2021/03/Guide-for-IS-facilitators_online.pdf

United Nations. (2016). *The Paris Agreement*. United Nations Framework Convention on Climate Change. https://unfccc.int/sites/default/files/resource/parisagreement_publication.pdf

Maud H. Schmiedeknecht is a Professor of Corporate Governance and Corporate Social Responsibility at the ESB Business School of the Reutlingen University in Germany and a Business Design Coach. What drives her is supporting and empowering business leaders and business school students to develop skills and learn methods to (re)design sustainable, circular and regenerative business models. Maud previously worked as a consultant at the global management consulting firm Kearney in Germany and abroad. As a research assistant at the Konstanz Institute for Intercultural Management, Values and Communication of Prof. Dr. Josef Wieland, she undertook consulting and research projects in the fields of corporate social responsibility and compliance and participated in national and international committees, e.g. the German CSR-Forum and ISO 26000 Working Group on Social Responsibility. She was an associate of the Stiftung Neue Verantwortung (now interface), a think tank for digital technologies, politics and society, on the topic of global economic ethic. Her dissertation on the governance of multistakeholder dialogues was awarded the Max Weber Prize for Business Ethics. Maud's current teaching, research and consulting focus is on corporate sustainability management, (social) entrepreneurship and sustainable business model innovation, business transformation and circular economy.

Part II
Theoretical Perspectives of Circular Economy in Global Supply Chains

Circular Economy and Sustainable Supply Chain Management

Elisabeth Fröhlich and Kristina Steinbiß

1 Introduction

Efficient supply chains are characterised by the fact that all processes are coordinated to the minute and interlock in the best possible way. Methods have been developed to further optimise these value creation networks by, for example, reducing inventories or increasing globalisation to fully exhaust price differences in terms of labour, energy or raw material costs. Nevertheless, the focus is on customer satisfaction. Unfortunately, the goals of further optimising and streamlining supply chains are not necessarily in harmony with the goal of meeting customer needs in the best possible manner (Werner, 2020).

2 Fundamentals

Before defining sustainable supply chains in the following section, the fundamental concept of supply chain management must first be discussed. Efficient supply chains are now being replaced by "robust" supply chains. Companies are trying to make these supply chains more resilient and minimise the risks that arise from them (Werner, 2020). The following section will show that sustainability risks are responsible for the largely dysfunctional supply chains. This highlights the importance of the topic of this chapter.

E. Fröhlich (✉)
ispira Think Tank for Sustainable Supply Chains, Hürth, Germany
e-mail: lisa.froehlich@ispira-thinktank.com

K. Steinbiß
ESB Business School, Reutlingen University, Reutlingen, Germany

E. Fröhlich et al. (eds.), *Circular Economy in Sustainable Supply Chains*, CSR, Sustainability, Ethics & Governance,
https://doi.org/10.1007/978-3-031-70749-0_14

2.1 Definition of Supply Chain Management

The fundamental task of supply chain management is to fulfil the orders acquired on the sales market in the agreed quantity, by the agreed date and to the agreed delivery location in the respective supply chain and production networks. There are many possible definitions in the literature. This article uses the definition by Wiendahl (2020) and the CSCMP (2013):

> *Supply chain management* describes an integrated, process-orientated planning and control of all flows of goods, information and money along the entire value chain.

The focus here is on cooperation and coordination with sales partners, suppliers and end customers, as well as intermediaries and third-party service providers, considering all relevant restrictions and capacities. In simple terms, supply chain management takes control of the entire supply and demand within a company and beyond company boundaries. This increasing autonomisation of the individual supply chain stakeholders increases the importance of trust, control and power aspects. For this reason, it is necessary to closely examine the barriers and success factors of sustainable supply chains in Sect. 3.2.

Based on the above definition, a supply chain consists of various stages and actors. Which and how many stages a supply chain comprises depends, on the one hand, on the type of product and, on the other hand, on whether the product is distributed directly or indirectly via intermediaries. As Fig. 1 illustrates, the "classic" supply chain comprises raw material and component suppliers, the manufacturing company, distributors and retailers, as well as the end consumer (Chopra & Meindl, 2016).

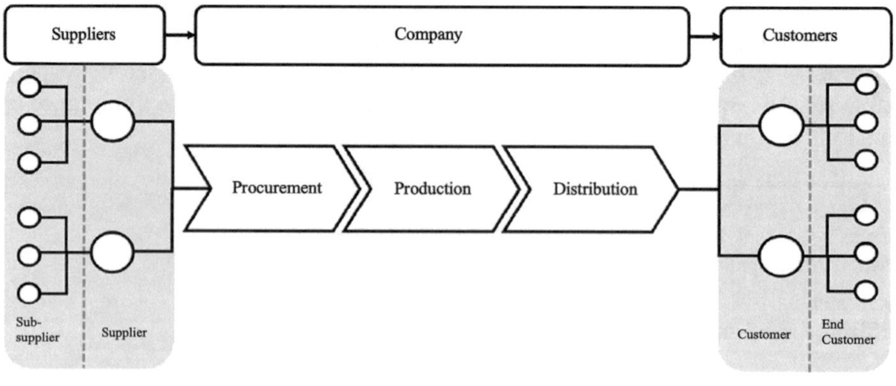

Fig. 1 A classic supply chain (Chopra & Meindl, 2016, p. 15)

2.2 Risks in Global Supply Chains

Risks are defined as negative and positive deviations from an expected target value. In business practice, however, the focus is usually on the negative deviation from the target. This means that risks in supply chains affect not only the manufacturing company but all actors in the value network (Fig. 1). Therefore, these negative deviations can lead to damage within the company as well as for all supply chain actors (Kersten et al., 2008).

The risks discussed below are taken from a not yet published master's thesis on "Sustainable supply chain set-up: What are the characteristics of resilient supply chains? How can sustainable and resilient supply chains be built?" (Schuhmacher, 2022). By analysing the global risk reports from 2020, 2021 and 2022 (WEF, 2020, 2021, 2022), four key risk groups were identified that can lead to volatile supply chains: the impacts of the COVID-19 pandemic, associated global economic and political dependencies, global warming and digitalisation.

Many experts see the cause of the *COVID-19 pandemic* as the unsustainable manner of our business action (Sarkis et al., 2020). Lockdowns, port closures and border closures have not only led to the partial paralysis of large parts of supply chains due to a global shortage of materials but were, ultimately, caused by our unsustainable economic system. The "war for talent", which was already frequently cited before the pandemic, has become a veritable war for qualified employees. The critical situation of the labour market has still not eased and continues to make supply chains less robust. In particular, the *economic divergences between industrialised and developing countries* (especially the problems of the global south) and the associated differences in medical care affect the availability of labour. Low vaccination rates in key procurement markets are promoting disruptions to global supply chains and thus affecting global consumption. The World Economic Forum (WEF, 2021) warned companies of possible commodity shocks, price instabilities and debt crises following an economic recovery. Russia's attack on the Ukraine, which began in 2022, made these predicted risks a sad reality while also leading the global economy straight into the next recession.

The pandemic has not only led to economic disruptions but also to new *political conflicts and dependencies*. At this point, a reference to the intensification of the trade war between the USA and China or the Taiwan crisis should suffice.

Due to *climate change* and the accompanying destruction of nature, it is becoming increasingly difficult for companies to do business successfully. Only 5% of Fortune 500 companies communicate concrete targets and KPIs on biodiversity loss through their business activities. Heat stress will lead to the loss of around 80 million full-time jobs in 2030 (WEF, 2020). This destructive force of corporate behaviour can continue as desired and is one of the main reasons for the numerous disruptions and perceived price distortions in global supply chains.

Digitalisation as a key enabler for sustainable supply chains also harbours major risks due to cyber vulnerabilities. Global supply chain operators are dependent on technology companies and, therefore, third parties. Companies without strong corporate governance in the area of cyber security also suffer considerable reputational damage in the event of a security breach (WEF, 2022).

3 Sustainable Supply Chains: Definition, Meaning and Models

The above statements demonstrate that social and environmental sustainability are among the greatest challenges facing global supply chains. The following section defines sustainable supply chains and discusses their main barriers and drivers. There are countless models of sustainable supply chains, only two of which are highlighted below. The supply chain operational reference (SCOR) model is the most widely cited in the context of sustainable supply chains. The second model is based on a research project by Fröhlich and Steinbiß (2019) and prioritises the integration of the United Nation's Sustainable Development Goals (SDGs).

3.1 Definition of Sustainable Supply Chains

The discussion of potential risks to global supply chains in Sect. 2.2 was deliberately based on the risk categories of the World Economic Forum. This made it clear that the topic of environmental and social challenges must also be anchored in global supply chains long term because, without sustainability, disruption-free and thus resilient supply chains cannot be realised (EY, 2016).

Carter and Rogers (2008) define *sustainable supply chains* as the strategic, transparent integration and achievement of an organisation's social, environmental and economic goals and the systemic coordination of all key intercompany processes to improve the long-term economic performance of the company and its supply chains.

The definition is based on the triple bottom line (Elkington, 1998) and operates with three cornerstones of sustainability, which can be defined as follows (Sadowski, 2010):
As part of the *economic bottom line*, the individual business processes must be optimised to ensure the company's competitiveness. A balance between community and individual interests must be realised by safeguarding knowledge.
Ecological aspects comprise the conservation of natural resources and thus the goal of minimising the use of materials. Climate protection focuses on the use of renewable energies and the avoidance of fossil fuels. Compliance with environmental

protection requirements forms the core of ecological activities in corporate behaviour.
The *social dimension* of sustainable supply chains is currently the focus of legal regulations. For example, compliance with minimum human rights standards is one of the greatest challenges facing global supply chains (see also Sect. 4.1.2).

3.2 Reasons for Establishing Sustainable Supply Chains

Following the brief definition of sustainable supply chains, we must now discuss the *opportunities and drivers*, or rather barriers, that need to be considered when implementing ecologically and socially compliant supply chains.

3.2.1 Opportunities and Drivers of Sustainable Supply Chains

Sustainable supply chain management offers companies a wide range of opportunities: regulatory certainty, reduction of supply chain disruptions, fulfilment of increasing stakeholder expectations, protection and improvement of reputation as well as positive effects on economic performance and competitiveness (Müller & Siakala, 2020).

By implementing environmental and social sustainability measures, companies can increase their competitiveness by reducing costs, improving quality, increasing efficiency and thus increasing productivity. However, the positive effects of a solid sustainability performance throughout the supply chain on the company's image, the willingness of end consumers to buy and the willingness of external investors to invest should also not be forgotten (Müller & Siakala, 2020).

These previously discussed opportunities for establishing a sustainable, global supply chain can be summarised into three key drivers: managing risks, increasing efficiency gains and offering more sustainable products. In order to successfully integrate these drivers, a *culture of responsibility* is required within the company. The joint derivation of sustainability strategies and goals with all relevant stakeholders in the supply chain is a key success factor. These endeavours should be supplemented by training measures for the company's employees (e.g. procurement staff) and the most important strategic suppliers. Training is considered a "release from liability" in the context of the German Act on Corporate Due Diligence (German translation "Lieferkettensorgfaltspflichtengesetz" (LkSG) - term used in this article) in Supply Chains (BMAS, 2022). Its importance for procuring companies should thus not be underestimated. To better *manage risks* in supply chains in the future, a realignment of cooperative supplier-buyer relationships is required above all (see Sect. 4.1.3). In particular, the fulfilment of minimum social standards, such as the avoidance of child labour or compliance with safety standards in production facilities, requires trusting cooperation with suppliers throughout the entire supply chain. One way to realise *efficiency gains* is through the aforementioned joint definition of targets with strategic suppliers to counter the current energy crisis by reducing energy consumption. Organising training courses for suppliers to identify energy-saving opportunities or setting up energy management systems can also help increase efficiency potential. The *offering of sustainable products* can be achieved, among other things, through the procurement of raw materials that explicitly consider social and ecological effects. For example, biological plastics emit relatively fewer greenhouse gases throughout their entire life cycle (BSR, 2010).

3.2.2 Internal and External Barriers

Even though there are many positive drivers and reasons for implementing sustainable supply chains, it should not be forgotten that companies have to contend with internal and external challenges that are difficult to overcome, especially in times of crisis.

The main *internal barriers* are financial hurdles, communication gaps and a lack of know-how. For example, the lack of concrete *knowledge* about the implementation of "clean technologies" inevitably leads to misunderstandings about the supposedly higher costs and risks associated with these investments. They are often seen as a "necessary evil" in order to comply with environmental directives. An additional problem is that machines that can be used beyond their amortisation period contribute significantly to profitability. For this reason, companies are extremely reluctant to invest in new, greener technologies. Ironically, this unwillingness is exacerbated by the *financial performance pressure*, the keyword here being shareholder value. As a result, the profit potential of sustainable technologies often remains untapped. Therefore, the ecological added value of an investment is often overlooked in favour of possible dividend payouts based on the relative strength index (RSI), which measures past financial performance and ignores future earnings or dynamics. Inter-organisational rivalry and *miscommunication* are counterproductive in most cases. If different departments and employees only compete for power and influence instead of contributing to the common goal of embedding sustainability in the organisation through the evaluation and implementation of new and existing technologies, many of the opportunities discussed earlier will remain untapped. This trend is intensified if middle management is not sufficiently involved in sustainability communication (Achillas et al., 2018).

Additionally, one should not disregard *external barriers*. These are often the negative effects of the aforementioned opportunities and drivers if the incentives set do not lead to the desired results. For instance, the legal requirements that force companies to act more sustainably, such as the LkSG, are often less effective than desired. Laws lead to more bureaucracy and, therefore, less flexibility in global supply chains, which can be fatal for companies, especially in times of crisis. Depending on local and global circumstances, public subsidy policies can also have a counterproductive effect on corporate environmental performance. *Energy costs* are probably the best example of this mechanism. Since particularly fossil energy sources are still financed and partially subsidised by the government, companies can purchase energy at unrealistically low costs. This promotes economic growth at the expense of our environment. A company's ability to access external funding is crucial for steady growth. Although small and medium-sized enterprises make up a significant proportion of the global economy, it is difficult for them to access these external sources of finance (Achillas et al., 2018).

3.3 *Models of Sustainable Supply Chain Management*

The first approach is based on the most frequently used model from supply chain management—the SCOR model from APICS (2017)—which is expanded to include the dimension of sustainability. The second approach originates from the two authors of this article (Fröhlich & Steinbiß, 2019). The focus of this general management model for supply chains is on the practical integration of the SDGs into global supply chains. This is based on the United Nation's supply chain management model (UN, 2015).

3.3.1 The SCOR Model Extended to Include Sustainability

The fundamental idea behind the SCOR model (Fig. 2) is to *standardise* the processes within a supply chain. The approach comes from the USA and was launched by the Supply Chain Council (Werner, 2020).

The process reference model covers all stages of a supply chain, whereby the processes can be configured from the source of supply to the point of consumption.

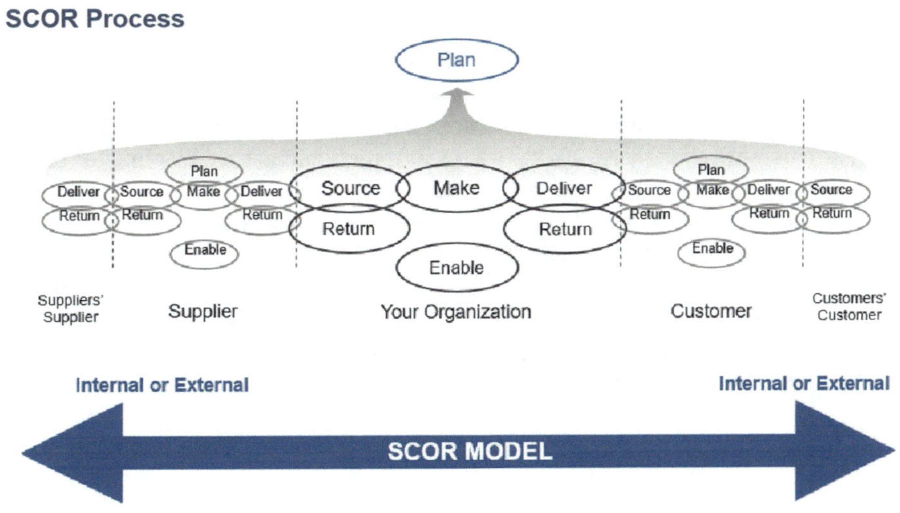

Fig. 2 The SCOR model extended by the "enable" dimension (APICS, 2017, p. 5)

By offering various alternatives for similar processes, a standardised language is developed for internal and external communication processes within the supply chain. This approach is one of the model's key strengths, as it enables a performance comparison between all partners in the value chain (Werner, 2020).

The SCOR model comprises four hierarchical levels: top, configuration, process, and implementation level. A brief explanation of the top level is sufficient here (Werner, 2020):

- **Plan** At this stage, the primary task is to plan the supply and anticipated demand structures.
- **Source** This core process involves identifying and evaluating possible suppliers in order to ensure a sufficient supply. Achieving climate neutrality requires the identification of suitable suppliers to reduce CO_2 emissions and be able to offset them in the next step.
- **Make** Manufacturing processes must be optimised in the context of coordination with all interfaces as well as customer satisfaction.
- **Deliver** This includes all necessary processes to satisfy customer demand.
- **Return** This final core process focuses on all administrative processes for returning raw materials, defective products or no longer required items.

Another core process has been added to the latest version of the SCOR model: "*enable*". This includes supporting processes, such as supply chain risk management or supply chain technology management (APICS, 2017). The importance of establishing sustainable supply chains for greater robustness in global value chains has already been addressed in the previous sections. Therefore, it seems sensible to also relate "enable" to the topic of sustainability, e.g. by focusing on the *triple bottom line* (Elkington, 1998 and Sect. 3.1). This allows social and environmental sustainability to be integrated into all core processes at the top level. Efficient planning of global supply chains leads, among other things, to the avoidance of waste.

The *embedding of minimum social standards* is only possible with the inclusion of suppliers and the introduction of joint codes of conduct in the "*source*" process. In the core "*make*" process, it is particularly important to increase the use of renewable energy sources in order to counteract what is currently the biggest problem of climate change (Schöneberg, 2022). The serious problems caused by climate change in supply chains have already been highlighted elsewhere.

The "*deliver*" and "*return*" processes are of crucial importance for the development of circular business models. Electrification of distribution processes or the expansion of collection concepts for recycling activities are two concrete examples of the sustainable consideration of these two process stages. It would also be conceivable to expand the previously integrated core processes, e.g. to include design, disposition or utilisation of the products by the end consumer.

3.3.2 A Supply Chain Model for Implementing the SDGs

This model (Fig. 3) attempts to combine the different contents that need to be considered in order to build sustainable supply chains. The original model was focused on the procurement case but can be transferred to *any functional area of business*. The 2030 Agenda (BMZ, 2022) provides a global, binding sustainable target framework for companies. Nevertheless, most companies develop their sustainable action plan according to the logic of the triple bottom line (Elkington, 1998). For this reason, it is essential to bring these two systematisation approaches together. The aim is to identify the most important SDGs that can be realised through the company's sustainability activities.

In a second step, the "*behaviour change model*" (4E model by Berry, 2011) should be applied to "swear in" all stakeholders in the global supply chain to the 2030 Agenda. Berry (2011) distinguishes between four dimensions: enable, exemplify, engage and encourage. "*Enable*" focuses on generating shared knowledge in order to encourage the stakeholders involved to improve their sustainable behaviour. Providing resources, offering joint training programmes or conducting audits are possible measures. "*Exemplify*" encompasses the role model function. The company can only expect sustainable behaviour from its stakeholders, such as customers, suppliers or the media, if it has taken its own employees on this sustainability journey. "*Engage*" is one of the most important dimensions in the 4E model. The company's expectations of the sustainability agenda must be clearly communicated to all stakeholders. Based on this, sustainability goals and strategies must be developed together. As a result, this dimension leads to stakeholder "enthusiasm" for the 2030 Agenda. In the context of the sales function, for example, this is referred to as customer delight. The fourth and final dimension, "*encourage*", is about rewarding or punishing stakeholders based on sustainable action. Examples include bonus systems for consumers, supplier awards or financial reward mechanisms or subsidies for the use of environmentally friendly technologies.

The various sustainability measures developed by applying the 4E model in the company must now be specified, depending on the industry, company size, global production sites or the relevant value creation stage:

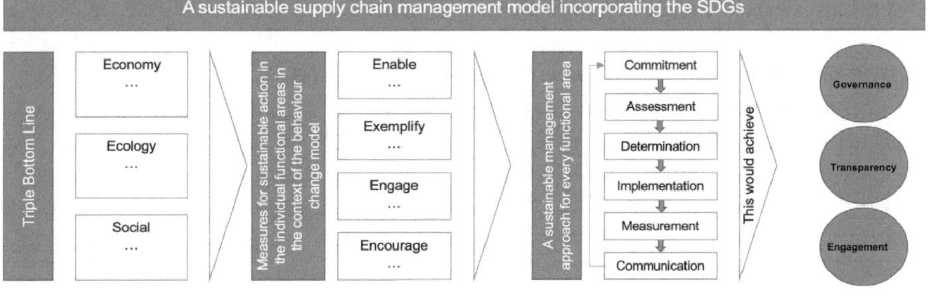

Fig. 3 A supply chain model for implementing the Sustainable Development Goals (translated from Fröhlich & Steinbiß, 2019, p. 46)

- **Commitment** Guidelines must be implemented to ensure the integration of the SDGs into the corporate sustainability strategy. Building on this, measures must be derived that contribute to the achievement of social goals.
- **Assessment** As not all 17 SDGs are equally relevant to the corporate sustainability agenda, they must be assessed and prioritised. In addition, the risks and opportunities that could arise from the individual SDGs must be analysed across the entire global supply chain wherever possible.
- **Determination** The outcome of this step is the generation of a key performance indicator (KPI) pool that ensures good performance along the global supply chain based on the results of the previously completed SDG assessment. The UN SDG Compass (2022) assists in deriving possible KPIs.
- **Implementation** The next step is embedding the relevant SDGs throughout the entire supply chain. To realise this, all stakeholders must have the necessary knowledge and willingness to support this process. In the context of procurement, this can be achieved, for example, by establishing sustainable supplier management (see Sect. 4.1.3). In the sales function, the aim is to implement sustainability in consumer behaviour. To this end, customer insight must first be gained in order to design products sustainably and utilise communication along the entire customer journey (Steinbiß & Fröhlich, 2022).
- **Measurement** The impact of the measures implemented along the supply chain—based on the prioritised SDGs—must be reviewed at this stage of the process. Here too, the approach of the UN SDG Compass (2022) can be used, which provides concrete instructions for the implementation of a monitoring system.
- **Communication** The results of this process must be communicated to all stakeholders, both internal and external to the company. This approach builds trust and motivates all stakeholders along the global supply chain to further optimise their contribution to shaping a sustainable supply chain. In the context of the 4E model, the two dimensions "encourage" and "engage" are addressed.

Ultimately, companies will be obliged to consider three meta-aspects in the context of sustainable, global supply chains in the future (Fröhlich & Steinbiß, 2019). The management approach discussed previously supports the realisation of these principles. "*Governance*" refers to the regulatory framework for the management and monitoring of companies, which is now being expanded to include the implementation of the SDGs. *Transparency* in global supply chains is one of the most important requirements of sustainable supply chains. This is induced by the tightening of national and international legislation. However, it also serves to build trust in order to counteract potential damage to the image of the company or brand. The obligation of all stakeholders involved in a global supply chain to continuously improve the level of sustainability in this chain leads to the successful implementation of the last principle of "*engagement*" through open and comprehensible communication.

4 Challenges in Sustainable Supply Chain Management

This section focuses on two areas. Firstly, the majority of environmental and social problems arise in the part of the supply chain concerning the procurement market. At the other end of the supply chain, it is also essential to consider the needs of the end consumer. The customer must be understood and provided with relevant information. Only then will they align their behaviour with sustainability.

4.1 The Role of Procurement in Sustainable Supply Chains

This section explains why purchasing plays such a central role in sustainable supply chains. This is explained based on empirical findings as well as the introduction of stricter legal requirements. These new challenges can only be met by involving strategically relevant suppliers through the development of cooperative supplier-buyer relationships. Relationship management in purchasing must be rethought.

4.1.1 Procurement as the Missing Element of Sustainable Supply Chains

Initial empirical studies show that sustainable supply chains can only be implemented by realigning procurement towards more sustainability and digitalisation: procurement is the missing link in sustainable supply chains!

As part of a qualitative study, Villena and Gioia (2020) identified the following steps for promoting the social and environmental responsibility of suppliers:

- The buyer and supplier define a *joint sustainability strategy* and long-term sustainability goals.
- Building on this, first-tier suppliers formulate their own long-term *sustainability goals*.
- Lower-tier suppliers are integrated into the joint sustainability strategy by the first-tier suppliers.
- All employees in the purchasing department of the procuring company have the task of extending the sustainability programme to first- and second-tier suppliers. A so-called responsible "point person" is appointed for this purpose.

Villena and Gioia (2020) propose the following four approaches to successfully implement these steps in a procurement company and thus create a basis for the fulfilment of the LkSG:

- **Direct Approach** First-tier suppliers are involved in the establishment of a joint sustainability strategy based on common sustainability goals and standards.
- **Indirect Approach** Responsibility for the sustainability management of lower-tier suppliers is transferred to first-tier suppliers. To this end, training is offered

to suppliers and incentives are created for the implementation of sustainable measures.

- **Global Approach** Working with international organisations or non-governmental organisations (NGOs) that share the buyers' sustainability goals enables companies to realise their sustainability vision for global supply chains.
- **Collective Approach** The challenges can only be overcome if buyers work together with their competitors and key suppliers to develop industry-wide sustainability standards.

4.1.2 The German Supply Chain Due Diligence Act in the Context of Sustainable Supply Chains

Now that initial empirical considerations have demonstrated the high relevance of the procurement function for the realisation of sustainable supply chains, this section will look at the stricter legal regulations. The German Federal Government adopted the National Action Plan for Business and Human Rights (NAP) in 2016 and launched the annual monitoring programme to review implementation in companies in 2018. As the sustainability goals set by the German government were only poorly implemented by the German economy, the first draft of the LkSG was created in 2021 (Fedato et al., 2021).

The Federal Council passed the law on the 25th of June 2021. The LkSG defines the requirements for responsible management of global supply chains. It sets out binding rules on the *due diligence obligations* that a company must fulfil. Improving the protection of fundamental human rights and integrating relevant environmental concerns into global supply chains is at the centre of the LkSG. The law contains the following key regulations, whereby the responsibility of companies extends to the entire supply chain (KMU Kompass, 2022):

- How companies analyse their environmental and human rights risks.
- How companies implement appropriate risk management.
- Which preventive and remedial measures companies take based on the analysed risks.
- How companies set up an internal complaint management system.
- How companies document their efforts to fulfil the aforementioned requirements of the LkSG.

To conclude this section, some critical comments on the LkSG will be discussed. These are based on presentations and content from the initiative "Economists Demand Supply Chain Law" (Initiative Lieferkettengesetz, 2021). The introduction of the LkSG should not be seen purely as a cost driver for German companies but as a perfectly justifiable *investment in the future viability* of a company. Therefore, it is vital to change the communication surrounding this law. It is more important to use "lighthouse examples" to show companies the added value that compliance with human rights and environmental requirements can have for the success of their company than to focus on the financial and liability consequences (penalties). The law is preventative in nature: companies are not to be penalised for their past actions, but rather a general action framework should be created that avoids the so-called "free rider positions" in the future and thus significantly improves the social and ecological conditions of global supply chains. Therefore, the LkSG creates a *fair framework for global supply chain management*. The law's restrictions based on the number of employees and turnover make little sense. This leads to a "false sense of security" for small and medium-sized supplier companies. In fact, these suppliers are the basis for ensuring that companies with more than 3000 employees that are subject to reporting requirements in 2023 can implement the LkSG requirements at all. Similar size restrictions can also be found in the draft of the CSDDD (Corporate Sustainability Due Diligence Directive), which is expected to take effect from 2028. Against this backdrop, it can be concluded that German companies should align themselves directly with the CSDDD, as it sets significantly higher requirements, especially in the area of environmental sustainability (BMUV, 2024). As explained at the beginning of this chapter, sustainability and resilience in supply chains go hand in hand. This should be incentive enough to steer entrepreneurial behaviour in new directions and jointly develop a "new normal" for our economy. The LkSG is an important first step on this path!

4.1.3 Designing Sustainable Supplier-Buyer Relationships

As explained previously, strategic partnerships with suppliers are essential for the successful integration of sustainability practices in the supply chain and the achievement of corresponding corporate goals (Kumar & Rahman, 2015). If the cooperation between buyer and supplier is understood as part of a joint organisation, this is referred to as the so-called relational view of relationship management. This moves the relationship away from the level of purely contractual control mechanisms and creates the opportunity to build a collaboration supported by both partners (Bai et al., 2021).

The *TrueSRM framework* (Schuh et al., 2014) serves as the basis for the model discussed below. Nine different types of supplier relationships are derived from the two dimensions of "performance" and "strategic potential". In order to further develop this approach into a sustainable supplier management model, the two dimensions are expanded to include the contents of the triple bottom line (Elkington, 1998). In addition to the basic supplier evaluation parameters of cost, time and quality (Schuh et al., 2014), "*performance*" includes economic (e.g. fair competition or the inclusion of local suppliers), social (e.g. job security or training opportunities) and environmental aspects (e.g. KPIs such as CO_2 footprint, water footprint or waste minimisation). Schuh et al. (2014) use the assessment parameters of growth, innovation, scope and cooperation to position suppliers along the "*strategic potential*" dimension:

- *Growth* is about the supplier's ability to generate added value for the procuring company. In the context of sustainable supply chain management, this ability of the supplier extends to making the supply chain more sustainable for the procuring company and upstream suppliers.

- *Innovation* is focused on the proactive advancement of the procuring company's sustainability performance.
- *Scope of action* is understood as the importance of the supplier for different business areas of the procuring company. In the context of sustainability, this includes the supplier's ability to influence society and the environment.
- *Collaboration* describes the alignment of the "mindsets" of the supplier and the procuring company. As explained previously, it is particularly important for the implementation of more sustainable supply chains that the sustainable procurement strategy and goals are supported by the supplier company and that close cooperation is thus established between the buyer and supplier.

Figure 4 provides an overview of the TrueSRM model expanded to include the component of sustainable management. It should be noted that only 5% of suppliers have a very high or very low performance level. 90% are in the medium range. The same applies to strategic potential: only 10% provide high added value for the procuring company, while the remaining 90% can only be categorised as having low or medium strategic potential. The application of the TrueSRM framework to sustainability is taken from an article by Kux (2024).

The "improve", "sustain" and "harvest" types belong to the group of "*ordinary*" suppliers. This group forms the foundation of a supplier portfolio. Due to their medium to high performance and low to medium strategic potential, they require only minimal management support from the procuring company, as the number of

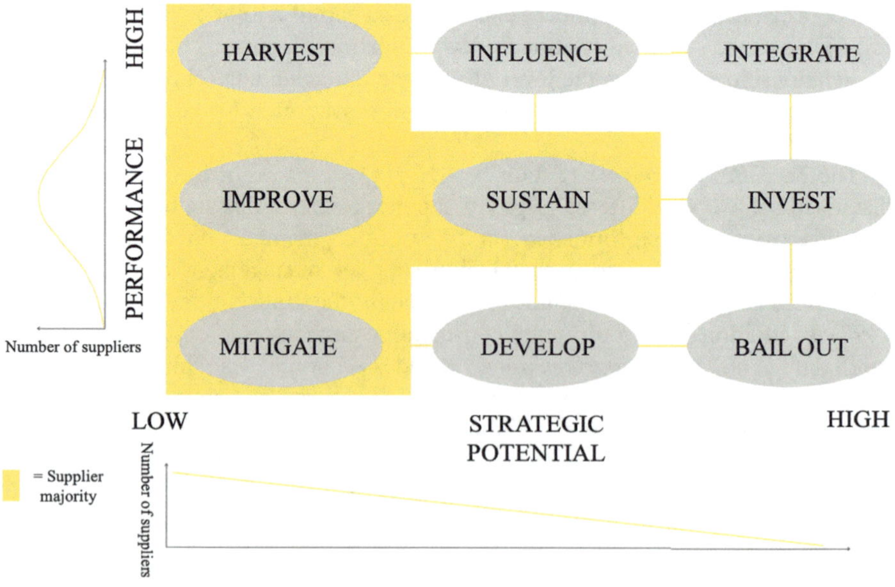

Fig. 4 The TrueSRM framework (Schuh et al., 2014, p. 30)

suppliers available in this group is relatively high. The group of "problematic" suppliers comprises the types "mitigate", "develop" and "bail out". This group can harbour a high risk potential for achieving the sustainability goals of the procuring company. In principle, however, attempts should be made to develop this supplier group, as establishing new relationships with suppliers in the context of sustainability is always very resource-intensive.

The "influence", "invest" and "integrate" types form the group of the so-called "*critical*" *suppliers*. The contribution of this group to achieving the sustainability strategy and goals is, as the name of this group implies, of great importance. A sustainable relationship between buyer and supplier is created through the development of a bond that is intensified through joint investments and increases the joint strategic potential of this relationship. The sustainable performance level of suppliers categorised as the "*influence*" type is optimal. In contrast, the innovation potential of these suppliers does not make a significant contribution to the capability development of the procuring company. Building a trusting relationship is particularly important here, as the large customer base makes it difficult to establish an intensive bond with the supplier company. Leveraging the strategic advantages on both sides can be a promising differentiator for a relationship with this type. The opposite is true for the "*invest*" type. The suppliers in this category have performance gaps in a few sustainability areas but can significantly influence the achievement of the procuring company's sustainability goals through their expertise, innovative strength and the associated large scope of action. Engaging with the needs of the other party and the joint development of a transparent and proactive action plan are seen as central to this type of supplier. Suppliers of the "*integrate*" type meet with the procuring company on an equal footing. Both parties are equally important for the long-term success of this relationship, which is based on cooperative proactivity and trust. The supplier recognises the growth opportunities of this relationship through the use of their own innovations concerning social and environmental sustainability and thus becomes an integral part of the procuring company's production process. Joint investments lead to a high degree of sustainability in the global supply chain designed by the supplier and buyer.

4.2 The Role of Sales in Sustainable Supply Chains

In addition to procurement, the role of sales in sustainable supply chains will also be analysed. Two focal points were identified for this purpose: on the one hand, the relevance of different motives or influencing factors on individual purchasing behaviour will be outlined, and on the other hand, the possible discrepancy between consumer intention and behaviour will be clarified. The basic principle here is that the sustainable supply chain must always be orientated towards the sales market and thus the consumer as the end user.

The literature contains a multitude of models that analyse why customers act sustainably. One of the most comprehensive is the *SHIFT model* (White et al.,

2019). The acronym "SHIFT" stands for the groups of criteria that have been identified as behaviourally relevant in relation to sustainable consumption: social influence, habits, the individual self, feelings and cognition and tangibility. Figure 5 shows these different groups of criteria.

The SHIFT model identifies five categories of influence on sustainable consumer behaviour:

- **Social Influence** Consumers generally base their behaviour on what is socially desirable, social norms or the expectations of others. The aim is to make a positive impression on others and to act in the same way as others do.
- **Habits** Sustainable behaviour is promoted through habits. The easier sustainable actions are made for consumers in their daily routines, the more firmly they can be embedded.
- **The Individual Self** Besides self-worth, consistent behaviour, an interest in one's own physical and mental well-being, and the impact of one's own actions contribute to this influencing factor.
- **Emotions and Cognition** Sustainable action can be ensured by negative or positive emotions as well as by building up relevant knowledge during the decision-making process.
- **Tangibility** Another influencing factor is convincing consumers that environmentally friendly actions are not abstract, vague and far-removed from their own experience. They must be directly linked to their actions and made tangible for consumers.

The sales function in sustainable supply chains now has the task of addressing consumers precisely on the product-specific or consumption-critical influencing factor that is relevant to them. The primary undertaking here is to define the relevance of the influences on the individual consumption motive, which varies in importance depending on the situation and product (Steinbiß & Fröhlich, 2022).

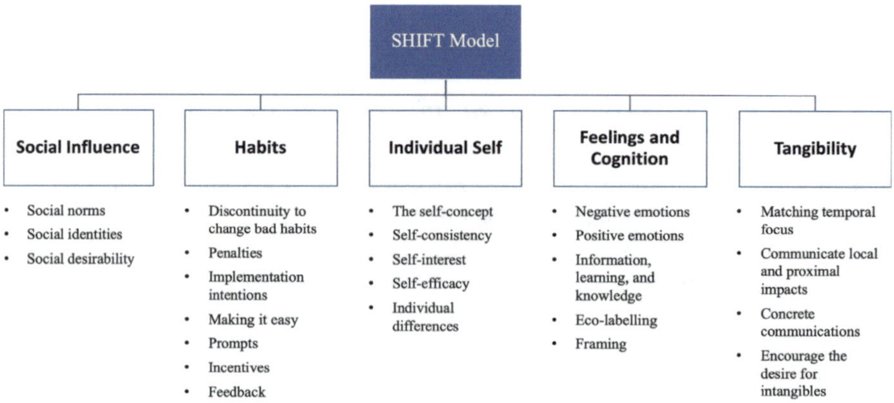

Fig. 5 The SHIFT model of sustainable consumer behaviour (White et al., 2019, p. 24)

A recurring phenomenon in this context—that needs to be overcome in order to successfully design a sustainable supply chain from a sales perspective—is the *intention-behaviour gap*. This describes the disparity between what a consumer intends to do and what they actually do (Auger & Devinney, 2007). Closing this gap is the goal of sustainable supply chains. Changing behavioural patterns is of central importance in this regard, extending across the entire consumption cycle from product selection to disposal or recycling of the product after use.

To close the intention-behaviour gap, Capgemini's attitude-based cycle can be adapted (Capgemini, 2022). This consists of four fields of action that cluster the sustainability activities of companies in a customer-oriented manner: recognise, grasp, enable and emotionalise. Transparency across the entire supply chain forms the basis on which, in a second step, an awareness of the responsibility of finite resources can be established on the customer side. Particularly, this aspect is currently not being actively addressed to the extent that would be desirable in order to design sustainable supply chains. Customers must also be incentivised to become actively involved and gain the understanding that this involvement can influence the establishment of a sustainable supply chain. The higher the level of personal involvement and the clearer the consequences of sustainable behaviour for the individual, the more likely it is that corporate activities relating to sustainability will be successful.

5 Conclusion and Outlook

This chapter presented action-orientated concepts and models for sustainable supply chain management. A particular focus was placed on robust supply chains as the basis for sustainable supply chains. Readers were made aware of the key role that suppliers and end consumers play in successfully designing and implementing sustainable global supply chains. In the future, these findings must be applied to supply chains in different industries in order to realise the sustainable transformation of the economy.

This chapter was previously published in German with additional examples of sustainability in sports and culture: Fröhlich, E., & Steinbiß, K. (2023). Nachhaltiges Supply Chain Management. In A. Bühler & G. Nufer (Eds.), Nachhaltigkeits management in Sport und Kultur (pp. 123–146). Erich Schmidt Verlag.

References

Achillas, C., Bochtis, D. D., Aidonis, D., & Folinas, D. (2018). *Green Supply Chain Management*. Routledge.

APICS. (2017). *Supply Chain Operations Reference Model SCOR*. https://www.apics.org/docs/default-source/scor-training/scor-v12-0-framework-introduction.pdf?sfvrsn=2

Auger, P., & Devinney, T. M. (2007). Do what consumers say matter? The misalignment of preferences with unconstrained ethical intentions. *Journal of Business Ethics, 76*, 361–383. https://doi.org/10.1007/s10551-006-9287-y

Bai, C., Kusi-Sarpong, S., Khan, S. A., & Vazquez-Brust, D. (2021). Sustainable buyer-supplier relationship capability development: A relational framework and visualization methodology. *Annals of Operations Research, 304*, 1–34. https://doi.org/10.1007/s10479-021-04090-4

Berry, C. (2011). *The sustainable procurement guide: procuring sustainability using BS 8903*. BSI British Standards Institution.

BMAS. (2022). *Sorgfaltspflichtengesetz*. https://www.bmas.de/DE/Service/Gesetze-und-Gesetzesvorhaben/gesetz-unternehmerische-sorgfaltspflichten-lieferketten.html

BMUV. (2024). Europäische Lieferkettenrichtlinie (CSDDD). https://www.bmuv.de/themen/nachhaltigkeit/wirtschaft/lieferketten/europaeische-lieferkettenrichtlinie-csddd

BMZ. (2022). *Die globalen Ziele für nachhaltige Entwicklung*. https://www.bmz.de/de/agenda-2030

BSR. (2010). *The Business Case for Sustainable Supply Chain Management: A Brief for Business Leaders*. https://www.bsr.org/reports/Beyond_Monitoring_Business_Case_Brief_Final.pdf

Capgemini. (2022). *Attitude-Behaviour Gap: Warum Konsumenten vorgeben nachhaltig leben zu wollen – es aber nicht tun*. https://www.capgemini.com/de-de/insights/blog/attitude-behaviour-gap-warum-konsumenten-vorgeben-nachhaltig-leben-zu-wollen-es-aber-nicht-tun/

Carter, C. R., & Rogers, D. S. (2008). A framework of sustainable supply chain management: moving toward new theory. *International Journal of Physical Distribution & Logistics Management, 38*(5), 360–387. https://doi.org/10.1108/09600030810882816

Chopra, S., & Meindl, P. (2016). *Supply chain management: Strategy, planning, and operation* (6th ed.). Pearson.

CSCMP. (2013). *Supply chain management terms and glossary*. https://cscmp.org/CSCMP/Educate/SCM_Definitions_and_Glossary_of_Terms.aspx

Elkington, J. (1998). Accounting for the triple bottom line. *Measuring Business Excellence, 2*(3), 18–22. https://doi.org/10.1108/eb025539

EY. (2016). *The State of Sustainable Supply Chains: Building responsible and resilient supply chains*. https://unglobalcompact.org/library/4451

Fedato, C. et al. (2021). Das deutsche Lieferkettengesetz ist da! Wuppertal

Fröhlich, E., & Steinbiß, K. (2019). Integration der "Sustainable Development Goals" in eine nachhaltige Supply Chain: Der "Nachhaltige Beschaffungs-Case". In W. Wellbrock & D. Ludin (Eds.), *Nachhaltiges Beschaffungsmanagement* (pp. 37–54). Springer.

Initiative Lieferkettengesetz. (2021). *Initiative Lieferkettengesetz*. https://lieferkettengesetz.de/

Kersten, W., Hohrath, P., & Winter, M. (2008). Risikomanagement in Wertschöpfungsnetzwerken – Status Quo und aktuelle Herausforderungen. *Supply Chain Risk Management, 8*, 7–21. https://www.fh-vie.ac.at/uploads/pages/forschung/SR-2008-08-Web.pdf

KMU Kompass. (2022). *Der KMU Kompass: Lieferketten nachhaltiger managen*. https://kompass.wirtschaft-entwicklung.de/

Kumar, D., & Rahman, Z. (2015). Sustainability adoption through buyer supplier relationship across supply chain: A literature review and conceptual framework. *International Strategic Management Review, 3*(1–2), 110–127. https://doi.org/10.1016/j.ism.2015.04.002

Kux, C. (2024). Ein Lieferantenbeziehungsmanagementmodellentwurf zur Förderung der nachhaltigen Beschaffung. In E. Fröhlich & Y. Jamal (Eds.), *CSR und Beschaffung* (2nd ed.) (pp. 149–180). Springer.

Müller, M., & Siakala, S. (2020). *Nachhaltiges Lieferkettenmanagement: Von der Strategie zur Umsetzung*. De Gruyter Oldenbourg.

Sadowski, P. (2010). *Grüne Logistik: Grundlagen, Ansätze und Hintergründe zur Optimierung der Energieeffizienz in der Logistik*. Verlag Dr. Müller.

Sarkis, J., Cohen, M. J., Dewick, P., & Schröder, P. (2020). A brave new world: Lessons from the COVID-19 pandemic for transitioning to sustainable supply and production.

Resources, Conservation & Recycling, 159, Article 104894. https://doi.org/10.1016/j.resconrec.2020.104894
Schöneberg, K. (2022, November 20). Klimakatastrophe auf Kurs. *Tageszeitung*. https://taz.de/Reaktionen-auf-COP-27/!5893264/
Schuh, C., Strohmer, M. F., Easton, S., Hales, M., & Triplat, A. (2014). *Supplier Relationship Management: How to maximize supplier value and opportunity*. Apress.
Schuhmacher, M. (2022). *Nachhaltiges Lieferketten Set-Up: Was sind die Merkmale von widerstandsfähigen Lieferketten? Wie lassen sich nachhaltige und widerstandsfähige Lieferketten aufbauen?* [Unpublished master's thesis]. CBS International Business School.
Steinbiß, K., & Fröhlich, E. (2022). Die Verpackung als Enabler für nachhaltige Kundenerlebnisse. *PraxisWISSEN Marketing, 7*(1), 21–32. https://www.econstor.eu/bitstream/10419/266163/1/1820742245.pdf
UN. (2015). *Supply Chain Sustainability: A Guide for Continuous Improvement, Second Edition*. https://unglobalcompact.org/library/205
UN SDG Compass. (2022). *SDG Compass*. https://sdgcompass.org/
Villena, V. H., & Gioia, D. A. (2020, March–April). A more sustainable supply chain. *Harvard Business Review*. https://hbr.org/2020/03/a-more-sustainable-supply-chain
WEF. (2020). *The Global Risks Report 2020*. https://www.weforum.org/publications/the-global-risks-report-2020/
WEF. (2021). *The Global Risks Report 2021*. https://www.weforum.org/publications/the-global-risks-report-2021/
WEF. (2022). *Global Risks Report 2022*. https://www.weforum.org/publications/global-risks-report-2022/
Werner, H. (2020). *Supply Chain Management: Grundlagen, Strategien, Instrumente und Controlling* (7th ed.). Springer.
White, K., Habib, R., & Hardisty, D. J. (2019). How to SHIFT consumer behaviors to be more sustainable: A literature review and guiding framework. *Journal of Marketing, 83*(3), 22–49. https://doi.org/10.1177/0022242919825649
Wiendahl, H.-H. (2020). Auftragsmanagement. In T. Bauernhansl (Ed.), *Fabrikbetriebslehre 1: Management in der Produktion* (pp. 193–294). Springer.

Elisabeth Fröhlich is the founder of ispira Think Tank for Sustainable Supply Chains. She invented an innovative business coach concept to prepare companies for the challenges of the different European sustainability laws. Her major aim is to bridge the gap between research and practice in this field to enable companies to implement the sustainable transformation of global supply chains successfully. Prof. Dr. habil. Elisabeth Fröhlich was president of the CBS International Business School in Germany until 2022. She holds a full professorship in Sustainable Procurement and Supply Chains and is an internationally recognised expert in sustainable supply chain management. She is also highly engaged in the field of responsible management education. She serves as a board member of PRME, chairs the PRME Nomination and Governance Committee and is the PRME DACH Chapter Chair. Additionally, she leads the AOM Management Education Division Ambassador Program as vice-chair, leads the Board of JARO Institute and offers online training in the field of sustainable procurement. Her research focuses on sustainable supply chain management and green procurement, qualification in purchasing, Procurement 4.0 and strategic supplier relationship management. Innovative teaching formats of responsible management education are further main areas of her research. She has published several books and articles on the above-mentioned topics and supports several journals as an external reviewer.

Kristina Steinbiß holds a professorship in Management with a focus on Marketing at the ESB Business School of the Reutlingen University in Germany, where she teaches primarily in the Industrial Engineering programmes. Her fundamental teaching goal is to enable her students to think and act sustainably while adopting an integrative perspective combining business and technology. Kristina's current research focuses on implementing new business models to accomplish a company's sustainability goals, as well as on influencing and nudging consumers towards sustainable behaviour. She is the author of numerous publications in the above-mentioned fields, including the textbook *Marketing*, which is already available in its second edition. Next to her research activities, she is also enthusiastic about the Theory of Inventive Problem Solving, the TIPS innovation methodology. As a TIPS trainer and coach, she likes to combine this methodology with her marketing approach and adapt it to implement sustainable solutions in day-to-day business.

Circular Procurement in Global Supply Chains

Chiara Bernd

1 Global Supply Chains: An Overview

Global supply chains are the backbone of the modern economy and play a crucial role in the production and distribution of goods and services worldwide. They represent complex networks that connect companies, suppliers, manufacturers, logistics service providers and distribution channels in order to bring products from their point of origin to the end consumer. These supply chains can extend across different countries and continents and are therefore considered a central component of the global economy (Chopra & Meindl, 2020, p. 5).

Supply chains are a multi-layered construct that encompasses several key areas, as visualised in Fig. 1. The *procurement* of raw materials often marks the point of origin of a supply chain. This is where companies identify potential suppliers, negotiate contracts and acquire resources from them, often from different parts of the world (Krajewski et al., 2016, p. 570). Once the materials have been procured, the *production and manufacturing* phase follows. The acquired materials are further processed in factories or production facilities to create finished products or components. This can involve different processes and several intermediate products (Kummer et al., 2009, p. 24). The subsequent step focuses on *logistics, transport and storage*. After production, the finished products must be transported to their destinations. Transport can take place in various ways, whether by land, sea or air. Logistics companies and sophisticated logistics systems play an important role in ensuring the smooth, safe and efficient transport of goods. During transport and before final distribution, products are often temporarily stored in warehouses.

C. Bernd (✉)
Cologne, Germany
e-mail: chiarabernd@gmx.de

E. Fröhlich et al. (eds.), *Circular Economy in Sustainable Supply Chains*, CSR, Sustainability, Ethics & Governance,
https://doi.org/10.1007/978-3-031-70749-0_15

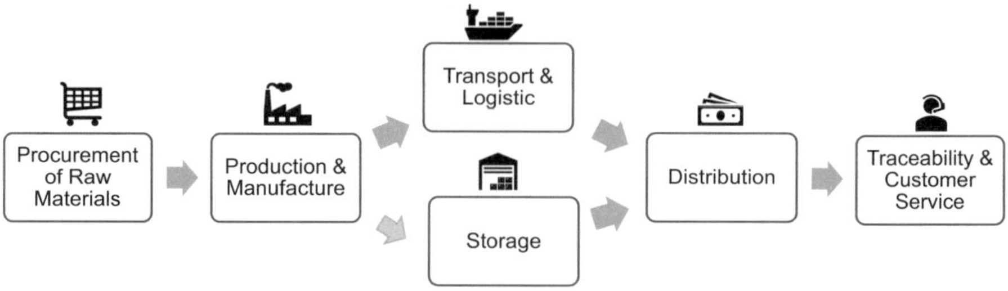

Fig. 1 Supply chain structure (own illustration)

Warehouses are used to sort and manage the products safely and efficiently before they are sent on (Kummer et al., 2009, p. 81).

Finally, the products are *distributed*. Various distribution channels play a role here, be it bricks-and-mortar retail, online shops or other points of sale through which the product finds its way to the consumer (Kummer et al., 2009, p. 24). As the last link in the supply chain, companies often offer *after-sales services* for customers and may also track products to enable recalls or to gather information for improving products and processes.

Sustainability is playing a progressively more important role in global supply chains and is influencing them in a variety of ways. Companies are increasingly recognising that sustainable practices not only offer environmental benefits but also have economic and social impacts. This realisation has led to sustainability becoming an important factor in the design and management of supply chains (Carter & Easton, 2011, p. 48).

One of the main aspects is environmental sustainability. Companies are increasingly endeavouring to reduce their CO_2 emissions, use resources more efficiently and introduce more environmentally friendly production processes. This includes, for example, the use of renewable energies, the reduction of waste and the implementation of recycling and circular economy concepts. These measures directly impact the entire product life cycle within the supply chain described above. Social sustainability also plays a crucial role. Companies are increasingly striving to ensure and implement ethical standards along the supply chain. This includes fair working conditions, appropriate remuneration, respect for human rights and the promotion of social justice. The involvement of suppliers who are committed to these principles is becoming progressively more important (Zimmermann, 2016, pp. 8–9). Another point of influence is economic sustainability. The search for long-term, stable and reliable suppliers and the reduction of risks that can arise from ecological or social problems are the focus here. Companies recognise the value of long-term partnerships to build a sustainable supply chain and simultaneously strengthen their resilience to external influences (Gartner, 2024).

All these aspects of sustainability directly impact supply chain management and influence how companies organise their procurement strategies. Sustainability therefore becomes a key concern when selecting suppliers, designing procurement processes and ensuring ethically and environmentally responsible procurement. Companies that take a sustainable approach to their procurement help make the entire supply chain more sustainable and achieve long-term positive effects on the environment, society and the economy.

2 Sustainable Procurement Management and Circular Procurement

Sustainable procurement management refers to a strategic approach to the planning, organisation and implementation of procurement activities in companies or organisations, considering a wide range of environmental, social and economic aspects. This approach aims not only to achieve short-term goals, such as cost savings and efficiency, but also to consider long-term effects on the environment, society and the economy (Wellbrock & Ludin, 2019, pp. 3–6).

At its core, it is about organising supply chains and purchasing processes in such a way that they are more environmentally friendly, ethical and socially responsible. Sustainable procurement management encompasses a variety of measures, including the selection of suppliers, products and services based on environmental standards, the promotion of fair working conditions along the entire supply chain, the reduction of resource consumption, the use of renewable energy and the support of recycling and reutilisation processes for materials.

In addition, sustainable procurement management involves integrating sustainability criteria into procurement strategies and decisions, working with suppliers to improve environmental and social standards, complying with legal requirements and developing long-term partnerships with suppliers who share the responsibility for sustainability. The overall approach of sustainable procurement management aims not only to achieve short-term operational goals but also to help reduce environmental impacts, promote social justice and ensure long-term economic stability. It is an integrative approach that pursues a balanced consideration of economic, environmental and social objectives in order to establish and promote sustainable and responsible procurement practices in the long term (Fröhlich, 2015, pp. 20–21).

Circular procurement is an addition to the sustainable procurement approach. To date, there is no standardised dictionary definition. However, the term circular procurement refers to sustainable and responsible procurement that manufactures and recycles resources or products according to the principle of the circular economy. These are products that are "designed for durability, repairability and recycling and can at the end of their life cycle be broken down into components, materials or raw materials, which can then be used again in the production chain" (Jones et al., 2017, p. 1). What makes circular procurement special is not just buying circularly manufactured products or recycled raw materials but rather using these products and raw materials circularly and ensuring that the cycle continues after consumption (van Oppen et al., 2018, p. 21).

The relationship between the two approaches shows that sustainable procurement management, together with circular procurement, provides a comprehensive framework that promotes environmental sustainability as well as maximises social and economic benefits. Both approaches complement each other by emphasising the reuse of resources while ensuring that procurement practices are ethical, socially responsible and economically viable in the long term. By combining these practices, companies can make their supply chains more sustainable and make a significant contribution to protecting the environment while strengthening their operational efficiency and market position.

2.1 The Importance of Sustainable Procurement and Circular Procurement for the Sustainable Transformation

The integration of sustainable procurement practices in companies plays a central role in today's globalised world and contributes significantly to sustainable transformation. This transformation refers to a comprehensive change that considers ecological, social and economic aspects in order to promote long-term ecological stability, social justice and economic prosperity (Fröhlich, 2015, p. 4; Wellbrock & Ludin, 2019, pp. 3–4).

Environmentally, it promotes environmental protection by minimising the use of resources, favouring renewable energy sources and reducing the environmental impact over the life cycle of products. These practices are core components of circular procurement, which aims to maximise the reusability of products and minimise waste (Wellbrock & Ludin, 2019, p. 8).

Socially, sustainable procurement attaches great importance to fair working conditions, ethical standards and the creation of a transparent supply chain. These aspects are particularly important for circular procurement, as it promotes the responsible use of resources and a fair distribution of the resulting benefits (Fröhlich, 2015, p. 33; Wellbrock & Ludin, 2019, p. 8).

Economically, sustainable procurement offers significant advantages by reducing costs and risks in the supply chain and promoting innovation for sustainable products and services. These economic aspects are crucial for the viability and competitiveness of companies and are further strengthened by circular procurement models that support the longevity of products and the efficiency of resource utilisation (Wellbrock & Ludin, p. 9).

In the context of new legal provisions and regulatory requirements, such as the German Act on Corporate Due Diligence Obligations in Supply Chains, sustainable procurement is becoming increasingly important. Companies must comply with these regulations to avoid legal conflicts and contribute to sustainable development (Fröhlich, 2015, pp. 57–58).

Overall, sustainable and circular procurement act as key catalysts for companies to play their role as drivers of change towards a more sustainable future. By integrating environmental, economic and social issues into procurement decisions, it enables companies to have a positive impact on their supply chains while making a significant contribution to conserving natural resources and promoting a fairer and more stable society.

2.2 *The Impact of Procurement on the SDGs*

The United Nations Sustainable Development Goals (SDGs), also known as the 2030 Agenda, were adopted in 2015 and comprise 17 overarching sustainability goals (Fig. 2), which are broken down further into 169 detailed sub-goals (BMUV, 2023). Their main objective is to promote sustainable development at an ecological, economic and social level worldwide by 2030. This includes combating poverty and hunger, ensuring education and gender equality, protecting the climate and promoting sustainable cities and communities (Die Bundesregierung, 2023).

Procurement, as a strategic process within an organisation for the acquisition of goods and services, plays a significant role in sustainability. The decisions made in the procurement process have a direct impact on various areas, such as the environment, resource consumption, social conditions, innovation and compliance with guidelines and regulations (HEC Paris, 2017). Through the selection of suppliers, materials and production processes, procurement can make a significant contribution to minimising a company's impact on the environment. This includes the reduction of greenhouse gas emissions, the promotion of renewable energies and the use of sustainable materials. It also contributes to the promotion of fair working

Fig. 2 The Sustainable Development Goals (World Economic Forum, 2020)

conditions and compliance with social standards in supply chains. The close link between sustainable procurement and the SDGs is reflected in several goals. For example, a more conscious use of resources and materials in the procurement process can contribute directly to realising the goal of *responsible production and consumption* (SDG 12). Promoting fair working conditions along the supply chain can contribute to achieving the goal of *decent work and economic growth* (SDG 8). Other aspects, such as ensuring supply chain resilience, promoting innovation processes for sustainable products and working with suppliers to develop common sustainability goals, directly impact the realisation of various SDGs.

The possibilities for influencing sustainable procurement measures are presented in more detail using SDG 12. The aim of SDG 12 is to promote sustainable consumption and production patterns by using resources efficiently, avoiding waste and developing sustainable consumption patterns (BMZ, 2023). One way to achieve this goal is through procurement, focusing on environmentally friendly, durable and recyclable products and materials. In this way, procurement can help reduce environmental impact and waste volumes, which in turn helps reduce greenhouse gas emissions and promote circular economy approaches. The selection of suppliers that already pursue sustainable production patterns and circular economy approaches can further support the implementation of SDG 12. Overall, the targeted selection of products and suppliers through procurement can help achieve the objectives of this SDG and thus make a significant contribution to environmental protection and the promotion of sustainability. In this SDG, seven sub-goals can be addressed through procurement.

Procurement plays a decisive role in the implementation of various sub-goals (12.2; 12.3; 12.4; 12.5; 12.6; 12.7; 12.a) of SDG 12. In the context of SDG 12.2, the *sustainable management and efficient use of natural resources*, procurement can make a positive contribution through the targeted selection of products and materials manufactured with environmental aspects in mind. This also includes a preference for suppliers that use sustainable cultivation methods and environmentally friendly production processes to minimise the impact on natural resources.

Furthermore, in accordance with SDG 12.3—*halve global per capita food waste*—procurement contributes to counteracting food waste and losses. By strategically purchasing food in appropriate quantities and considering its shelf life, procurement can help reduce food waste. Collaboration with suppliers also plays an important role in promoting efficient supply chains and storage practices to minimise food loss.

Regarding SDG 12.4, the *environmentally sound use of chemicals*, procurement can contribute through the targeted selection of products that use environmentally friendly and safer chemical substances. Cooperation with suppliers who are committed to environmentally friendly manufacturing processes and the use of safe chemicals is of crucial importance here.

The *reduction of waste generation*, as set out in SDG 12.5, can also be achieved through sustainable procurement practices. This includes selecting durable products that are less prone to wear and tear and waste, as well as promoting recycling initiatives by selecting suppliers that use recyclable materials and implement waste reduction measures.

SDG 12.6, the *encouragement of sustainable practices and sustainability reporting*, can also be supported by procurement. This includes selecting suppliers that implement sustainable production processes and practices and supporting suppliers in the introduction of transparent sustainability reporting mechanisms.

In addition, procurement following SDG 12.7, the *promotion of sustainable public procurement practices*, can help raise awareness of sustainable developments and lifestyles. The targeted selection of suppliers that actively contribute to raising awareness of sustainable development and the integration of training programmes and information initiatives into procurement practices can play a key role in this.

Finally, in accordance with SDG 12.a—*develop and implement tools to monitor sustainable tourism*—procurement can support developing countries in terms of scientific and technological capacities. This can be done through partnerships with suppliers from developing countries to promote the exchange of scientific and technological know-how and by prioritising suppliers that support initiatives to promote research and technology development in developing countries.

Overall, it can be stated that procurement plays a key role in realising the goals of SDG 12. The targeted selection of environmentally friendly products and suppliers can promote resource efficiency, waste reduction and sustainable management. Procurement also contributes to the fight against food waste, the environmentally friendly use of chemicals and the creation of awareness for sustainable developments. Cooperation with suppliers from developing countries supports their scientific and technological capacities. Overall, sustainable procurement is a key lever for a comprehensive transformation towards sustainable consumption and production patterns in line with global sustainability goals.

The detailed analysis (Fig. 3) of the influence of procurement on the implementation of the SDGs using the example of SDG 6 (access to clean water and sanitation) makes it clear that companies can actively contribute to achieving

SDG 6 – Clean Water and Sanitation

Sub-Goal	Description	Contribution of Sustainable Procurement
6.1	Access to safe and affordable drinking water for all	Selection of suppliers or partner companies that demonstrably promote access to drinking water in developing countries and, for example, carry out regular water tests there.
6.2	Access to equitable sanitation for all	Selection of suppliers or partner companies that demonstrably promote access to sanitary facilities in developing countries and, for example, install sanitary facilities there and thus counteract infectious diseases.
6.3	Improve water quality by reducing pollution	Selection of suppliers or partner companies that avoid harmful chemicals in their production processes, promote sustainable water treatment processes and reduce water pollution, e.g. by utilising the circular economy.
6.4	Increase water-use efficiency and counteract water scarcity	Selection of suppliers or partner companies that use resource-conserving processes and technologies, promote the reuse of water and simultaneously reduce water consumption, e.g. by utilising the circular economy.
6.5	Integrated water resources management at all levels	Application of a circular economy to keep the resource of water in the use cycle as long and efficiently as possible.
6.6	Protection of water-related ecosystems	Procurement of resources or products that are biodegradable, can be produced using renewable energy and can be used in a circular way to protect ecosystems.
6.a	Support developing countries for access to better water supply and sanitation	Selection of suppliers or partner companies that demonstrably promote access to sanitary facilities and water in developing countries and create healthier living conditions there.

Fig. 3 Influence of procurement using the example of SDG 6 (own illustration)

the global sustainability goals through targeted and sustainable procurement strategies.

3 Circular Economy in Global Supply Chains

The concept of the 7Rs in the circular economy is an extended approach that builds on the original concept of three Rs: reduce, reuse and recycle (bifa Umweltinstitut, 2011, p. 1). Over the course of the approach's development, new Rs were added again and again, so that modern models can comprise up to ten Rs: refuse, rethink, reduce, reuse, rebuild, refurbish, remanufacture, repurpose, recycle and recover (Burri, 2022). This extension of the original concept offers a broader range of options for minimising resource consumption and making products more sustainable. The 7R approach, illustrated in Fig. 4, aims to integrate circular thinking and

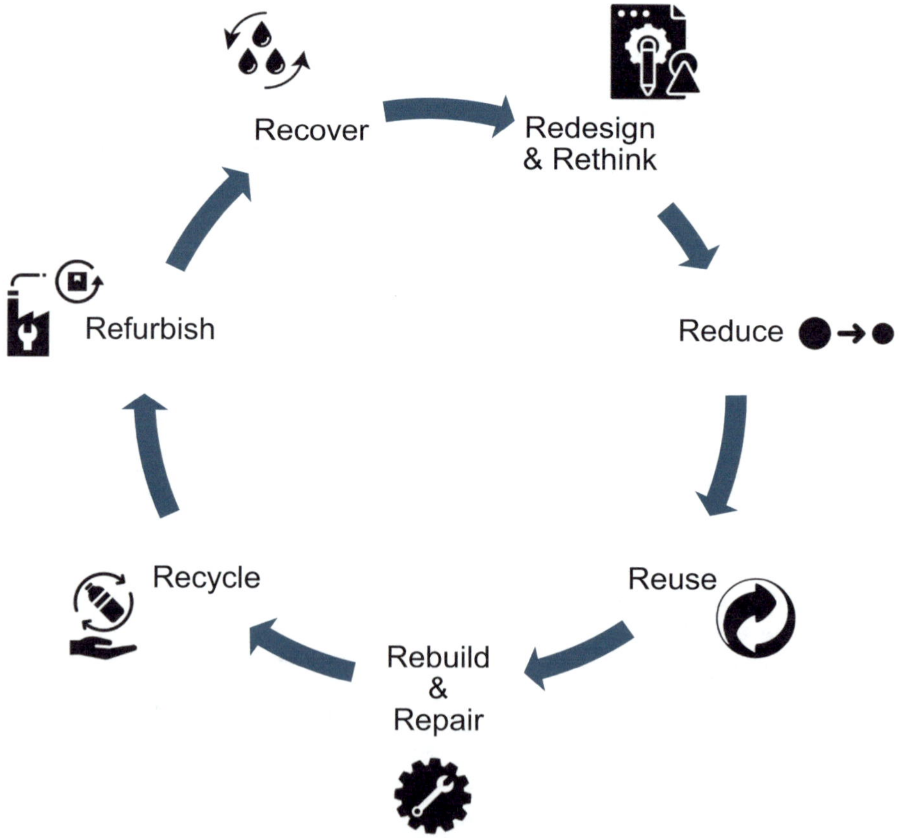

Fig. 4 The 7R model (own illustration)

production approaches into the economy and counteract the extent of current waste volumes. The core task is to actively keep materials in the cycle and thus maximise the service life of products. This involves implementing seven measures that companies need to consider in order to fully integrate circularity. Implementing these measures enables companies to reduce their environmental impact, increase their resource efficiency and promote sustainable business practices (Münger, 2023).

Redesign & Rethink. Fundamental changes to existing systems, designs and processes are essential to create closed loops. This change begins with the fundamental consideration of purchasing requirements and the resources needed. Economic and environmental benefits can be achieved simply by avoiding unnecessary resources and excessive quantities. The focus of rethink and redesign is on manufacturing products that are suitable for circulation and can be used without restriction in closed-loop systems. These products must be made from environmentally friendly resources and be durable, recyclable, reusable, repairable and fully degradable. Optimised product design contributes significantly to durability, reusability and reparability and thus reduces the need for new products. It is of great importance to promote processes and systems that implement the concept of the circular economy and take a holistic view of the entire life cycle of resources and products (Münger, 2023; Umwelt Bundesamt, 2020).

Reduce. The reduce measure aims to promote a circular economy by reducing purchase and order quantities and the use of environmentally harmful resources. This can be achieved through improved demand planning, in which the need for products or resources is carefully analysed and minimised. In addition, the reduction can be achieved by using recyclable and durable raw materials instead of environmentally harmful resources. The use of such raw materials helps reduce the environmental impact and optimise the consumption of resources (Umwelt Bundesamt, 2020, pp. 24–25).

Reuse. The reuse measure in the circular economy is concerned with the multiple uses of materials and products. The aim is to use materials or products as often as possible for the same or other purposes until they reach the end of their life cycle. This can be done either by directly reusing an entire product or by dismantling discarded products into individual parts that can then be used to manufacture new products. The main focus of this measure is to maximise reuse and thus extend the life of products (ZWIA, 2022).

Rebuild & Repair. Rebuild and repair aims to repair materials and products to extend their service life and conserve resources—a key aspect of a sustainable circular economy. This measure includes the repair of defective products and the use of repaired individual parts for the manufacture of new products. By carrying out repairs early in the product life cycle, functionality is maintained, and value is maximised. This not only helps reduce the need for new products but also cuts resource consumption and reduces waste (Dick, 2023).

Recycle. Recycling endeavours to obtain new, reusable raw materials from used products and materials. Thereby, waste is collected, sorted and reprocessed accordingly in order to keep resources in the cycle and reduce the need for new products and resources. The main objective is to extend the life cycle of resources and reduce the use of natural resources (BZ, 2019).

Refurbish. Refurbish aims to recondition used products to make them attractive and functional again. Refurbishing creates a new life cycle for the products by dismantling them and replacing non-functional parts, which significantly extends their lifespan. The aim is to conserve resources and reduce the amount of waste by refurbishing products rather than throwing them away (Ambros, 2023). This is supported by the previous measure of redesign and rethink, which aims to design products in such a way that they can be more easily recycled or reused later.

Recover. This measure aims to recover reusable materials at the end of the life cycle of disposed and defective products, which occurs when all other Rs are no longer feasible to recover raw materials or energy from the products (Swiss Recycling, 2023). In an ideal circular economy, this is rare, as resources should move continuously in the cycle. If resources can no longer be recycled or reused, the last individual parts are saved here, or energy is recovered from disposal, which flows back into the cycle.

In summary, the 7R model of the circular economy is a comprehensive strategy that aims to minimise the consumption of natural resources and reduce waste. The various measures of this model offer numerous approaches to conserve resources, minimise waste and simultaneously promote a stable economy. Procurement plays an important role in the implementation of this circular economy model, as it contributes significantly to achieving sustainability goals by focusing on the selection of environmentally friendly materials and suppliers that support the principles of the circular economy right at the start of the supply chain.

3.1 The Impact of the 7Rs on the SDGs

The circular economy concept is anchored in some of the United Nations' SDGs. The question arises to what extent this circular economy model can have an impact on achieving the SDGs. These potential impacts are explained in more detail below, with the argumentation based on the principle of plausibility. The explanations are based on existing findings and recognised facts.

Rethinking and redesigning products and production processes is an essential step in bringing about sustainable changes in companies and simultaneously promoting environmental protection. These efforts to create a more sustainable world have many positive effects on the 17 SDGs. As the first step towards a more sustainable future is rethinking and redesigning, *rethink* influences every SDG.

Through redesigning microfinance approaches and innovative financial models, rethink contributes to reducing poverty (SDG 1) by creating economic opportunities for low-income population groups and improving their access to education and health services. In agriculture, the rethinking of farming methods and the use of agroecological approaches promotes food security and protects the environment, which helps combat hunger (SDG 2). Using telemedicine and the development of less harmful products, rethink supports public health and well-being (SDG 3). In education, rethinking teaching methods and integrating digital solutions enables better access to quality education (SDG 4). Rethinking business strategies to implement women's empowerment programmes contributes to gender equality (SDG 5), among other things. New, innovative water filtration systems and recycling technologies optimise water consumption and minimise waste, which supports clean water and sanitation (SDG 6). Rethinking the energy sector promotes the use of renewable energy and more efficient energy consumption patterns, which contributes to energy efficiency and cleaner energy (SDG 7). By promoting socially responsible business practices and creating fair labour conditions, rethink supports decent work and economic growth (SDG 8). The introduction of sustainable manufacturing processes and the revision of infrastructure projects contribute to the sustainable development of industry and infrastructure (SDG 9). Rethinking social and human resource policies can reduce inequalities and promote inclusive employment practices, which contributes to reducing social inequalities (SDG 10). By considering sustainable architecture and urban planning innovations, rethink promotes sustainable cities and communities (SDG 11). Additionally, rethinking consumption and production patterns contributes to sustainable consumption patterns by promoting the circular economy and environmentally friendly packaging (SDG 12). Promoting carbon capture technologies and sustainable lifestyles reduces greenhouse gas emissions and supports climate action (SDG 13). Rethinking fishing practices and introducing new protected areas helps prevent overfishing and conserve marine biodiversity, which protects life underwater (SDG 14). By rethinking agricultural methods and land use practices, rethink promotes environmentally sustainable practices and protects biodiversity on land (SDG 15). Rethinking justice systems and implementing transparent institutions contributes to the promotion of peace and justice and improves access to justice (SDG 16). Finally, rethinking business practices promotes global partnerships and ethical supply chains, which contributes to the achievement of the SDGs through increased cooperation between governments, businesses and civil society (SDG 17).

By reducing the use of resources and the amount of waste, *reduce* contributes directly to the promotion of sustainable consumption and production patterns (SDG 12). At the same time, it supports climate protection (SDG 13) by reducing greenhouse gas emissions and contributes to the protection of marine biodiversity (SDG 14) and biodiversity on land (SDG 15). The positive effects also extend to clean water (SDG 6) and the promotion of sustainable industrial and infrastructure development (SDG 9). In addition, reduce influences the design of sustainable cities (SDG 11) and contributes to the promotion of clean energy (SDG 7). Overall, reduce makes a fundamental contribution to achieving various SDGs by aiming to use resources efficiently and minimise environmental impacts.

Reuse significantly contributes to promoting sustainable consumption and production patterns (SDG 12) by reducing the amount of waste generated. Simultaneously, it supports decent work and sustainable economic growth (SDG 8) by creating jobs in the recycling industry. The promotion of innovation and sustainable infrastructure development (SDG 9) is strengthened by the focus on durable and reusable materials. In urban environments, reuse helps promote sustainable practices and create sustainable cities (SDG 11). In addition, it reduces resource consumption in water treatment plants, supports clean water (SDG 6) and facilitates the transition to renewable energy (SDG 7). The principle of reuse also minimises the ecological footprint of production, contributes to climate protection (SDG 13) and supports the protection of rural ecosystems (SDG 15). Overall, reuse promotes a circular economy and helps achieve the SDGs by minimising resource consumption and environmental impact.

By extending the product life cycle, *repair* contributes directly to the promotion of sustainable consumption and production patterns (SDG 12), as the need for new production is reduced. The creation of jobs in the repair industry supports the goals for decent work and sustainable economic growth (SDG 8). Promoting innovation in the development of durable and repairable products contributes to a more sustainable industry and infrastructure (SDG 9). In addition, the repair of products supports the development of sustainable urban environments (SDG 11) by reducing waste and utilising resources efficiently. On a global level, repair contributes to the reduction of resource consumption and greenhouse gas emissions, which promotes the SDGs concerning climate protection (SDG 13) and sustainable land use (SDG 15). Overall, the repair principle contributes to the implementation of various SDGs by promoting the principles of the circular economy and enabling a more sustainable use of resources.

Through closing the material loop, *recycle* significantly contributes to promoting sustainable consumption and production patterns (SDG 12) by reducing the need for new raw materials. At the same time, the recycling industry creates jobs and supports the goals for decent work and sustainable economic growth (SDG 8). Promoting innovation in recycling technologies contributes to the sustainable development of industry and infrastructure (SDG 9). By introducing recycling programmes in cities, recycle promotes sustainable waste management and supports the development of sustainable urban communities (SDG 11). Globally, recycle contributes to the reduction of greenhouse gas emissions (SDG 13) and the protection of marine (SDG 14) and terrestrial ecosystems (SDG 15). Overall, the recycle principle promotes several SDGs by driving sustainable practices and making a positive contribution to the environment.

Refurbish contributes to the promotion of sustainable consumption and production patterns (SDG 12) by extending the product life cycle and reducing the need for new production. The refurbishing industry creates jobs and supports the goals for decent work and sustainable economic growth (SDG 8). Promoting innovation in the development of remanufacturing technologies contributes to the sustainable development of industry and infrastructure (SDG 9). The renovation of buildings and urban infrastructure promotes sustainable urban living spaces (SDG 11). By refurbishing electronic equipment, refurbish also supports the sustainable use of energy (SDG 7) and reduces greenhouse gas emissions associated with the production of new products (SDG 13). By making a positive contribution to the environment, refurbish promotes the achievement of various SDGs.

SDG 6 - Access to clean water and sanitation

1. Rethink & Redesign	2. Reduce	3. Reuse	4. Repair
New technologies can encourage water savings	Reduce can minimise water consumption, e.g. in agriculture and industry	Reuse of products leads to fewer new productions and thus less water consumption	Repairing products leads to fewer new productions and thus less water consumption

5. Recycle	6. Refurbish	7. Recover
Recycling reduces water pollution from industrial waste and simultaneously reduces the amount of water required for production	Refurbishment of sanitary facilities and water supply systems contributes to more efficient use of water	Recovering water as a raw material enables the resource to be used again

Fig. 5 Impacts of the 7Rs on SDG 6 (own illustration)

Recover helps promote sustainable consumption and production patterns (SDG 12) by enabling the efficient use and reuse of resources. The recovery of energy from waste supports the goals for affordable and clean energy (SDG 7) and contributes to the reduction of greenhouse gas emissions (SDG 13). By recovering water, recover promotes the sustainable and efficient use of water resources (SDG 6). It contributes to sustainable land use by promoting methods for reforestation and soil rehabilitation (SDG 15). Promoting innovation in the recycling industry by recovering materials contributes to the sustainable development of industry and infrastructure (SDG 9). In addition, recover supports sustainable waste management practices and promotes the development of sustainable urban communities (SDG 11). Through the safe recovery and disposal of medical waste, it also contributes to the health and well-being of the population (SDG 3). Overall, recover promotes the implementation of various SDGs by targeting the efficient use of resources and supporting sustainable practices in various sectors.

Detailed influences of each R on the SDGs can be found in the appendix. An example of the impact of the Rs on SDG 6 (access to clean water and sanitation) can be found in Fig. 5.

4 The Impact of Circular Procurement on the SDGs

As part of an extensive analysis, based on findings from my unpublished master's thesis and my professional experience, twelve SDGs were identified that can be influenced by one or more Rs of the circular economy model in addition to procurement (Fig. 6). Of these, nine SDGs are of particular importance (SDGs 6, 7, 8, 9, 11, 12, 13, 14 and 15), as all Rs of the circular economy have a direct influence alongside procurement.

SDG		Impact of Procurement Managment	Impact of Circular Economy (7R Model)						
			1. Rethink/ Redesign	2. Reduce	3. Reuse	4. Repair	5. Recycle	6. Refurbish	7. Recover
1	No Poverty	X	X					X	
2	Zero Hunger	X	X					X	
3	Good Health and Well-Being	X	X	X	X	X			
4	Quality Education	X	X						
5	Gender Equality	X	X						
6	Clean Water and Sanitation	X	X	X	X	X	X	X	X
7	Affordable and Clean Energy	X	X	X	X	X	X	X	X
8	Decent Work and Economic Growth	X	X	X	X	X	X	X	X
9	Industry, Innovation and Infrastructure	X	X	X	X	X	X	X	X
10	Reduced Inequalities	X	X						
11	Sustainable Cities and Communities	X	X	X	X	X	X	X	X
12	Responsible Consumption and Production	X	X	X	X	X	X	X	X
13	Climate Action	X	X	X	X	X	X	X	X
14	Life Below Water	X	X	X	X	X	X	X	X
15	Life on Land	X	X	X	X	X	X	X	X
16	Peace, Justice and Strong Institutions	X	X						
17	Partnerships for the Goals	X	X						

Fig. 6 SDGs influenced by procurement and the 7R model (own illustration)

This analysis shows that the implementation of circular economy approaches, especially procurement approaches, has a differentiated impact on the various SDGs, depending on which of the 7R strategies is applied. In particular, *rethink/ redesign*, which should be seen as a fundamental step in the value chain, has the potential to directly influence each of the 17 SDGs.

The example of SDG 6, access to clean water and sanitation, shows how essential the combination of the circular economy principles and procurement are for achieving this goal. Circular procurement plays a central role by supporting sustainable technologies and processes in the rethink and redesign process, which enables water savings in various sectors, such as agriculture and industry. For example, companies can define in their procurement policy to only work with suppliers that have a proven track record of using water-saving technologies, such as drip irrigation systems in agriculture or water-efficient cooling processes in industry.

By *reducing* water consumption in procurement—e.g. by purchasing highly efficient machines and systems—resources can be conserved and the availability of clean water increased. Conscious purchasing of raw materials that require less water-intensive production processes also plays a role here.

Furthermore, the *reuse* of products contributes to the reduction of new production, which in turn reduces water consumption. In procurement, this can be promoted by selecting suppliers who offer modular and easily repairable products to extend their service life and reduce the need for new production.

The *repair* of products is also supported by circular procurement strategies, for example, by companies signing contracts with service providers specialising in the repair and maintenance of industrial plants and machinery, which ultimately reduces the need for new production and, therefore, water consumption.

Recycling plays a dual role: procurement can support this by sourcing secondary raw materials made from recycled materials. Such materials often require significantly less water in the production phase, which not only reduces pollution from industrial waste but also minimises water consumption.

The renovation and modernisation of sanitary facilities and water supply systems (*refurbish*) can also be supported by circular procurement measures. Companies can favour special technologies that are known for their efficient water use, for example, by purchasing advanced sanitary technology that reduces water consumption.

Finally, the *recover* principle enables the recovery and reuse of water. Circular procurement can support this by investing in treatment technologies that enable contaminated water to be purified and returned to the production cycle. This contributes significantly to the sustainability of the water cycle and the development of sustainable water management.

The conclusion from analysing the circular economy and its principles is that circularity has a comprehensive and profound impact on the achievement of the SDGs. By promoting reuse, repair, recycle and refurbish, circularity not only contributes to economic efficiency and the creation of new jobs (SDG 1), but it also influences responsible consumption and production patterns (SDG 12) by increasing the efficient use of resources and reducing waste. It supports climate protection measures (SDG 13) by lowering energy consumption and reducing greenhouse gases while improving access to clean water (SDG 6) through innovative recycling technologies. In urban areas, it promotes sustainable growth and reduces resource consumption (SDG 11) whilst contributing to energy efficiency (SDG 7) by recovering energy from waste materials.

In addition, the circular economy indirectly improves public health (SDG 3) by reducing pollution, promotes quality education (SDG 4) and supports gender equality (SDG 5) and the reduction of inequalities (SDG 10). It contributes to protecting the oceans (SDG 14), promoting strong institutions (SDG 16) and strengthening global partnerships (SDG 17).

Overall, it is clear that the circular economy contributes significantly to achieving sustainable development through its circular approaches and is, therefore, a key building block for a fairer, healthier and more resilient world.

Procurement is a decisive factor in the realisation of the SDGs. Every aspect of procurement, from supplier selection to production methods, has the potential to influence each of the SDGs. Particular attention is paid to procurement management, which sets direct priorities by selecting suitable, innovative and responsible suppliers and trading partners. These decisions not only communicate future interests to the outside world and can reshape existing business relationships, but they also make it attractive to switch to new suppliers who are committed to the principles of sustainability.

The type and quantity of procurement are central to promoting the circular economy principles, which aim to use resources efficiently and minimise waste. This can be illustrated using the example of SDG 8 (decent work and economic growth). For example, a conscious choice of resources (8.4) and the improvement of productivity through technology (8.2) support the circular economy and thus directly promote goals such as sustainable production patterns and efficient resource consumption.

In addition, circular production processes that are structured according to the 7Rs can directly impact the SDG sub-goals and indirectly influence other SDGs, such as SDG 10, which aims to reduce inequalities. A well-established circular economy not only creates secure jobs but also promotes equal opportunities and self-determination in companies, which in turn counteracts inequalities.

Overall, the integration of circular procurement in companies leads to a shift towards a circular economy. This change can have both direct and indirect positive effects on the environment and the people in the supply chains. For example, a large electronics company can reduce the need for virgin plastic by sourcing recycled plastic for its products, significantly reducing water and energy consumption in the production chain. Another scenario is a clothing company that has switched to biodegradable materials that require less water in production and ensure that fewer harmful microfibres are released into the water when the products are washed.

There are also companies in the construction industry that save costs and significantly reduce the carbon footprint of their projects by sourcing recycled concrete and reused steel. The use of such materials helps reduce the extraction of natural resources and supports, among other things, the goal of sustainable cities and communities (SDG 11).

Nevertheless, the exact interactions between the SDGs influenced by circular procurement are not yet fully understood and can vary depending on the industry or company. For instance, a car manufacturer could not only reduce resource consumption by using recycled materials in car parts but also reduce emissions in the supply chain. However, further research and adjustments to corporate strategies are needed to fully utilise the potential of circular procurement and have a sustainable, positive impact on all SDGs.

In theory, circular procurement is almost exclusively associated with advantages, such as cost reductions or more environmentally friendly production (Schaller, 2021). These advantages are reflected in increased security of supply, which reduces dependence on fluctuating external sources of supply. Employees trained in circular economy contribute to increasing the value of the company through their expanded expertise and develop it into a centre of excellence. A company's market value can increase through such positioning while simultaneously responding to market fluctuations and improving the customer experience by taking back and refurbishing products. All of this contributes to greater customer loyalty and, at the same time, offers employees new incentives and more responsibility. In addition, procurement becomes more flexible and can react better to price fluctuations. Innovative strength increases through the search for and implementation of circular processes and materials, which also reduces the carbon footprint. Cost savings are also achieved by reducing transport routes. Despite these benefits, companies face challenges, such as the complexity of legal frameworks, which can hinder recovery approaches under certain circumstances. The switch to circular models is also time-consuming and cost-intensive, and the relationship between supply and demand is shifting, resulting in a shortage of recycled goods and a drastic increase in prices (Bernd, in press).

5 Outlook

Circular procurement plays a crucial role in promoting the SDGs. The principle of reusing and repurposing resources can directly impact goals, such as responsible consumption and production (SDG 12), by paving the way for more efficient and less wasteful production and consumption processes. It also contributes to combating climate change (SDG 13), as CO_2 emissions can be significantly reduced through reduced raw material consumption and waste production. Strengthening the resilience of infrastructures and promoting inclusive and sustainable industrialisation (SDG 9) are further important potentials of circular procurement. By creating closed loops, global supply chains are fundamentally changed, which creates resilience and stability and reduces dependencies on international markets. This requires coordinated cooperation and the creation of effective take-back and recycling processes that fulfil international standards. In procurement, management supports the principles of the circular economy by reducing material consumption, promoting reuse and recycling, recovering resources, repairing products and rethinking business models. This comprehensive approach not only makes companies more environmentally friendly but also promotes sustainable economic stability.

Appendix

	SDG	1. Rethink & Redesign	2. Reduce	3. Reuse	4. Repair	5. Recycle	6. Refurbish	7. Recover
1	**No Poverty**	– Rethinking financial models and enabling economic opportunities for low-income population groups through innovative financial design					– Refurbishment gives people on low incomes access to otherwise expensive products – Provides opportunities to purchase products they could not normally afford – Reduces financial barriers to certain products through refurbished options	
2	**Zero Hunger**	– Introduction of sustainable agricultural practices and cultivation of food with higher nutritional value to combat hunger and improve food security					– New refurbishment work processes in the circular value chain promote more sustainable agriculture – Supports more ecological methods and practices in agriculture – Contributes to the reduction of waste and resource wastage in the agricultural industry	
3	**Good Health and Well-Being**	– Redesigning products and processes can improve public health – Utilisation of telemedicine and innovative healthcare technologies	– Reducing the use of harmful raw materials directly improves public health	– Reuse can indirectly contribute to public health by reducing environmental pollution and the associated health risks (e.g. less single-use plastic)	– Repair can indirectly contribute to public health by reducing environmental pollution and the associated health risks (e.g. better air quality due to less waste incineration)			

	SDG	1. Rethink & Redesign	2. Reduce	3. Reuse	4. Repair	5. Recycle	6. Refurbish	7. Recover
4	**Quality Education**	– New learning methods and digital platforms can make education accessible worldwide						
5	**Gender Equality**	– Redesigning corporate strategies that specifically promote women in management positions						
6	**Clean Water and Sanitation**	– New technologies can encourage water savings	– Reduce can minimise water consumption, e.g. in agriculture and industry	– Reuse of products leads to less new production and therefore less water consumption	– Repair of products leads to less new production and therefore less water consumption	– Recycling reduces water pollution from industrial waste and at the same time reduces water consumption in production	– Refurbishment of sanitary facilities and water supply systems contributes to more efficient use of water	– Recovering water as a raw material enables the resource to be used again

	SDG	1. Rethink & Redesign	2. Reduce	3. Reuse	4. Repair	5. Recycle	6. Refurbish	7. Recover
7	**Affordable and Clean Energy**	– Designing more efficient products can accelerate the transition to cleaner energy sources	– Reducing energy consumption through more efficient transportation/ buildings reduces CO2 emissions	– Reuse of products leads to less new productions and therefore less energy consumption	– Repair of products leads to less new productions and therefore less energy consumption	– Recycling can improve energy efficiency and contribute to the utilisation of waste as an energy source	– Refurbish can reduce energy consumption in the product life cycle, especially for energy-intensive products such as electrical goods	– Recovered energy from discarded products or materials is used, for example, as self-generated renewable energy for new production processes
8	**Decent Work and Economic Growth**	– New circular processes create better and safer jobs – These jobs are characterised by fair pay and a pleasant working environment	– Reduce promotes decent jobs through new circular economy processes – Supports economic growth through these new labour processes – Ensures fair pay and working conditions	– Reuse promotes decent jobs through new circular economy processes – Supports economic growth through these new labour processes – Ensures fair pay and working conditions	– Repair promotes decent jobs in the circular economy – Supports economic growth through new labour processes – Ensures fair pay and working conditions	– Recycle promotes decent jobs in the circular economy – Supports economic growth by introducing new labour processes – Ensures fair pay and working conditions	– Refurbishment enables innovative work processes that create new jobs – These jobs are characterised by fair pay – A high level of occupational safety is guaranteed – There is a pleasant working environment	– Recover creates new work steps and processes that are characterised by fair pay, a high level of occupational safety and a pleasant working environment

	SDG	1. Rethink & Redesign	2. Reduce	3. Reuse	4. Repair	5. Recycle	6. Refurbish	7. Recover
9	**Industry, Innovation and Infrastructure**	– Redesigning products or processes promotes sustainable growth – Enables companies to expand into new business areas – Supports innovation and development of new market segments	– Reduce promotes innovation in products and processes – Supports sustainable growth for companies – Enables the development of new business areas through technological progress	– Reuse promotes innovation in products and processes – Supports sustainable growth for companies – Enables the development of new business areas through technological progress	– Repair promotes innovation in products and processes – Supports sustainable growth for companies – Enables the development of new business areas through technological progress	– Recycle promotes innovation in products and processes – Supports sustainable growth for companies – Enables the development of new business areas through technological progress	– Refurbishment technologies increase technological demands on industry and infrastructure – Drives sustainable growth – Promotes innovative solutions in various sectors	– New disposal and recovery methods increase technological demands on industry and infrastructure – Drives sustainable growth in various sectors – Promotes innovative solutions to environmental challenges
10	**Reduced Inequalities**	– Rethink/ redesign contributes to reducing social inequalities by promoting inclusive and equitable solutions for society through the rethinking of social and HR policies						

	SDG	1. Rethink & Redesign	2. Reduce	3. Reuse	4. Repair	5. Recycle	6. Refurbish	7. Recover
11	**Sustainable Cities and Communities**	– Redesigning urban spaces/ buildings with green roofs, for example, can reduce emissions and improve insulation	– Reducing waste promotes clean cities. Waste is reduced in advance through reduce	– Waste reduction contributes to the creation of clean cities – Longer product life cycles through reuse lead to less waste – Reduced waste improves urban cleanliness and environmental quality	– Waste reduction contributes to the creation of clean cities – Longer product life cycles through repair lead to less waste – Reduced waste improves urban cleanliness and environmental quality	– Waste reduction contributes to the creation of clean cities – Longer product life cycles through recycle lead to less waste – Reduced waste improves urban cleanliness and environmental quality	– Waste reduction contributes to the creation of clean cities – Longer product life cycles through refurbish lead to less waste – Reduced waste improves urban cleanliness and environmental quality	– By utilising the last remnants through recover, waste is reduced and cities are therefore cleaner

	SDG	1. Rethink & Redesign	2. Reduce	3. Reuse	4. Repair	5. Recycle	6. Refurbish	7. Recover
12	**Responsible Consumption and Production**	– Product design changes influence production patterns – Rethought production processes adapt to changing consumer patterns – Changes in production patterns are reflected in changing consumer preferences	– Waste reduction and circular economy minimise environmental impact – Creates new production routes and longer product life cycles – Increases resource efficiency and reduces greenhouse gas emissions – Fulfils growing customer demands on companies	– Waste reduction and circular economy minimise environmental impact – Creates new production routes and longer product life cycles – Increases resource efficiency and reduces greenhouse gas emissions – Fulfils growing customer demands on companies	– Waste reduction and circular economy minimise environmental impact – Creates new production routes and longer product life cycles – Increases resource efficiency and reduces greenhouse gas emissions – Fulfils growing customer demands on companies	– Waste reduction and circular economy minimise environmental impact – Creates new production routes and longer product life cycles – Increases resource efficiency and reduces greenhouse gas emissions – Fulfils growing customer demands on companies	– Waste reduction and circular economy minimise environmental impact – Creates new production routes and longer product life cycles – Increases resource efficiency and reduces greenhouse gas emissions – Fulfils growing customer demands on companies	– Waste reduction and circular economy minimise environmental impact – Creates new production routes and longer product life cycles – Increases resource efficiency and reduces greenhouse gas emissions – Fulfils growing customer demands on companies

	SDG	1. Rethink & Redesign	2. Reduce	3. Reuse	4. Repair	5. Recycle	6. Refurbish	7. Recover
13	**Climate Action**	– Redesigning products and processes to reduce waste – Development of resource-saving products for the circular economy – New designs can reduce CO2 emissions, which is positive for the climate – Promotes environmental protection by reducing waste and CO2 emissions	– Less waste = less greenhouse gas emissions in production, use and disposal – Active help in the fight against climate change through reduction – Reduce promotes a circular economy and extends product life cycles – A significant step in climate protection and for the environment	– Less waste = less greenhouse gas emissions in production, use and disposal – Active help in the fight against climate change through reduction – Reuse promotes a circular economy and extends product life cycles – A significant step in climate protection and for the environment	– Less waste = less greenhouse gas emissions in production, use and disposal – Active help in the fight against climate change through reduction – Repair promotes a circular economy and extends product life cycles – A significant step in climate protection and for the environment	– Less waste = less greenhouse gas emissions in production, use and disposal – Active help in the fight against climate change through reduction – Recycle promotes a circular economy and extends product life cycles – A significant step in climate protection and for the environment	– Less waste = less greenhouse gas emissions in production, use and disposal – Active help in the fight against climate change through reduction – Refurbish promotes a circular economy and extends product life cycles – A significant step in climate protection and for the environment	– Waste reduction supports the reduction of greenhouse gas emissions in the fight against climate change – Recover maximises the potential of products and materials by recovering valuable materials and energy – Utilising all resources to avoid waste and increase efficiency

	SDG	1. Rethink & Redesign	2. Reduce	3. Reuse	4. Repair	5. Recycle	6. Refurbish	7. Recover
14	**Life Below Water**	– Reduction of waste to preserve marine ecosystems and biodiversity – Product concepts focusing on longevity and low waste production – Preservation of marine ecosystems through sustainable product concepts	– Reduced use of plastics in production leads to less waste that can end up in the oceans	– Longer product life cycles through reuse reduce marine litter – Protection of marine ecosystems through reduced waste input – Conservation of biodiversity through less pollution of the oceans	– Longer product life cycles through repair reduce marine litter – Protection of marine ecosystems through reduced waste input – Conservation of biodiversity through less pollution of the oceans	– Longer product life cycles through recycle reduce marine litter – Protection of marine ecosystems through reduced waste input – Conservation of biodiversity through less pollution of the oceans	– Longer product life cycles through refurbish reduce marine litter – Protection of marine ecosystems through reduced waste input – Conservation of biodiversity through less pollution of the oceans	– Recover reduces the need to extract new raw materials and thus contributes to the protection of marine ecosystems

	SDG	1. Rethink & Redesign	2. Reduce	3. Reuse	4. Repair	5. Recycle	6. Refurbish	7. Recover
15	**Life on Land**	– Product concepts focusing on longevity for waste reduction. – Preserving marine ecosystems and biodiversity by reducing waste. – Goal: minimising waste through durable product concepts	– Reduced use of natural resources (e.g. wood) counteracts deforestation and protects biodiversity	– Longer product life cycles through reuse reduce waste – Reduces environmental impact and protects continental ecosystems – Supports biodiversity conservation and sustainable protection of animals and plants	– Longer product life cycles through repair reduce waste – Reduces environmental impact and protects continental ecosystems – Supports biodiversity conservation and sustainable protection of animals and plants	– Longer product life cycles through recycle reduce waste – Reduces environmental impact and protects continental ecosystems – Supports biodiversity conservation and sustainable protection of animals and plants	– Longer product life cycles through refurbish reduce waste – Reduces environmental impact and protects continental ecosystems – Supports biodiversity conservation and sustainable protection of animals and plants	– Recover reduces the need to extract new raw materials and thus contributes to the protection of terrestrial ecosystems

	SDG	1. Rethink & Redesign	2. Reduce	3. Reuse	4. Repair	5. Recycle	6. Refurbish	7. Recover
16	**Peace, Justice and Strong Institutions**	– Rethinking justice systems to promote justice and peace – Develop designs that promote transparency in the supply chain to actively counter corruption						
17	**Partnerships for the Goals**	– Redesign is crucial for achieving the Rs – Serves as the first instance for meeting and exchange (e.g. on innovations, standards) – Promotes exchange for the achievement of the Sustainable Development Goals (SDGs)						

References

Ambros, F. (2023). *Refurbishment als Beispiel für zirkuläre Geschäftsmodelle.* Haufe. https://www.haufe.de/sustainability/umwelt/kreislaufwirtschaft-und-circular-economy/refurbishment-als-beispiel-fuer-zirkulaere-geschaeftsmodelle_575774_579654.html

Bernd, C. (in press). Circular Procurement – Die neue nachhaltige Herausforderung im Einkauf. In E. Fröhlich & Y. Jamal (Eds.), *CSR und Beschaffung* (2nd ed.). Springer.

bifa Umweltinstitut. (2011). *Abschlussbericht: 3R-Studie.* https://www.bmuv.de/fileadmin/Daten_BMU/Download_PDF/Abfallwirtschaft/3r_abschlussbericht_bf.pdf

BMUV – Bundesministerium für Umwelt, Naturschutz, Nukleare Sicherheit und Verbraucherschutz. (2023). *17 Nachhaltigkeitsziele – SDGs.* https://www.bmuv.de/themen/nachhaltigkeit/nachhaltigkeitsziele-sdgs

BMZ – Bundesministerium für Wirtschaftliche Zusammenarbeit und Entwicklung. (2023). *SDG 12: Nachhaltige/r Konsum und Produktion.* https://www.bmz.de/de/agenda-2030/sdg-12

Burri, S. (2022). *Die "10 R" der Kreislaufwirtschaft.* Kreislaufwirtschaft Bau. https://kreislaufwirtschaft-bau.ch/2022/06/23/10r-der-kreislaufwirtschaft/

BZ. (2019, June 12). Wie funktioniert Recycling und wie gut ist das System? *B.Z. – Die Stimme Berlins.* https://www.bz-berlin.de/archiv-artikel/wie-funktioniert-recycling-und-wie-gut-ist-das-system

Carter, C. R., & Easton, P. L. (2011). Sustainable supply chain management: Evolution and future directions. *International Journal of Physical Distribution & Logistics Management, 41*(1), 46–62. https://doi.org/10.1108/09600031111101420

Chopra, S., & Meindl, P. (2020). *Supply Chain Management. Strategy, Planning, and Operation* (7th ed.). Pearson.

Dick, A. (2023, March 15). Kreislaufwirtschaft stärken: Reparieren vor Recyceln. *DAV Panorama.* https://magazin.alpenverein.de/artikel/kreislaufwirtschaft-staerken_616eb7db-fb4d-4a67-8bcf-49a47c57f90a

Die Bundesregierung. (2023). *Die 17 globalen Nachhaltigkeitsziele verständlich erklärt.* https://www.bundesregierung.de/breg-de/themen/nachhaltigkeitspolitik/nachhaltigkeitsziele-erklaert-232174

Fröhlich, E. (2015). Corporate Social Responsibility in der Beschaffung: Theoretisch wie Praktische Implikationen. In E. Fröhlich (Ed.), *CSR und Beschaffung. Theoretische wie praktische Implikationen eines nachhaltigen Beschaffungsprozessmodells* (1st ed.). Springer Gabler.

Gartner. (2024). *Nachhaltigkeit der Lieferkette.* https://www.gartner.de/de/supply-chain/insights/nachhaltigkeit-der-lieferkette

HEC Paris. (2017). *The State of Sustainable Procurement: Latest Ecovadis/HEC Paris Barometer is Released.* https://www.hec.edu/en/overview/news/state-sustainable-procurement-latest-ecovadis/hec-paris-barometer-released

Jones, M., Kinch Sohn, I., & Lysemose Bendsen, A.-M. (2017). *Circular Procurement. Best Practice Report.* ICLEI – Local Governments for Sustainability, European Secretariat. https://circulars.iclei.org/wp-content/uploads/2021/01/Circular_Procurement_Best_Practice_Report.pdf

Krajewski, L. J., Malhotra, M. K., & Ritzman, L. P. (2016). *Operations management: Processes and supply chains* (11th ed.). Pearson.

Kummer, S., Gruen, O., Jammernegg, W. (2009). Grundzüge der Beschaffung, Produktion und Logistik. (2nd. ed.). Pearson.

Münger, A. (2023). *Definition: Design Circular.* Haufe. https://www.haufe.de/sustainability/umwelt/kreislaufwirtschaft-und-circular-economy/definition-design-circular_575774_577298.html

Schaller, S. (2021). *Die strategische Bedeutung zirkulärer Beschaffung.* Cosinex Blog. https://blog.cosinex.de/2021/06/08/zirkulaere-beschaffung/

Swiss Recycling. (2023). *Die 10 "Re-" der Kreislaufwirtschaft - von Rethink über Reuse bis Recover.* https://swissrecycle.ch/de/aktuell/detail/die-10-re-der-kreislaufwirtschaft-von-rethink-ueber-reuse-bis-recover

Umwelt Bundesamt. (2020). *Leitsätze einer Kreislaufwirtschaft*. https://www.umweltbundesamt.de/sites/default/files/medien/1410/publikationen/2020_04_27_leitlinie_kreislaufwirtschaft_bf.pdf

van Oppen, C., Croon, G., & de Vroe, D. B. (2018). *Circular Procurement in 8 Steps*. Copper8. https://www.copper8.com/wp-content/uploads/2018/10/Circular-Procurement-in-8-steps-Ebook.pdf

Wellbrock, W., & Ludin, D. (2019). *Nachhaltiges Beschaffungsmanagement. Strategien-Praxisbeispiele-Digitalisierung*. Springer.

World Economic Forum. (2020). *Want to help the world meet the UN's Sustainable Development Goals? Here's where to start*. https://www.weforum.org/agenda/2020/09/sustainable-development-impact-summit-un-sustainable-development-goals/

Zimmermann, F. M. (2016). *Nachhaltigkeit wofür? Von Chancen und Herausforderungen für einen nachhaltige Zukunft*. Springer Spektrum.

ZWIA – Zero Waste International Alliance. (2022). *Zero Waste Hierarchy of Highest and Best Use 8.0*. https://zwia.org/zwh/

Chiara Bernd is a management consultant for supply chain management, procurement and sustainability. She completed her master's degree in Sustainable Management at the CBS—International Business School in Cologne and started to focus on research in the field of circular economy, with a special interest in circular procurement. In addition to her professional activities, she is also a PhD candidate in the field of sustainable transformation and circular economy. She is also passionate about sustainable education, which is why she is a PRME (Principles for Responsible Management Education) Sustainability Ambassador and a lecturer in sustainability management.

Circular Economy: Contested Definitions and an Evolving Concept

Roelof Vogel, Tharaka de Vass, Glen Croy, and Amrik Sohal

1 Introduction

Many nations have committed to achieving net zero carbon emissions by the year 2050. These commitments require systemic changes to the economic system, particularly imposed upon supply chains. The dominant economic system, characterised by the "take-make-use-dispose" linear supply chain model, needs to be transformed. The two necessary transformations are decoupling the economic system from (i) a dependence on extraction of finite resources and (ii) an inherent generation of waste streams.

The proposed solution, and gradually being adopted by nations and businesses, is the circular economy (CE) model. CE is based upon the principle that there are so many resources (materials, components and products) already in the economic system, that instead of extracting more, we can better utilise those already available. This approach also inherently reduces waste through the reorientation of value and specific efforts to capture resource value instead of allowing waste.

Hence, there is an emphasis on the adoption of a restorative and regenerative CE model that promotes iterative circulation of resources within the supply chain (Mishra et al., 2022). Transformation towards circularity requires enterprise remodelling of industrial supply chains, including the extraction, production, distribution and consumption process at the levels of design, implementation and operations (Khalifa et al., 2022; Sohal & de Vass, 2022). Minimising resource inputs and emissions, slowing, closing and narrowing resource loops, and promoting upcycling to sustain a regenerative system require collaboration of all enterprises and

R. Vogel (✉) · T. de Vass · G. Croy · A. Sohal
Monash Business School, Monash University, Caulfield East, VIC, Australia
e-mail: roelof.vogel@monash.edu; tharaka.devass@monash.edu; glen.croy@monash.edu; amrik.sohal@monash.edu

E. Fröhlich et al. (eds.), *Circular Economy in Sustainable Supply Chains*, CSR, Sustainability, Ethics & Governance,
https://doi.org/10.1007/978-3-031-70749-0_16

stakeholders in supply networks (de Vass et al., 2023). Therefore, a better understanding of how the CE concept progressed may stimulate behavioural changes, generate coordinated policies and promote innovative circular practices and technologies to accelerate circularity within the ecosystem.

A persistent challenge, however, is that the CE concept has evolved and branched into a diversity of understandings. As such, especially for those new to engaging in supply chain circularity discussions, appreciating the CE lexicon and even what CE is can be very confusing. That is, what some define and practice as CE has diversified to being something quite different to what others define and practice as CE.

In this context, we aim to explore and demonstrate the divergence of CE definitions into two definitional streams, and how this has affected the conceptualisation and then implementation of circularity in supply chains. To achieve the aim, we present three objectives. First, to trace the plethora of CE definitions, providing a grounding in the origins of CE principles. Second, to assess CE conceptualisation in relation to (i) coherence, (ii) permanence and (iii) construct collapse potential (Blomsma & Brennan, 2017). Third, to identify the relevance and implications of the definitional divergence and conceptualisation for supply chains.

The chapter is therefore presented in five parts. First, we interrogate CE definition narratives and the bifurcation into two streams. Second, we briefly introduce the genesis of the CE concept. Third, we present CE's conceptual evolution, demonstrating the growing sophistication and breadth of scope, and explore the current stage of evolution. Fourth, we focus on the implications for supply chains. Finally, we briefly conclude the chapter and its contributions.

2 Contested Circular Economy Definitions

Blomsma and Brennan (2017, p. 606) refer to CE as an umbrella concept, described as "a broad concept or idea used loosely to encompass and account for a set of diverse phenomena". However, as Chizaryfard et al. (2021) conclude, while there is general agreement on the paradigmatic nature of CE, it is essentially a contested concept. This contestation is founded in what Corvellec et al. (2022, p. 422) term a "definitional quagmire".

The Ellen MacArthur Foundation's (EMF) first major publication presents a CE definition which proved to be an important milestone in the CE conceptualisation (EMF, 2013, p. 24):

> A circular economy is an industrial system that is restorative or regenerative by intention and design. It replaces the 'end-of-life' concept with restoration, shifts towards the use of renewable energy, eliminates the use of toxic chemicals, which impair reuse, and aims for the elimination of waste through the superior design of materials, products, systems, and, within this, business models.

Nevertheless, this did not stop the variety of definitions being presented. For example, in a review of 114 CE publications, Kirchherr et al. (2017) find 95 different definitions of CE. The most used is the EMF definition, contributing 15 of the 114. Of the remaining definitions, only two others are used more than once, highlighting the emergent nature of CE theory and praxis, as well as on-going contests.

Key to the definitional contest is a broad or tight scope of CE to address the environmental, social and economic pillars of sustainability. Two main definitional streams have emerged. First, those with a philosophical position that CE needs to be holistic and clearly focus on supporting all sustainability outcomes (e.g. Kirchherr et al., 2017). Second, is the stream that argues that CE does not and even cannot address the three pillars of sustainability, particularly that of the social dimension (e.g. Murray et al., 2017).

In the first stream, authors explicitly note that the aim of CE is to address the three sustainability pillars. For example, "a circular economy describes an economic system … with the aim to accomplish sustainable development, which implies creating environmental quality, economic prosperity and social equity, to the benefit of current and future generations" (Kirchherr et al., 2017, p. 224), and "successful circular economy contributes to all three dimensions of sustainable development" (Korhonen et al., 2018, p. 37). This first stream considers CE "as a tool to organise resources for the purpose of maintaining or enhancing social well-being, environmental quality and economic prosperity" (Velenturf & Purnell, 2021, p. 1447).

However, even with the same general conceptualisation of CE, the second stream limits the scope of aims. Here, the aim of CE is to maximise resource efficiency and effectiveness. Murray et al. (2017) criticise the first stream, arguing that CE cannot address the three pillars of sustainability (also see Corvellec et al., 2022). They specify that "it is unclear how the concept of the circular economy will lead to greater social equality, in terms of inter- and intra-generational equity, gender, racial and religious equality and other diversity, financial equality, or in terms of equality of social opportunity. These are important moral and ethical issues which are missing from the construct" (Murray et al., 2017, p. 376).

Continuing this stream of a tighter scope, for example, Kalmykova et al. (2018) identify the aim of CE as decoupling the economic system from resource extraction, instead arguing sufficient resources already in the system can be better utilised (also see Demirel & Danisman, 2019). Furthering this stream, common principles are prioritised: (i) to preserve and enhance natural capital, (ii) to optimise resource yields and (iii) to foster system effectiveness (Acerbi & Taisch, 2020; EMF, 2013, 2015). The definitions emerging from this second stream reflect that presented by the European Union (EU, 2018 cited in Sarja et al., 2021, p. 2): "CE is a model where the value of products, materials, and resources is maintained in the economy for as long as possible, and the generation of waste is minimised".

Within this array of CE definitions, there emerges a clear bifurcation between those definitions that (i) seek to directly address the three pillars of sustainable development and (ii) those that remain more narrowly focused on resource utilisation and greater efficiencies in resource use. In the second stream, there is the possibility of incidental sustainability outcomes, though these are not inherent within the approach and aims. It is of note that each definitional stream appears to adopt a common conceptualisation of CE; nevertheless, the primary tension arises from the extent to which CE has the capacity to replace current socio-economic-political systems and influence the full breadth of economic, environmental, social governance and outcome needs of society.

We take the position aligning with the second stream that CE has a primary aim to maximise resource effectiveness and efficiencies. Subsequently, CE may and will likely contribute to wider sustainability goals and pillars, though we do not see this as the primary aim.

3 The Circular Economy Conceptual Framework

Despite the definitional variety and contest in stated CE aims, as noted, there appears to be a relatively common conceptual framework. The genesis of the CE concept is frequently attributed to Boulding (1966), e.g. by Blomsma and Brennan (2017) and Murray et al. (2017). Boulding (1966, p. 9), when critiquing the consumptive economic system and Earth's limits for resource extraction and pollution absorption, coined the phrase "a cyclical ecological system". His core argument for the cyclic ecological system was an economy that extracted resources within the limits of regeneration and polluted within the Earth's ability to absorb it. Despite this, by the end of the 1960s we had exceeded Earth's capacity; we are now using 1.75 earths, and this is expected to be 2 earths by 2030 (The World Counts, 2023).

All the same, in efforts to address the over-consumptive and wasteful traditional economic model and to live within the limits of what Boulding (1966) called "spaceship earth", others have proposed a more specific CE model. In these propositions, the same core argument persists; however, the focus has changed to recycling the resources (materials, components and products) in the economic system. It is the work of many scholars that has contributed to the introduction and refinement of CE definitions, and we review these chronologically.

Stahel and Reday-Mulvey (1976, cited by Stahel, 2019) focus on the relative costs of resource transformation into finished products, as well as the importance of labour as a factor of production. They identified the transformative process of value creation. This is a precursor to the importance of maintaining resources at their highest possible value (peak value). They additionally introduce the vision of an "economy in loops", with a focus on job creation, economic competitiveness, resource savings and waste minimisation (Stahel & Reday-Mulvey, 1976, p. 31).

Mollison (1979) contributes the concepts of resilient and regenerative biological systems to the still emerging CE definition. Building upon Boulding (1966), this is the genesis for the CE principle to regenerate natural systems. Indeed, the ecological perspective persists in the work of Frosch and Gallopoulos (1989, p. 144), but they viewed the economic system as an "industrial ecosystem". They define the industrial ecosystem in which "the consumption of energy and materials is optimised; waste generation is minimised and the effluents of one process … serve as the raw material for another process". This extends Boulding's (1966) minimisation by introducing resource circularity processes. Nonetheless, it was Pearce and Turner (1990) who first prominently used the term CE.

Extending the circular industrial ecosystem discussion, Lyle (1994, p. 10) focused on "regenerative design", specifically designing for resource regeneration opportunities. He provides a practical guide to the design and construction of regenerative energy and water systems, which are foundational precursors to the CE principles of (i) designing out waste and pollution and (ii) keeping products and materials in use (Webster, 2015). Even so, Lowe and Evans (1995) provide a clearer circularity in the industrial ecosystem. They specifically note the central goal to move from a linear to a closed-loop system in all realms of production and consumption, including material and energy flows within industrial systems.

Braungart and McDonough (2002), in their work *Cradle to cradle: Remaking the way we make things,* reinforce the already emerging CE principles of peak value (with reference to a four-stage Rs hierarchy: reduce, reuse, recycle and regulate), eco-effective design, maximising renewable energy and celebrating diversity. They exemplify closed-loop systems, proposing the elimination of the concept of waste. Instead of waste, they argue residual resources are valuable resources for other processes. Braungart and McDonough (2002) also divide material flows into biological and technological categories.

Pauli (2010) further adds the concept of cascading systems of resources. As well as reinforcing the desire of maintaining resource peak value, he emphasises the simple and multiple opportunities to capture and recreate value.

Through these and other works, core CE conceptual frames emerged. Table 1 provides a summary of the key contributions to the CE conceptual framework (see also Webster, 2015).

In summary, the conceptual framework for CE began to slowly emerge in the 1960s. However, the uptake of the CE concept started to accelerate in the 2010s, as evidenced by the emergence of a new wave of thought leadership and global promotion of CE (EMF, 2013), increasing academic attention and publications (Kirchherr et al., 2017) and public policy and governance (Alonso-Almeida & Rodríguez-Antón, 2020).

Table 1 Origins of the circular economy conceptual framework

Conceptual frame	Conceptual contribution	Key Author/s
Earth's finite resources and absorptive capacity	Finite limit of resources; reckless and exploitative open systems are an issue	Boulding (1966)
Value creation and peak value	Labour in the industrial economy; closed-loop approach to production processes; product life extension; long-life goals; reconditioning; waste prevention	Stahel and Reday-Mulvey (1976); Stahel (2006, 2013)
Regenerative systems	Design of productive agricultural ecosystems that have the diversity, stability and resilience of natural ecosystems	Mollison (1979)
Waste as an input resource	Waste from one industrial process as the raw material input for another, so reducing the impact of industry on the environment	Frosch and Gallopoulos (1989)
Circular economy term	Natural resource extraction and the environment; circular economy terminology and concept to recognise resource limits	Pearce and Turner (1990)
Regenerative design	Regenerative design can be applied beyond agriculture	Lyle (1994)
Closed-loop systems	Move from a linear to a closed-loop system in all realms of human production and consumption	Lowe and Evans (1995)
Cradle to cradle Residual resources (rather than waste) Biological and technological material flows	Waste equals food; renewable energy; celebrate diversity	Braungart and McDonough (2002)
Value capture and re-creation	Challenge and rethink the total economic model; use resources available in cascading systems	Pauli (2010)

4 The Phased Evolution of the Circular Economy Concept

The emergence of the CE conceptual framework reflects the staged transformative evolution of the concept itself. Various authors have proposed to explain evolutionary stages of CE (e.g. Blomsma & Brennan, 2017; Geissdoerfer et al., 2020; Kirchherr et al., 2017; Reike et al., 2018).

Blomsma and Brennan (2017) identify three periods, namely, *The preamble period* (1960–1985), *The excitement period* (1985–2013) and *The validity challenge period* (2013–present). Similarly, Reike et al. (2018) define the transformation periods as CE 1.0 (1970–1990s) *Dealing with waste,* CE 2.0 (1990s–2010) *Connecting input and output in strategies for eco-efficiency* and CE 3.0 (2010–Present) *Maximising value retention in the age of resource depletion.* Taking a slightly different division in the evolutionary phases, Kirchherr et al. (2017) and Geissdoerfer et al. (2020) provide two-step transformations with the second step occurring after 2010.

What follows is a description of these works by transformation period and their collective contribution to an understanding of the CE concept transformation.

The Blomsma and Brennan (2017) *Preamble period* (1960–1985) is characterised by a primary focus on waste handling and the polluting effects of waste streams. The resource life extending strategies (RLES) highlighted during the preamble period primarily relate to end-of-life (EoL) processes. Waste handling strategies at EoL were prominent, including cleaner incineration, waste-to-energy, composting and recycling. Blomsma and Brennan (2017) emphasise two important CE concept evolutions during the preamble period. First, the need for responsible management of natural resources. Second, the progress in the academic fields contributing to CE thinking as well as the emergence of multi-disciplinary approaches and new disciplines. Reike et al. (2018) define their first period as CE 1.0 (1970–1990s) *Dealing with waste*, and the focus is on the "output side" of processes and waste. The 3R concept of reduce, reuse and recycle gains in prominence.

The Blomsma and Brennan (2017) *Excitement period* (1985–2013) started with the emergence of valuing waste as an important resource. They highlight the five important developments during this second period as (i) extension of life-cycle thinking through recycling, urban mining and product service, (ii) a wider discussion on sustainable development framed as CE opportunity, (iii) recycling explicitly positioned as a source of raw materials, (iv) the emergence of two groups of CE frameworks, the first from policymakers in framing CE legislation and the second from business, and (v) the ascendancy of CE as the cognitive unit for the RLES movement and as the dominant umbrella concept for holistic resource management. Reike et al. (2018) identify CE 2.0 (1990s–2010) *Connecting input and output in strategies for eco-efficiency* with the focus turning to a stronger connection between preventive measures and outputs. Economic opportunity and systems thinking grows and the CE concept gains some prominence. Both Kirchherr et al. (2017) and Geissdoerfer et al. (2020) broadly review the first two periods discussed as a single transformation period during which CE definition, scope and theory is under development.

The Blomsma and Brennan (2017) *Validity challenge period* (2013–Present) is characterised by a new level of engagement with CE as the umbrella cognitive unit and a new discursive space to further develop CE knowledge. Many new facets of the CE concept emerge, and there is a lack of unified definitions, paradigms and theories. Blomsma and Brennan (2017) highlight the work ahead as first, the theoretical development of the CE concept, and second, the development of strategies and knowledge for implementation of CE. Reike et al. (2018) identify CE 3.0 (2010–present) *Maximising value retention in the age of resource depletion* as the period in which CE emerges as a solution for decoupling resource use and economic growth. The key complementary contribution of Kirchherr et al. (2017) to the understanding of the transforming CE concept in this period is that a "CE systems" perspective is recognised as a common theme. Geissdoerfer et al. (2020) extend the CE concept development to include the emergence of circular business models (CBMs) in this third period.

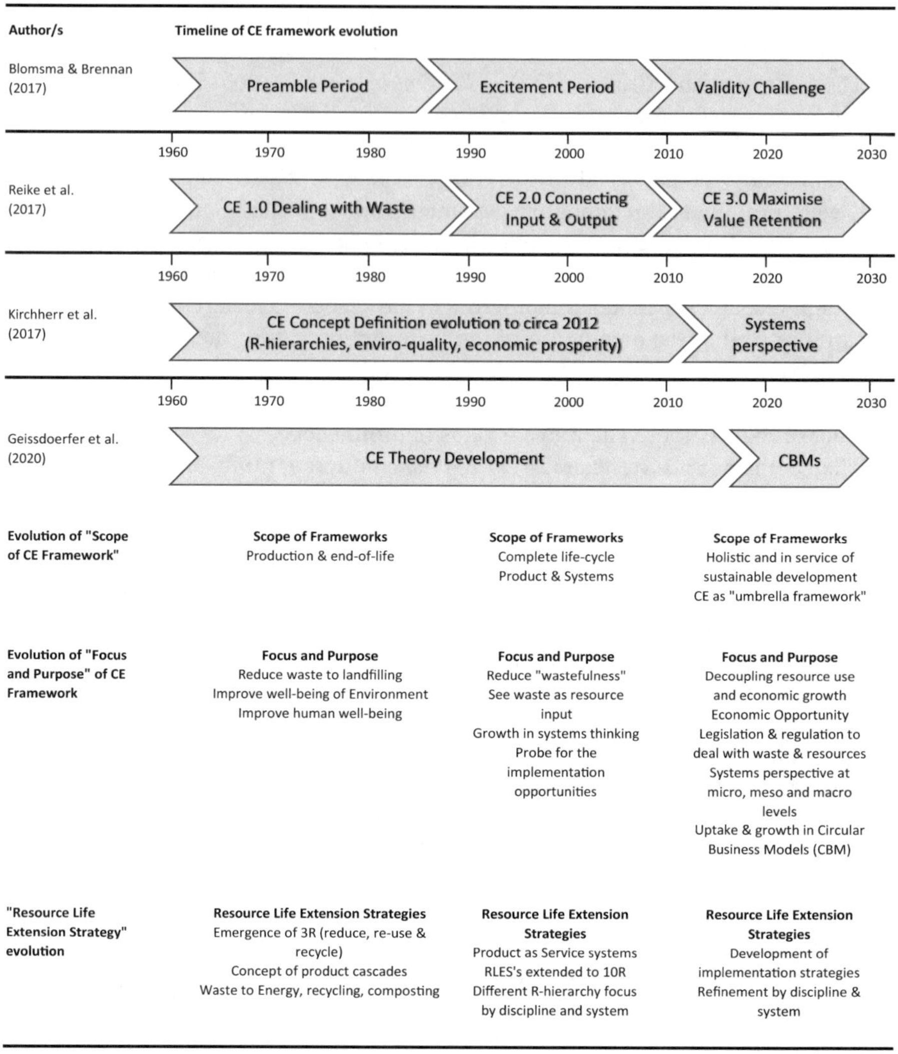

Fig. 1 Timeline of the circular economy concept evolution (own illustration)

Figure 1 provides a summary of the CE concept development periods discussed above. Below the timelines in Fig. 1, the key common themes of the CE concept framework (following the format of Blomsma & Brennan, 2017) are analysed under three headings: (i) scope of framework, (ii) focus and purpose and (iii) RLES.

In summary, the CE concept has undergone a transformation from the 1960s to the present. While the early focus was more confined to waste minimisation, resource utilisation and a narrower concept of RLES, a more holistic focus is emerging, advocating that CE should play a greater role in a systems approach and on business models.

Reflecting the evolution of the CE concept, its demonstration in practice is similarly evidenced. That is, for some businesses, cities, regions and nations we see a focus on waste management and minimisation. However, in others there are demonstrations of maximising value retention and systems perspectives. For example, demonstrating their advanced efforts, some governments have introduced CE policies and legislation at national, regional and local (city) level to advance the adoption of CE innovation and solutions within their precincts. The first major national level CE policy legislation was enacted in China in 2008. In 2005, the China State Council issued "Opinions on accelerating the development of the circular economy", which proposed a policy framework, CE principles, objectives, key tasks and policy measures to promote CE. The "Circular Economy Promotion Law" was published in 2008, with effect from 2009, and introduced the 3R strategy of reduce, reuse and recycle. This law was revised and updated in 2018 (Bleischwitz et al., 2022). The European Union (EU) CE initiatives started with the 2011 white paper "Roadmap to a resource efficient Europe", followed by the 2015 action plan "Closing the loop: An EU action plan for the circular economy" (Alonso-Almeida & Rodríguez-Antón, 2020).

5 The Validity Challenge Period

In this section, we seek to further explore the current *validity challenge period* and the progress to resolve the challenges (Blomsma & Brennan, 2017). With growing policy initiatives at all three levels of government in many developed and developing economies, and a systemic focus on macro, meso and micro levels, Blomsma and Brennan (2017) identified that issues for the CE would be in the areas of (i) coherence, (ii) permanence and (iii) construct collapse. We see four interlinked issues in the current global "polycrisis" and discuss their interaction in relation to CE:

1. Climate change mitigation and adaptation
2. SDG progress lag
3. Circular economy adoption
4. Hegemony of corporate capitalism

The evidence of accelerating climate change events is no longer in dispute. In part, this is due to a dramatic rise in climate related insurance claims and the global insurance industry exposure (GIMAR, 2022). Kahl (2023) reports that prior to 2005, insured losses for natural disaster events were never greater than US$50 billion, but that the "new normal" is for insured losses greater than US$100 billion. Indeed, the average annual global insured losses for the five-year period 2018–2022 were an estimated US$97 billion, while average annual global total losses (i.e. including uninsured) amounted to an estimated US$270 billion (NatCatServices, 2022). Global efforts to slow the rate of CO2-equivalent (CO2e) emissions into the atmosphere have been well below the targets identified by the United Nations

International Panel on Climate Change (IPCC) since its inception. The first IPCC report (1989) identified that anthropomorphic climate change was a result of excessive CO2e emissions. In 1994, the United Nations Framework Convention on Climate Change (UNFCCC) was created, and today it has near-universal membership of 198 member countries. The UNFCCC established the annual Conference of the Parties (COP) meetings to maintain a steady focus on targets and progress. Despite these efforts, more than 50% of cumulative industrialisation CO2e emissions has occurred in the 30 years since the first report by the IPCC (Saxifrage, 2019). We are witnessing that the fossil fuel driven economic growth is far too great a driving force for the COP global leadership to influence and achieve the rate of change that is needed (Nyberg et al., 2022). We should heed these lessons in our approach to the implementation of CE.

The 2023 report of the United Nations Secretary-General on the progress of Sustainable Development Goal (SDG) implementation (UN-RSG, 2023) marks the midpoint in time to the 2030 goals agreed by the signatories. The Secretary-General states, "It is time to sound the alarm. At the midpoint on our way to 2030, the Sustainable Development Goals are in deep trouble. A preliminary assessment of the roughly 140 targets for which data is available shows that only about 12 per cent are on track" (UN-RSG, 2023, p. 2). We argue that whilst CE will not address all SDGs, there are very real opportunities to contribute directly to a significant number.

Understanding the concept of the hegemony of corporate capitalism is as important for CE as it is deemed to be for climate change mitigation and adaptation (Nyberg et al., 2022). The theory of hegemony is described as meaning "political leadership based on the consent of the led, a consent which is secured by the diffusion and popularisation of the world view of the ruling class" (Bates, 1975, p. 352). Hegemony, therefore, involves winning the active and passive consent of a nation through intellectual and moral leadership and a construction of "common sense". Nyberg et al. (2022, p. 22) refer to the "vice-like grip" that corporate capitalism exerts on most nations through the establishment of the "natural order" of economic activity and cite this phenomenon as the major reason for the slow rate of appropriate response to climate mitigation.

There is a view that CE and sustainable development are on divergent pathways (Velenturf & Purnell, 2021). But we maintain that while CE primarily focuses on slowing and decoupling resources from economic growth and keeping goods in circulation at the highest value for as long as possible, this does not negate sustainability outcomes. For example, the 2022 interim report by the Netherlands Environmental Assessment Agency (PBL) makes a very clear case for the impact of CE on societal issues and policy themes to address them (PBL, 2022). Key amongst the societal issues identified are climate change, biodiversity loss, environmental pollution and supply risks. There is evidence of third tier (local) governments declaring a climate emergency and seeking ways to mitigate climate change through the implementation of practical CE action plans. The City of Glen Eira, within the Melbourne metropolitan area, is one such local governments that has published "Our climate emergency response strategy 2021–2025" (CGE, 2021), followed by the "Glen Eira circular economy plan 2022–2026" (CGE, 2022). We cite this example, as CE is identified in the climate emergency document as one of four innovation areas that will be implemented as part of its climate emergency response action plan.

In summary, we assess that the criterion of coherence sought by Blomsma and Brennan (2017) has not yet been met. A recent survey by Kirchherr et al. (2023, p. 30) of 221 CE definitions identified in articles since the Kirchherr (2017) survey finds that "… the development of a 'final' and consensus definition of CE is elusive". The criteria of "permanence" identified by Blomsma and Brennan (2017) can be measured by the rate at which relevant stakeholders are adopting and implementing CE principles and CBMs. Whilst increasing numbers of governments, at all tiers, are adopting CE strategies (e.g. CGE, 2022), only 6% of businesses in the Netherlands have adopted CE principles and CBMs (Hanemaaijer et al., 2023, p. 9). The low uptake indicates that CE is not sufficiently mainstream and of enough impact, in our view, to meet the "permanence" criterion. The third criterion contemplated by Blomsma and Brennan (2017), that of "construct collapse", has become of growing interest to researchers, as evidenced by Kirchherr et al. (2023) who find that "… scholars now increasingly insist that CE necessitates a fundamental systemic shift, particularly with respect to existing supply chains". We posit that without a systemic shift in major market sectors, driven by targeted policy intervention, the CE construct as a primary solution to resource over-consumption will collapse.

The literature and organisations tracking the progress of CE adoption suggest that at a global level we are not making progress with total global circularity, currently receding annually since it was first measured at 9.1% by Circle Economy in 2018 (Circle Economy, 2023). Whilst the limitations of the measurement and reporting arise from the focus of the measure on the use of secondary (recycled) materials, the areas identified by the report to reverse the resource consumption overshoot are of importance through the lens of supply chains. The four key areas identified for focus are (i) food systems, (ii) built environment, (iii) manufactured goods and consumables and (iv) mobility and transport. Each of these has its own unique complexities and often long supply chains that demand attention at the macro, meso and micro levels for CE advances to be made.

6 Supply Chain Implications

Global repercussions of resource overutilisation, pollution and waste generation highlight the drawbacks of the linear economy. The linear economy model is based upon mass material extraction, mass manufacturing products and discarding products after use, resulting in depletion of finite resources and growing waste that could be reused (Bag & Rahman, 2021). Implicit in the linear economy is a continual supply chain absolved from responsibility of the inefficiencies and waste created. It is, however, of note that much supply chain effort over the past decades has focused on growing efficiencies and effectiveness (e.g. lean supply chains). From the inception of the CE concept by Boulding (1966), it has been theorised that if this cyclical system is to come to fruition, a systematic approach with a synchronised commitment is required. In current supply chain terms, this is an integrated closed-loop supply chain. As such, the implications for supply chains in a CE are much more, targeting resource efficiency through the regenerative cycling of biological materials and the restorative cycling of technical materials (Zhang et al., 2022).

This integration has given rise to a spectrum of sustainability-focused supply chain frameworks, progressing from conventional (linear) supply chains to sustainable supply chains, closed-loop supply chains, open-loop supply chains, and finally, circular supply chains. Each of these additions to the spectrum aimed at enhancing value and waste reduction objectives and advancing the degrees of transformation and sustainability aspirations (Montag, 2023). Circular supply chain management (CSCM) integrates circular thinking into the realm of supply chain management. At a macroscopic level, the principles of CSCM offer countries the potential for accelerated advancements towards sustainable development objectives encompassing resource security, emissions mitigation and landfill utilisation (Theeraworawit et al., 2022). At an intermediary mesoscopic level, collaborative efforts within industries can mitigate resource scarcity, price instability and emissions while fostering community engagement to establish the critical mass necessary for economic viability. At a microscale, CSCM empowers businesses to position themselves strategically in global markets, cultivate a brand reputation, generate new revenue streams and mitigate business risks associated with inventory and supply constraints. Nevertheless, the conceptual development of CSCM remains in its initial and formative phase. Besides, CSCM research is dominated by recycling, limiting the representation of a broader view (Theeraworawit et al., 2022; Zhang et al., 2023).

To further advance CSCM, supply chain implications start with modified design to bolster financial and environmental performance at the material, component, product and organisational levels (Kirchherr et al., 2017). That is, CE functions as an economic paradigm, striving for optimal resource utilisation by every member of the supply chain while concurrently restoring the environment. Evident in this conceptualisation is the significant role of supply chain management, serving as the physical infrastructure and operational processes essential for the CE framework. The CE supply chain requires the integration and mobilisation of bidirectional physical, informational, financial and knowledge flows across entities, supply chain participants, as well as the physical and digital platforms used. By applying the circular flow principles along supply chains, a paradigm shift in material flow and process redesign occurs, establishing a viable ecological-economic relationship (Genovese et al., 2017). Thereby, the CE supply chain diminishes the requirement for natural resource extraction, wherein waste generated by one organisation is effectively repurposed as input material for others (Majiwala & Kant, 2022) while restoring biological nutrients to uphold the Earth's natural capital (Bag & Rahman, 2021).

CE is a systemic perspective that extends to the entire supply chain beyond the boundaries of production processes (Batista et al., 2023; Kirchherr et al., 2017). The contention is that products are embedded within organisations and organisations integrated within supply chains, and these hierarchical levels collectively represent a nested operational and supply chain management perspective (Braz & de Mello, 2023). As such, the implication is that to establish CE ecosystems where materials undergo repeated utilisation, it is imperative to engage vertical stakeholders in both upstream and downstream facets alongside collaboration among horizontal stakeholders (Genovese et al., 2017). Therefore, CE requires an overhaul of the entire supply chain (Kirchherr et al., 2017), where all levels of the supply chain prioritise environmental efficiency as a fundamental prerequisite for the successful adoption of the CE approach (Mirzaei & Shokouhyar, 2022). Each supplier thus has responsibility for post-consumption materials, components, and product reclamation and flows, including endeavours like reduction, reuse, remanufacturing, repurposing, recovery and recycling. Here, reciprocal role inversion between suppliers and buyers, as well as customers engaging in product return for EoL restoration, underscores the dual and multifaceted roles that supply chain participants can assume. Such supply chains with forward and reverse attributes are not new and have been functioning well in isolated cases, namely, pallets and containers or even scrap metal and glass in a more unorganised way. Those supply chains have been practical and economical but restricted to well-systematised industries. Therefore, the influence they had on scaling and spreading to other sectors is limited.

Combined, these implications translate to an integrated closed-loop supply chain or what is conceptualised as "supply chain integration". Supply chain integration promotes collaboration, trust and, more recently, real-time information sharing using innovative technologies at inter- and intra-organisational levels across supply networks for sustainable outcomes (de Vass et al., 2018). The implication is that successful macro-level adoption of CE requires remodelling of supply chains from design, implementation and operations, spanning all entities from extraction to consumption (de Vass et al., 2023; Mishra et al., 2022). Remodelling for CE necessitates the integration of collaborative efforts and systemic effectiveness management involving multiple agents and multi-tier supply chains, which conjoin the management of supply chains, ecosystems and business models (Braz & de Mello, 2023). Industry 5.0, grounded upon the evolution of information technologies, facilitates the integration of advanced manufacturing technologies and humans within distributed supply networks by utilising technologies such as the Internet of Things, big data and cloud computing to enable extensive automation (Aggarwal & Manaswi, 2022). Implementing such sophisticated technologies that transcend organisational boundaries and leverage machine intelligence enables more expeditious, flexible and resilient services (Liu et al., 2023).

Crucially, another implication is that especially prominent members of the supply chain have added responsibility to facilitate the adoption of the CE paradigm within their operational domain and throughout their supply chain (Zhang et al., 2022). The prominent members have the power and capacity to establish collaborations between upstream and downstream entities within circular and even closed-loop supply chains, aiding the recuperation and procurement of secondary raw materials. The realisation of circular supply chains necessitates intimate collaborative ties with partners both within and extending beyond the immediate industrial confines, encompassing suppliers, product designers, procurement professionals and regulatory bodies (Bag & Rahman, 2021). To help effectively recover and recycle products and packaging, further collaborative relationships need to be forged between shopping centres, retailers, waste collection cooperatives and recyclers (Batista et al., 2023).

In summary, the adoption of CE underscores the intrinsic connection with supply chain management, highlighting the reciprocal relationship and collaborative efforts required to achieve sustainable outcomes in a progressive economy. Supply chain management efforts have already made substantial efforts towards sustainable approaches; nonetheless, for CSCM further efforts are needed. Specifically, four core implications were identified. First, the necessary modified design of materials, components, products and organisations to optimise resource utilisation. For supply chains, the redesign changes both what is and how it is supplied. Second, the systemic perspective of the full supply chain, upstream, downstream and horizontally. In a CE, each member of the supply chain is responsible for sustainability and for providing opportunities to cycle resources. Third, the integrated supply chains need common tools to facilitate collaboration, trust and information sharing. Combining digital and technical advances with necessary human relationships, Industry 5.0 could facilitate these outcomes. Fourth, prominent supply chain members have responsibility to facilitate and lead the supply chain for CE adoption. Their power needs to be used to supportively influence other supply chain members' adoption of CE practices. Overall, CE necessitates a comprehensive transformation of the entire supply chain, involving all stakeholders, from product design to end-of-life management. The supply chain thus becomes the practical vehicle for executing CE principles and facilitating bidirectional physical, informational, financial and knowledge flows among various entities and platforms.

7 Conclusion

The plethora of definitions issued and used has continued unabated since CE was first introduced. We argue that while the definitions have a common conceptualisation, which has evolved, there are still two contesting definitional streams. One that suggests that CE has the opportunity and aim to address all three sustainability pillars. Another that has a tighter scope and indicates that CE only addresses resource efficacy and effectiveness. This is the primary tension in the debate over CE definitions. It arises from the extent to which CE has the inherent capacity to replace current socio-economic-politico systems and influence the full breadth of economic, environmental, social governance and outcome needs of society.

The process of making current economic activity more circular is crucial and one of adjusting governance, policy and regulation within each national economy. A necessary understanding of the emergent CE concept and its core frames guides the focus for adjusting strategies and practices at a business level, though also provides guidance for governance, policy and regulation at local, national and wider levels. This has also been demonstrated in the evolutionary phases of the concept, reflecting the adoption across nations. The current emphasis for many on waste management demonstrates the great step needed to move practice to match the current conceptualisation. The work of Blomsma and Brennan (2017) is particularly helpful in this regard, as it identifies the current phase as the *Validity challenge period*. The level of coherence of the CE concept is still a work-in-progress but has made progress, as evidenced by policy and regulation initiatives at all tiers of government. There is emerging coherence in the relationship between the adoption of CE principles and the policy themes at all levels of government to address climate change, biodiversity loss, environmental pollution and supply risks. Furthermore, the permanence of the CE concept is still under construction by individual nations and largely driven by governance, policy and regulation at the appropriate levels of government. However, the growing establishment of national, regional and city targets in relation to CE, as well as increasing rates of policy and regulation introduction, is indicating emerging permanence. We believe that the construct collapse of the CE concept will only come about if CE is pursued as a construct outside of the current consumption-driven capitalist economic model. We see a continuum between two hypothetical end states—namely, a purely linear economy and a purely circular economy—along which an economy is influenced to operate by the key actors in the unit of analysis.

We conclude that the adoption of CE at the rate required to deliver the outcomes needed is dependent upon supply chain actors. The supply chain actors need to craft reciprocal and collaborative efforts to (i) redesign materials, components, products and organisations to optimise resource utilisation, (ii) adopt a systemic perspective, (iii) integrate common tools to facilitate collaboration, trust and information sharing and (iv) ensure prominent supply chain members lead CE adoption. We appreciate that the implications are made in an existing context dominated by a hegemony of corporate capitalism that shapes national thinking. Therefore, the role of governance stakeholders is necessary to resolve this challenge at the required scale, so that the application of CE principles may deliver the essential benefits espoused by its thought leaders.

References

Acerbi, F., & Taisch, M. (2020). A literature review on circular economy adoption in the manufacturing sector. *Journal of Cleaner Production*, *273*, Article 123086. https://doi.org/10.1016/j.jclepro.2020.123086

Aggarwal, P., & Manaswi, K. (2022). Role of circular economy, Industry 4.0 and supply chain management for tribal economy: A systematic review. *Journal of the Anthropological Survey of India, 71*(2), 265–280. https://doi.org/10.1177/2277436x221125914

Alonso-Almeida, M. d. M., & Rodríguez-Antón, J. M. (2020). The role of institutional engagement at the macro level in pushing the circular economy in Spain and its regions. *International Journal of Environmental Research and Public Health*, *17*(6), Article 2086. https://doi.org/10.3390/ijerph17062086

Bag, S., & Rahman, M. S. (2021). The role of capabilities in shaping sustainable supply chain flexibility and enhancing circular economy-target performance: An empirical study. *Supply*

Chain Management: An International Journal, 28(1), 162–178. https://doi.org/10.1108/scm-05-2021-0246

Bates, T. R. (1975). Gramsci and the theory of hegemony. *Journal of the History of Ideas, 36*(2), 351–366. https://doi.org/10.2307/2708933

Batista, L., Seuring, S., Genovese, A., Sarkis, J., & Sohal, A. (2023). Theorising circular economy and sustainable operations and supply chain management: a sustainability-dominant logic. *International Journal of Operations & Production Management, 43*(4), 581–594. https://doi.org/10.1108/ijopm-12-2022-0765

Bleischwitz, R., Yang, M., Huang, B., Xiaozhen, X., Zhou, J., McDowall, W., Andrews-Speed, P., Liu, Z., & Yong, G. (2022). The circular economy in China: Achievements, challenges and potential implications for decarbonisation. *Resources, Conservation and Recycling, 183*, Article 106350. https://doi.org/10.1016/j.resconrec.2022.106350

Blomsma, F., & Brennan, G. (2017). The emergence of circular economy: a new framing around prolonging resource productivity. *Journal of Industrial Ecology, 21*(3), 603–614. https://doi.org/10.1111/jiec.12603

Boulding, K. E. (1966). The economics of the coming spaceship earth. In H. Jarrett (Ed.), *Environmental quality in a growing economy* (pp. 3–14). John Hopkins University Press. https://doi.org/10.4324/9781315064147-11

Braungart, M., & McDonough, W. (2002). *Cradle to cradle: Remaking the way we make things*. North Point Press.

Braz, A. C., & de Mello, A. M. (2023). Supply chain management strategies, types and tactics for circular economy transitions. *Management Review Quarterly*. https://doi.org/10.1007/s11301-023-00354-4

CGE. (2021). *Our climate emergency response strategy 2021-2025*. https://www.gleneira.vic.gov.au/media/9226/our-climate-emergency-response-strategy-2021-2025.pdf

CGE. (2022). *Glen Eira circular economy plan 2022-2026*. https://www.gleneira.vic.gov.au/services/rubbish-and-recycling/circular-economy

Chizaryfard, A., Trucco, P., & Nuur, C. (2021). The transformation to a circular economy: Framing an evolutionary view. *Journal of Evolutionary Economics, 31*(2), 475–504. https://doi.org/10.1007/s00191-020-00709-0

Circle Economy. (2023). *The circularity gap report 2023*. https://assets.website-files.com/5e185aa4d27bcf348400ed82/63ecb3ad94e12d3e5599cf54_CGR%202023%20-%20Report.pdf

Corvellec, H., Stowell, A. F., & Johansson, N. (2022). Critiques of the circular economy. *Journal of Industrial Ecology, 26*(2), 421–432. https://doi.org/10.1111/jiec.13187

de Vass, T., Nand, A. A., Bhattacharya, A., Prajogo, D., Croy, G., Sohal, A., & Rotaru, K. (2023). Transitioning to a circular economy: Lessons from the wood industry. *The International Journal of Logistics Management, 34*(3), 582–610. https://doi.org/10.1108/IJLM-04-2022-0200

de Vass, T., Shee, H., & Miah, S. J. (2018). The effect of "Internet of Things" on supply chain integration and performance: An organisational capability perspective. *Australasian Journal of Information Systems, 22*, 1–29. https://doi.org/10.3127/ajis.v22i0.1734

Demirel, P., & Danisman, G. O. (2019). Eco-innovation and firm growth in the circular economy: Evidence from European small-and medium-sized enterprises. *Business Strategy and the Environment, 28*(8), 1608–1618. https://doi.org/10.1002/bse.2336

EMF. (2013). *Towards the circular economy Vol. 1: An economic and business rationale for an accelerated transition*. https://ellenmacarthurfoundation.org/towards-the-circular-economy-vol-1-an-economic-and-business-rationale-for-an

EMF. (2015). *Towards a circular economy: Business rationale for an accelerated transition*. https://ellenmacarthurfoundation.org/towards-a-circular-economy-business-rationale-for-an-accelerated-transition

EU. (2018). *Communication from the Commission to the European Parliament, the Council, the European Economic and Social Committee and the Committee of the Regions on a Monitoring Framework for the Circular Economy* (52018DC0029). https://eur-lex.europa.eu/legal-content/EN/TXT/?uri=COM%3A2018%3A29%3AFIN

Frosch, R. A., & Gallopoulos, N. E. (1989). Strategies for manufacturing. *Scientific American, 261*(3), 144–153. https://doi.org/10.1038/scientificamerican0989-144

Geissdoerfer, M., Pieroni, M. P., Pigosso, D. C., & Soufani, K. (2020). Circular business models: A review. *Journal of Cleaner Production, 277*, Article 123741. https://doi.org/10.1016/j.jclepro.2020.123741

Genovese, A., Acquaye, A. A., Figueroa, A., & Lenny Koh, S. C. (2017). Sustainable supply chain management and the transition towards a circular economy: Evidence and some applications. *Omega, 66*(B), 344–357. https://doi.org/10.1016/j.omega.2015.05.015

GIMAR. (2022). *Global insurance market report: December 2022*. https://www.iaisweb.org/uploads/2022/12/GIMAR-2022.pdf

Hanemaaijer, A., Kishna, M., Koch, J., Lucas, P., Rood, T., Schotten, K., & van Sluisveld, M. (2023). *Integral Circular Economy Report 2023. Assessment for the Netherlands: Summary and Main Findings*. PBL Netherlands Environmental Assessment Agency. https://www.pbl.nl/en/publications/integral-circular-economy-report-2023-assessment-for-the-netherlands

Kahl, S. (2023, Jan 9). Insured losses hit $120 Billion as extreme weather spreads. *Bloomberg*. https://www.bloomberg.com/news/articles/2023-01-09/insured-losses-hit-120-billion-as-extreme-weather-upends-norms#xj4y7vzkg

Kalmykova, Y., Sadagopan, M., & Rosado, L. (2018). Circular economy–From review of theories and practices to development of implementation tools. *Resources, Conservation and Recycling, 135*, 190–201. https://doi.org/10.1016/j.resconrec.2017.10.034

Khalifa, A. A., Ibrahim, A.-J., Amhamed, A. I., & El-Naas, M. H. (2022). Accelerating the transition to a circular economy for net-zero emissions by 2050: A systematic review. *Sustainability, 14*(18), Article 11656. https://doi.org/10.3390/su141811656

Kirchherr, J., Reike, D., & Hekkert, M. (2017). Conceptualising the circular economy: An analysis of 114 definitions. *Resources, Conservation and Recycling, 127*, 221–232. https://doi.org/10.1016/j.resconrec.2017.09.005

Kirchherr, J., Yang, N. N., Schulze-Spüntrup, F., Heerink, M. J., & Hartley, K. (2023). Conceptualising the circular economy (revisited): An analysis of 221 definitions. *Resources, Conservation & Recycling, 194*, Article 107001. https://doi.org/10.1016/j.resconrec.2023.107001

Korhonen, J., Honkasalo, A., & Seppälä, J. (2018). Circular economy: the concept and its limitations. *Ecological Economics, 143*, 37–46. https://doi.org/10.1016/j.ecolecon.2017.06.041

Liu, L., Song, W., & Liu, Y. (2023). Leveraging digital capabilities toward a circular economy: Reinforcing sustainable supply chain management with Industry 4.0 technologies. *Computers & Industrial Engineering, 178*, Article 109113. https://doi.org/10.1016/j.cie.2023.109113

Lowe, E. A., & Evans, L. K. (1995). Industrial ecology and industrial ecosystems. *Journal of Cleaner Production, 3*(1–2), 47–53. https://doi.org/10.1016/0959-6526(95)00045-G

Lyle, J. T. (1994). *Regenerative design for sustainable development*. John Wiley & Sons.

Majiwala, H., & Kant, R. (2022). A state-of-art review of circular economy in the supply chain management: Scientometric mapping. *Management of Environmental Quality: An International Journal, 33*(5), 1226–1248. https://doi.org/10.1108/meq-12-2021-0291

Mirzaei, S., & Shokouhyar, S. (2022). Applying a thematic analysis in identifying the role of circular economy in sustainable supply chain practices. *Environment, Development and Sustainability, 25*(5), 4691–4722. https://doi.org/10.1007/s10668-022-02217-6

Mishra, R., Singh, R., & Govindan, K. (2022). Net-zero economy research in the field of supply chain management: a systematic literature review and future research agenda. *The International Journal of Logistics Management*. Advance online publication. https://doi.org/10.1108/IJLM-01-2022-0016

Mollison, B. (1979). *Perma culture two: Practical design for town and country in permanent agriculture*. Tagari Publications.

Montag, L. (2023). Circular economy and supply chains: Definitions, conceptualisations, and research agenda of the circular supply chain framework. *Circular Economy and Sustainability, 3*(1), 35–75. https://doi.org/10.1007/s43615-022-00172-y

Murray, A., Skene, K., & Haynes, K. (2017). The circular economy: An interdisciplinary exploration of the concept and application in a global context. *Journal of Business Ethics, 140*(3), 369–380. https://doi.org/10.1007/s10551-015-2693-2

NatCatServices. (2022). *Natural disasters in 2022*. https://www.munichre.com/content/dam/munichre/mrwebsitespressreleases/natcat_stats_2022_factsheet.pdf/_jcr_content/renditions/original./natcat_stats_2022_factsheet.pdf

Nyberg, D., Wright, C., & Bowden, V. (2022). *Organising response to climate change*. Cambridge University Press.

Pauli, G. A. (2010). *The blue economy: 10 years, 100 innovations, 100 million jobs*. Paradigm Publications.

PBL. (2022). *Circular Economy Progress Report 2022: English summary*. https://www.pbl.nl/en/publications/circular-economy-progress-report-2022

Pearce, D. W., & Turner, R. K. (1990). *Economics of natural resources and the environment*. Johns Hopkins University Press.

Reike, D., Vermeulen, W. J., & Witjes, S. (2018). The circular economy: new or refurbished as CE 3.0?—Exploring controversies in the conceptualisation of the circular economy through a focus on history and resource value retention options. *Resources, Conservation and Recycling, 135*, 246–264. https://doi.org/10.1016/j.resconrec.2017.08.027

Sarja, M., Onkila, T., & Mäkelä, M. (2021). A systematic literature review of the transition to the circular economy in business organisations: Obstacles, catalysts, and ambivalences. *Journal of Cleaner Production, 286*, Article 125492. https://doi.org/10.1016/j.jclepro.2020.125492

Saxifrage, B. (2019, July 31). Fossil fuel burning leaps to new record, crushing clean energy and climate efforts. *National Observer*. https://www.nationalobserver.com/2019/07/31/opinion/fossil-fuel-burning-leaps-new-record-crushing-clean-energy-and-climate-efforts

Sohal, A., & de Vass, T. (2022). Australian SME's experience in transitioning to circular economy. *Journal of Business Research, 142*, 594–604. https://doi.org/10.1016/j.jbusres.2021.12.070

Stahel, W. R. (2006). *The performance economy*. Palgrave Macmillan.

Stahel, W. R. (2013). Policy for material efficiency—Sustainable taxation as a departure from the throwaway society. *Philosophical Transactions of the Royal Society A, 371*(1986), Article 20110567. https://doi.org/10.1098/rsta.2011.0567

Stahel, W. R. (2019). *The circular economy: A user's guide*. Routledge.

Stahel, W. R., & Reday-Mulvey, G. (1976). *The potential for substituting manpower for energy (Research Contract No 76/13-V/343/78-EN)*. Commission of the European Communities.

The World Counts. (2023). *Number of planet Earths we need*. Retrieved April 29, 2023, from https://www.theworldcounts.com/challenges/planet-earth/state-of-the-planet/overuse-of-resources-on-earth

Theeraworawit, M., Suriyankietkaew, S., & Hallinger, P. (2022). Sustainable supply chain management in a circular economy: A bibliometric review. *Sustainability, 14*(15), Article 9304. https://doi.org/10.3390/su14159304

UN-RSG. (2023). *Progress towards the Sustainable Development Goals: towards a rescue plan for people and planet* (A/78/80-E/2023/64). https://unstats.un.org/sdgs/files/report/2023/secretary-general-sdg-report-2023%2D%2DEN.pdf

Velenturf, A. P., & Purnell, P. (2021). Principles for a sustainable circular economy. *Sustainable Production and Consumption, 27*, 1437–1457. https://doi.org/10.1016/j.spc.2021.02.018

Webster, K. (2015). *The circular economy: A wealth of flows* (2nd ed.). Ellen MacArthur Foundation.

Zhang, A., Hartley, J., Wang, Y., & Wang, S. (2022). Special issue Editorial: Logistics and supply chain management in an era of circular economy. *Transportation Research Part E: Logistics and Transportation Review, 166*, Article 102911. https://doi.org/10.1016/j.tre.2022.102911

Zhang, A., Seuring, S., & Hartley, J. L. (2023). Guest editorial: Empirically grounded research in logistics and supply chain management for a circular economy. *The International Journal of Logistics Management, 34*(3), 517–522. https://doi.org/10.1108/ijlm-05-2023-602

Roelof Vogel BSc (Mech) Eng, MSc (Eng), MBA (Cum Laude), M.Ind. Relations, worked in the global packaging industry for 35 years in roles that spanned Australia, New Zealand; Western, Central and Eastern Europe; and countries in Asia and Southern Africa. His corporate roles have been in operations, sales and marketing, and executive management. Roelof's new and developing interest lies in the concept and practice of circular economy, and he is a PhD candidate at the Monash Business School in this field. He now divides his time between his research and teaching activities and advisory work.

Tharaka de Vass is a Research Fellow at the Department of Management, Monash Business School, Monash University. His research primarily focuses on the critical themes of circular economy, Industry 5.0 and lean optimisation from operations and supply chain management perspectives. Tharaka has published in leading journals on the impact of the Internet of Things (IoT) on supply chain integration, lean thinking for healthcare operational optimisation, the adoption of circular economy in manufacturing, AI application in the Industry 5.0 era, project management performance domains and on how smart logistics facilitate smart cities.

Glen Croy is an Associate Professor at the Department of Management, Monash Business School, Monash University. He is an award-winning educator with expertise in the activation of training and learning for the adoption of new approaches to address emerging challenges. Glen has effectively demonstrated this in a range of business, community and education contexts, with participants achieving and retaining development objectives. He evidences this and researches in the areas of circular economy, higher education and tourism.

Amrik Sohal is a Professor at the Department of Management, Monash Business School, Monash University, where he has held several faculty level positions, including Associate Dean (Research). He has also held positions at the University of Melbourne and the University of Bradford. He has won multiple awards for his research excellence and postgraduate supervision. Amrik has impressive research experience both in Australia and overseas and has authored or co-authored over 200 papers published in refereed journals, as well as a number of book chapters. His current research interests are in supply chain management, circular economy and lean/agile production systems.

The Economic Geography of the Moral Supply Chain in a Circular Economy

Guli-Sanam Karimova and Stephen LeMay

1 Introduction

Supply chains are instruments with no inherent moral or ethical nature (Beske-Janssen et al., 2015; Mason & Lalwani, 2006). They can elevate the social and economic well-being of those who live and work along their paths or exploit, suppress, impoverish and even destroy those who touch them. To support a circular economy (CE), a supply chain must also be a circular supply chain (CSC).

Because supply chains often reach across national and cultural borders and have become more complex, they carry many moral implications. These include moral hazard (Guo et al., 2018; Lee et al., 2018), sustainability (Ha-Brookshire, 2017), disengagement (Eriksson, 2016) and human trafficking (Russell et al., 2017) amongst others. Supply chains are theoretically neutral, yet most real supply chains carry a broad range of social, ethical and moral issues.

The CE aims to eliminate waste and pollution by keeping materials and products in use as long as possible. The means to make this work includes product design, reuse, repair and recycling. A CE works only with a CSC, offering the chance to minimise or eliminate some ethical problems in a supply chain. For example, it could reduce pollution and create a more sustainable society. Before a CE can become a reality, it must overcome many ethical and practical barriers along the

G.-S. Karimova
Institute for Ethics in Management, Vienna University of Economics and Business, Vienna, Austria
e-mail: guli-sanam.karimova@wu.ac.at

S. LeMay (✉)
Institute for Supply Chain Logistics and Marketing, University of West Florida, Pensacola, FL, USA
e-mail: slemay@uwf.edu

E. Fröhlich et al. (eds.), *Circular Economy in Sustainable Supply Chains*, CSR, Sustainability, Ethics & Governance,
https://doi.org/10.1007/978-3-031-70749-0_17

supply chain, including the lack of circularity and focus on sustainability at the expense of labour (Montag, 2023).

We introduce the moral supply chain (TMSC) to analyse the ethical and practical issues of creating CSCs and a CE. Supply chains are instruments (Beske-Janssen et al., 2015; Mason & Lalwani, 2006), but they can also become the ground on which differing perspectives on ethics encounter one another. This chapter focuses on the moral dimensions of actions along a supply chain and how they affect a CE as well as the potential for circularity in supply chains and economies.

Supply chains and economies connect cultures, values and policies across national and other borders (Karimova & LeMay, 2021). They also connect geographies that may host a variety of ethical problems and differences in values. To reach circularity, at least some of those differences must be reconciled. A CSC may improve outcomes for society, reducing pollution and creating a more efficient and sustainable economy. However, to reach that point, we must address the ethical and practical issues for implementation. They are often related, and key actors along the supply chain must agree on at least some issues.

We argue that TMSC is a broader concept than corporate social responsibility (CSR), although it is not intended to supplant or dismiss it. CSR focuses, by definition, on the corporation, whereas TMSC incorporates a broader range of actors. To analyse a CSC, let alone a CE, CSR is insufficient.

2 The Moral Supply Chain: From Design Through Raw Materials and Finished Products

TMSC is the network through which values are conveyed, and key players interact in the context of getting, using, delivering and disposing of physical goods and services (LeMay et al., 2017). These interactions may be as neutral as the infrastructure that supports them. They require that the actors in each transaction agree on the key issues.

These issues include ethical sourcing of raw materials, labour practices, working conditions and environmental impact. The actors include consumers in developed countries, managers in developing countries, workers along the supply chain, multinational corporate officials and government officials at all levels (Karimova, 2020; Karimova et al., 2023). Who acts and where they act is a function of economic geography. This list is not exhaustive, but it goes far beyond CSR.

TMSC and CSR are two related but distinct concepts in the context of a CE. TMSC is a network of organisations that share a common ethical vision and values and that collaborate to create products or services that minimise environmental and social impacts, while maximising positive outcomes for all stakeholders (Karimova & LeMay, 2019; Milošević et al., 2023). CSR is a voluntary commitment by a company to integrate social, environmental and economic concerns into its business operations and interactions with its stakeholders (Hofstetter et al.,

2021). In a CE, both TMSC and CSR can contribute to the transition from a linear to a circular model of production and consumption, where resources are used more efficiently, and waste is reduced or eliminated (Ha-Brookshire, 2017). However, there are some differences between the two concepts.

TMSC focuses more on the inter-organisational level, while CSR focuses more on the intra-organisational level. This means that TMSC emphasises the collaboration and coordination among different actors in the value chain, while CSR emphasises the internal policies and practices of a single company (Carroll, 1999, 2008). TMSC is more normative, while CSR is more pragmatic (Karimova & LeMay, 2019). This means that TMSC is based on a shared moral framework and principles that guide the decisions and actions of the value chain partners, whereas CSR is based on a strategic approach that aligns the company's goals with the expectations and needs of its stakeholders. Additionally, TMSC is more holistic, while CSR is more selective. This means that TMSC considers the impacts and outcomes of the value chain, whereas CSR considers the impacts and outcomes of specific aspects or dimensions of the company's activities.

TMSC lens allows for many possibilities, but a CE does not. A CE requires that enough parties along the CSC agree on practical and ethical issues (Milošević et al., 2023). Under TMSC, this means that business officials agree to abide by government regulations, government officials enforce regulations and consumers value goods produced in this environment. It also means that workers do their jobs in this environment and that their work helps sustain them as it sustains the CE and CSC (Karimova & LeMay, 2019, 2022).

TMSC acknowledges that cultural values may clash, blend or bypass one another entirely. Take the example of Haribo candy production. Haribo altered its products for markets in the Middle East. In other parts of the world, Haribo gel candies contain pork ingredients (Neo, 2023). This offends consumers in Muslim countries, so they altered the product composition for that market. In this instance, the corporation aligned its product and thus the supply chain with consumer values.

From a consumer perspective, TMSC can offer several benefits, such as enhancing consumer trust and loyalty by demonstrating transparency, accountability and integrity in the supply chain practices; providing consumer satisfaction and value by delivering high-quality products or services that meet or exceed consumer expectations and needs; promoting consumer empowerment and participation by enabling consumers to make informed choices and voice their opinions and feedback on the supply chain performance; contributing to consumer well-being and social justice by ensuring fair trade, human rights, labour standards and environmental protection in supply chain operations (Karimova & LeMay, 2019, 2021).

However, TMSC also faces some challenges, such as balancing the trade-offs between economic efficiency and ethical responsibility in supply chain decisions or aligning the interests and incentives of different supply chain actors under the condition of moral hazard (Lee et al., 2018), where some parties may act opportunistically or dishonestly at the expense of others. Therefore, TMSC requires innovation, a multi-stakeholder approach and governance to overcome these challenges and achieve responsible, sustainable outcomes. TMSC also needs to adapt to the changing consumer preferences and demands, as well as the emerging trends and technologies in the market.

3 Ethics and Economic Geography of Supply Chains: The Spatial Dimensions of Ethical and Circularity Decisions

Economic geography is the spatial organisation of exchange, production, distribution and consumption over large areas (Pászto, 2020). A CE and CSCs rely on nodes located far apart, even if some supply chains compress to smaller geographic areas. Nearshoring and onshoring cannot move cobalt mines out of the Democratic Republic of the Congo (DRC). Electronic products will continue to use cobalt in lithium-ion batteries until there is a change in product design based on new technology (Niarchos, 2021).

Physical and cultural distance may make understanding differences in ethical and moral views more problematic. The consumer in New York may know little about labour laws and practices in Vietnam or the Philippines unless they become the focus of New York media.

Economic geography changes the availability and cost of raw materials, labour and land—the basics of manufacturing. It also changes the political landscape and brings into play differences in policy, practice and enforcement (Pászto, 2020). This includes labour regulations, sustainability and political corruption. Fairtrade practices, localisation of production and distribution, and regionalisation require widespread agreement among key decision-makers, including government officials, corporate officials, workers and consumers in the developed world.

We regard ethical and moral issues as central to implementing a CE (Karimova & LeMay, 2022). A CE needs aligned goals that originate in ethics and profitability; otherwise, it is unlikely to be achieved. Furthermore, the CE is not a new concept, but rather a combination and extension of existing concepts that share a common vision of closing material loops and enhancing resource efficiency. This is not a static concept but a dynamic and evolving one that adapts to changing contexts and challenges. The role of ethics in the CE is an important and complex topic.

Ethics can be understood as the moral principles and values that guide human behaviour and decision-making. The CE also aims to create value for all stakeholders, including customers, suppliers, society and the environment. Therefore, ethics can play a role in the CE in several ways. For example, ethics can help define the purpose and vision of a CE, such as what kind of society and environment we want to achieve through circularity, and help evaluate the impacts and trade-offs of a CE, such as how to balance the benefits and costs of circularity for different groups and generations. Additionally, ethics can help design and implement CE strategies and policies, such as how to ensure fairness, transparency, accountability and participation in circularity initiatives (Karimova & LeMay, 2022). Ethics can also help foster a culture and mindset of circularity, such as how to promote values such as responsibility, stewardship, collaboration and innovation among individuals and organisations.

3.1 *The Circular Economy: Transforming Waste into Wealth*

Economies and supply chains become circular when their design seeks to minimise waste and maximise resource use. The CE tackles global challenges such as climate change, biodiversity loss, waste and pollution by decoupling economic activity from the consumption of finite resources. In addition, the CE is based on three principles driven by design: eliminate waste and pollution, circulate products and materials (at their highest value) and regenerate nature. Their definition excludes current economic and social conditions.

Geisendorf and Pietrulla (2018) point out that neither profitability nor ethical issues are at the concept's core. Rather, circularity is when "[…] the value of products and materials is maintained, waste is avoided, and resources are kept within the economy when a product has reached the end of its life" (Geisendorf & Pietrulla, 2018, p. 779). They regard ethics and profits as framing issues. Thus, CSC management seeks to reduce waste, emissions and resource consumption by applying principles such as reuse, repair, remanufacturing, recycling and recovery to the supply chain processes. The CE is a resilient system that is good for business, people and the environment. It gives us the power to grow prosperity, jobs and resilience while cutting greenhouse gas emissions, waste and pollution. It also gives us the tools to tackle climate change and biodiversity loss while addressing important social needs.

3.2 *Scopes of Social Responsibility*

Dahlsrud (2008) identified five major dimensions of CSR: social, economic, environmental, stakeholder and voluntary. Each of these five dimensions can disrupt a CE by failing to provide the motive to implement or by placing barriers in its path. We treat the stakeholder dimension and voluntary dimensions differently because these are the dimensions in which TMSC and CSR differ (Karimova & LeMay, 2019, 2022).

Three conditions are likely to produce the kinds of abuses found along current supply chains: (1) a large sum of money is involved; (2) many people are involved and (3) activities are opaque, anything but transparent. What constitutes a large sum of money can vary by geography. A princely sum in one place may be pocket change in another. As supply chains become more complex, they also involve more people and become opaquer. This opacity may confound circularity, as circularity needs transparency. The following examples demonstrate the opportunities and limits of TMSC and CE.

3.2.1 Cobalt Mining in the Democratic Republic of the Congo

Cobalt mining in the DRC is a complex and controversial issue that involves various economic, social and environmental aspects. Cobalt is a key ingredient in the lithium-ion batteries that power electric vehicles, which are seen as essential for the transition to a low-carbon and CE and a wide range of other electronics. However, cobalt mining poses serious challenges and risks for the human rights, health and livelihoods of the miners and communities in the DRC, as well as for the conservation of nature and biodiversity. The same mines also yield alleged healing crystals, some sponsored or sold by prominent celebrities (Campbell, 2020; Thomsen & Hess, 2022).

Our first example shows the most common forms of ethical violations—poor working conditions, child labour, environmental damage and official corruption.

According to the World Economic Forum (2020), more than 70% of the global production of cobalt takes place in the DRC, of which 15–30% comes from the so-called artisanal and small-scale mines (ASMs). These are informal and often illegal operations that employ millions of people, mostly in poverty-stricken areas. ASMs are notorious for their unsafe and abusive working conditions, such as child labour, low wages, lack of protective equipment, exposure to toxic substances and frequent accidents or fatalities. ASMs also have negative impacts on the environment, such as deforestation, soil erosion, water pollution and wildlife poaching. A recent estimate suggested that over 40,000 children (about twice the seating capacity of Madison Square Garden) work in the DRC mines (Amnesty International, 2016). Officials in charge of policing child labour take bribes from the children in the mines, keeping them silent about the children working there. Few workers of any age have safety equipment. In addition, the mines damage the local environment, particularly with their water runoff.

Some companies in the cobalt supply chain have promised to stop sourcing from ASMs and instead get the mineral from large-scale industrial mines (LSMs), which are seen as a safer and more responsible option for both workers and corporate reputations. However, LSMs are not without problems either. The Guardian reports that some workers at LSMs—often employed through subcontractors—allege they are victims of severe exploitation, including wages as low as 30 cent an hour, precarious employment with no contracts and paltry food rations (Kelly, 2021). Some LSMs run by Chinese companies face allegations of discrimination and racism against Congolese workers (Kelly, 2021).

Another example includes the case of Freeport-McMoRan Gold and Copper, Inc., (FCX) merging with Phelps-Dodge for 56% control of the Tenke Fungurume mine in the DRC (Poole, 2016; Mining Technology, 2020). Freeport-McMoRan is a U.S.-based mining company that has been involved in the extraction of cobalt and copper in the DRC for several years. The company has faced criticism and controversy for its environmental and social impacts, as well as its business dealings with the Congolese government and other actors in the region. The company has also sold some of its assets in the DRC to China Molybdenum Co., Ltd. (CMOC), a

Chinese company that is another major player in the cobalt and copper market (Pinto & Thomas, 2016).

The relationship between Freeport-McMoRan and the mines of the DRC, as well as the concepts of TMSC and CE, is complex and multifaceted. On the one hand, Freeport-McMoRan has claimed that it has contributed to the economic development and social welfare of the DRC by providing jobs, taxes, royalties, infrastructure, health care, education and community development programmes. The company has also stated that it has implemented high standards of environmental management and human rights protection in its operations (Freeport-McMoRan Inc., 2020). Furthermore, the company has argued that its products are essential for the transition to a low-carbon and CE, as cobalt and copper are key ingredients in the batteries that power electric vehicles and other electronics. On the other hand, Freeport-McMoRan has been accused of causing or contributing to serious environmental and social harms in the DRC, such as land degradation, water pollution, greenhouse gas emissions, waste generation, deforestation, biodiversity loss, human rights violations, labour abuses, corruption, conflict and violence (Niarchos, 2021). The company has also been criticised for selling its assets to CMOC without consulting or compensating its local partner Gecamines, a state-owned mining company that holds a 20% stake in the Tenke Fungurume mine. Moreover, the company has been challenged for not doing enough to ensure the traceability and transparency of its supply chain or to support the formalisation and regulation of the artisanal and small-scale mining sector that employs millions of people in the DRC (Niarchos, 2021).

Therefore, it is difficult to assess whether Freeport-McMoRan and the mines of the DRC have been part of TMSC or CE. It seems that there are both positive and negative aspects of their activities, depending on the perspective and criteria used to evaluate them. However, there is room for improvement and innovation in their practices, as well as for collaboration and governance among all stakeholders involved in the cobalt and copper supply chain. This could help create more sustainable and ethical outcomes for businesses, society and the environment.

Thus, cobalt mining in the DRC poses a dilemma for TMSC and CE. While cobalt is essential for the development and deployment of electronics that can reduce greenhouse gas emissions and resource consumption, it also causes serious human rights violations and environmental degradation that undermine the principles and values of sustainability and ethics.

To address this dilemma, there is a need for innovation, collaboration, governance and education among all stakeholders involved in the cobalt supply chain. Some possible solutions include improving the traceability and transparency of the cobalt supply chain by using technologies such as blockchain or artificial intelligence; supporting the formalisation and regulation of ASMs by providing them with access to finance, technology, training, health care and social protection; implementing best practices and standards for LSMs by ensuring they comply with environmental and social laws and norms; promoting alternative sources of cobalt by recycling existing batteries or developing new battery technologies that use less or no cobalt; and raising awareness and engagement among consumers by informing them about the impacts and implications of their choices and actions.

As mentioned previously, the concept of TMSC shares a common ethical vision to collaborate and create products or services that minimise environmental and social impacts, while maximising positive outcomes for all stakeholders. TMSC holds everyone accountable for their actions and decisions in the supply chain. Rather than blaming or ignoring the problems or failures in the supply chain, TMSC puts individuals and their ethical decisions at the centre of the analysis. This means that every actor in the supply chain, from producers to consumers, must consider the moral implications and consequences of their choices and behaviours.

Many actors have responsibility in creating a CE and CSCs, because the CE is a system that aims to eliminate waste and keep resources in use for as long as possible by applying principles such as reducing, reusing, repairing, remanufacturing and recycling. The CE requires collaboration and coordination among different actors in the value chain, such as designers, manufacturers, distributors, retailers, consumers, recyclers, regulators and policymakers. Each actor has a role and duty to contribute to the transition from a linear to a circular model of production and consumption, where resources are used more efficiently, and waste is reduced or eliminated.

Corporations can influence the implementation of a CE and CSCs, but they cannot implement them alone. They operate in a context created by governments at all levels, consumers, multinational entities like the United Nations, other corporations and suppliers, and non-governmental organisations. Some of these supply chain stakeholders hold consistent views and apply known standards. If persuaded, they would likely commit themselves to creating a CE and CSCs. However, other stakeholders are fickle. Furthermore, some supply chain stakeholders are transparent, while others are opaque to the point of invisibility. Yet, transparency is key to creating a CE and CSCs.

3.2.2 The Moral Supply Chain and Scopes of Responsibility in a Circular Economy

The central starting point of the current discussion about the concept of CE is the critical question of whether waste production is an unavoidable evil of the current mode of capitalist production. Is a world without waste possible? Alternative approaches—such as a CE, zero waste, closing loops, resource efficiency, waste prevention, reuse, recovery and recycling—pursue the idea of responsible use of resources, products and the environment. For some years now, these concepts have been receiving more attention. However, a "world without waste" can only be achieved with a holistic approach that spans the whole product life cycle, including raw material extraction and energy use levels.

TMSC stands apart from specific ethical standards in supply chains. Supply chain ethics and morality must be judged from specific perspectives, whether it be individual or cultural. This holds true for CSCs as much as for other supply chains. Supply chain ethics come from norms set by governments, multinational organisations and other actors outside the supply chains. These often apply to labour and the environment (Karimova & LeMay, 2019, 2022), as well as the creation of CSCs.

Ethics concern the normative content of standards and define the responsibilities of consumers, corporations, states and other entities in supplier and buyer countries along the value chains (Schlegelmilch & Öberseder, 2007; Preiss, 2019; Davids, 1999). Circularity demands that some of these standards be applied more extensively and consistently. In many respects, circularity is an ethical standard. When we argue for a CSC, we argue for a supply chain that moves towards defined standards, especially regarding sustainability and environmental responsibility.

CSCs may also work towards economic equity, social justice and intergenerational responsibility. But those ethical goals are not essential to circularity. A CSC might still exploit labour or use child labour, even as it cuts waste, improves sustainability and raises economic efficiency (Geisendorf & Pietrulla, 2018).

Ethics orient the norms on how supply chains "ought to be". Circularity in economies and supply chains exemplifies this. In addition, ethics guide economic actors on principles, obligations and norms. A CE and CSCs make some of these standards and norms explicit. These ethics become morals when they are adopted by individuals, but key individuals must adopt the standards that lead to a CE and CSCs.

We distinguish between morals and ethics. "Moral" applies to an individual's sense of right and wrong. "Ethics" refers to principles from an outside source. These two terms are often used interchangeably, as in the example here, but we rely on the distinction.

"Morality" can be defined (Gert & Gert, 2017) in two ways:

1. Descriptively, morality refers to certain codes of conduct put forward by a society or group (such as a religion) or accepted by an individual for her own behaviour.
2. Normatively, morality refers to a code of conduct that, given specified conditions, would be put forward by all rational people.

The code of conduct put forth by the United Nations would qualify as ethics until it was accepted as moral by an individual—the second part of the first definition. The second definition refers to an individual's moral judgement, as that individual believes this norm should apply to all under the given conditions. For a CE and CSCs to work, they need widespread acceptance of their key principles and practices. Many individuals, especially those in power, must accept key principles and put them into practice. When we apply the concept of TMSC to specific situations, we find that culture may determine moral evaluations and behaviour in economic actors. CSCs present specific situations that lend themselves to this analysis. Supply chains require collaborations and interactions among varying cultures and levels of economic well-being. Changes in the structure of supply chains may alter a local culture, changing values and worldviews as well as buyer–supplier collaborations. With CSCs, changes in culture may need to precede implementation. So, in this sense, the ethics that are transmitted by the supply chain may become or influence the morals of people along the chain. By the same token, morals along the supply chain may encourage or restrict the realisation of a CSC and, consequently, CE.

We define morality as principles and values that guide human behaviour and decision-making. They can be influenced by various factors, such as culture, religion, education and personal experience. Values and principles can vary from person to person and from situation to situation. CSC is a supply chain management approach that incorporates circular thinking based on the CE framework. CSCs also aim to create value for all stakeholders, including customers, suppliers, society and the environment.

Morals along the supply chain may encourage or restrict the realisation of CSCs and a CE in several ways. For example, morality may influence the attitudes and behaviours of the actors along the supply chain, such as managers, workers, consumers and regulators. Morals may affect their motivations, preferences, choices and actions regarding CSC and CE practices. Additionally, morality may shape the expectations and demands of the stakeholders along the supply chain. Values and principles may determine what kind of CSC and CE activities and outcomes are desirable, acceptable or required by the stakeholders, and they may create opportunities or challenges for the implementation of CSCs and a CE. Moreover, moral values and principles may facilitate or hinder collaboration, communication, innovation and learning among the actors along the supply chain.

Therefore, TMSC can have a significant impact on the adoption and performance of CSCs and a CE by companies. TMSC can be a source of competitive advantage or disadvantage for companies that pursue CSC and CE goals.

Furthermore, TMSC incorporates descriptive and normative issues of supply chains. At its core, TMSC shows how morality is lived and how normativity is practiced throughout the supply chain. TMSC sets up encounters between what one group sees as what "is" and another group sees as what "ought to be". It can also be more than descriptive, incorporating elements of normative ethics and transmitting them directly and indirectly across national and cultural boundaries. TMSC looks on historically embedded traditions and transformations of cultural values and emphasises responsibilities throughout the supply chain.

This means that corporations become instruments of moral action, but they are not themselves moral actors. Corporations are artificial constructs, but they are populated at all levels by people who are moral actors. An international body might examine a corporation to see whether it complies with published ethical standards; however, the moral actions are taken by the people who operate the corporation. Regardless of legal standing, corporations do not make moral decisions; people do.

This makes TMSC essential to the creation of CSCs and a CE. The values must encounter one another along the supply chain. To create CSCs and a CE, key values must align. They cannot align in the absence of the encounter. TMSC is the network of actors and interactions that are guided by moral principles and values along the supply chain. Therefore, TMSC can affect the creation of CSCs and a CE, which are based on the CE framework that aims to reduce waste, emissions and resource consumption by applying principles such as reuse, repair, remanufacturing, recycling and recovery to the supply chain processes.

The values are the beliefs and preferences that influence the goals and actions of the actors along the supply chain. In addition, the values can be related to various aspects, such as economic, environmental, social, technological and institutional. The values can also vary from actor to actor and from context to context. TMSC is essential to the creation of CSCs and a CE because it enables the alignment of values and principles among the actors along the supply chain. The alignment of values means that the actors share a common vision and understanding of CSCs and a CE and agree on the strategies and policies to implement them. The alignment of values and principles can also foster trust, cooperation and innovation among the supply chain actors.

The statement also implies that the encounter is necessary for the alignment of values among the actors along the supply chain. The encounter allows the actors to express and exchange their values, and to learn from each other's perspectives and experiences. Furthermore, the encounter enables the actors to negotiate and resolve any conflicts or trade-offs that may arise from their different values. The encounter can also facilitate feedback and evaluation of the outcomes and impacts of CSCs and a CE.

Therefore, TMSC is a key factor for the success of CSCs and a CE. In particular, the encounter is a key mechanism for achieving this.

4 Dimensions of Morality Along the Geography of a Circular Supply Chain

We define a CSC as a supply chain that is geared around reusing its ostensible waste materials and returns. It aims to convert these materials and returns into new products that they can sell once more. Thus, it marks a shift for supply chains.

CSCs can improve the economic, social and environmental sustainability of firms. They can also create greater moral value for firms by integrating the business ecosystem, coordinating the forward and reverse supply chains and extending the life cycle of products, services, by-products and wastes (Montag, 2023).

The geography of a CSC refers to the physical locations of the supply chain's different stages. This includes the locations of suppliers, manufacturers, distributors and customers. The geography of a CSC can have an impact on its efficiency and effectiveness.

4.1 Descriptive Dimension of the Moral Supply Chain

Normative and descriptive ethics differ in the methods researchers use to evaluate moral issues in CSCs. Researchers can use TMSC to describe how people interact morally throughout the supply chain. For example, consider child labour in the supply chain. Consumers, corporations, governments and others may judge child labour differently, depending on where they are in the supply chain. In India, the caste system has historically influenced people's understanding of right and wrong. If children from lower castes work, both Indian consumers and corporations may find it morally acceptable. A German corporation may invest in India and harbour a different view of child labour, namely that child labour is wrong. That same German company would never employ a 13-year-old child in Germany; however, they might accept child labour in India if it does no harm to the local culture.

When consumers in developed countries discover the product was made by child labour, they may hold the German corporation responsible for unacceptable behaviour based on German values, not Indian values. These groups require agreement on product design and the need for circularity to create and sustain a CSC. The existence of the caste system in India relates to supply chain ethics but not to the circularity of the supply chain. Sympathy for the caste system may rise as circularity demands greater transparency throughout the supply chain. Some may even judge the system as completely morally wrong. It is important to note that caste discrimination is a unique form of discrimination in the supply chain. Businesses and their suppliers need to familiarise themselves with the issue in order to mitigate against the risks.

CSCs can be seen as a way of expressing and implementing some values and morals in supply chain management, such as environmental stewardship, social responsibility, resource efficiency and customer satisfaction. However, CSCs also face some challenges and trade-offs that may conflict with some values and morals, such as profitability, competitiveness, quality and convenience. Therefore, CSCs require a careful balance and alignment of different values and morals among the stakeholders involved in the supply chain. TMSC is, thus, an intellectual tool for describing and discussing moral interactions along the CSC. TMSC looks at morality in different geographical contexts, a product of the existence of a supply chain. The descriptive dimension of TMSC explains the ethical and moral constructs at play, who has them and how strongly they influence behaviour. Furthermore, TMSC also has a normative dimension.

4.2 Descriptive and Normative Dimensions of the Moral Supply Chain

CSCs can be influenced by cross-cultural ethics, as different cultures may have different views and expectations on how supply chains should operate and what outcomes they should achieve. For example, some cultures may value environmental protection more than others, or some cultures may have different standards of quality, safety, labour rights, etc. Therefore, CSCs require careful consideration and respect of the cultural diversity and sensitivity among the supply chain stakeholders. How people make moral judgements depends on circumstances along the supply chain. People often judge based on morality that was either influenced or determined by their socialisation. Values could be also transmitted to the people through national culture. In China, people may make judgements about fairness and justice in ways that differ from people in Germany or France. The Chinese share a collectivist cultural profile that contrasts with the more individualistic profile of the West (Karimova & LeMay, 2021). Chinese collectivism has its roots in Chinese normative philosophy and Confucianism. Being socialised in a Confucian culture may influence people's judgement in China.

The normative dimension in TSMC must lead to some agreement to foster CSCs. This implies that the normative dimension can influence the descriptive dimension. If Chinese people value hierarchy and authority over democracy and liberal values (Karimova & LeMay, 2021), they contribute uniquely to how people judge normativity and morality. The exchange of values and the influence of market values through capitalism can shape values like hedonism and materialism that Western cultures export to the rest of the world through the supply chain. However, they must first agree on issues like the importance of sustainability before a CE and CSCs can manifest themselves.

TMSC is where the encounter between values will take place. The values that lead to CSCs must dominate, or CSCs will not exist. These norms must carry through what consumers demand, designers create, manufacturers produce and supply chains deliver. That means designing for reuse, repurposing or recycling and being willing to pay for that design. The circular ethical code must conquer alternative views. Otherwise, a CE and CSCs are ideals instead of realities.

4.3 The Dimension of Culture

CSC management incorporates circular thinking—based on the CE paradigm—into supply chain management. The incorporation of sustainability-related topics into supply chain management has evolved into a fast-growing and high-volume research stream (Montag, 2023).

The geography of supply chains involves economic transactions across national borders and, with them, cultural borders. Each country involved in a supply chain imbues it with cultural values. Culture is the set of shared values, beliefs, norms and practices that shape the behaviour and identity of a group of people. These cultural values change as countries develop in a historical progression. We base our understanding of these aspects of culture on work by Taylor (1999), including a sense of cultural modernism.

According to Taylor (1999, p. 73), culture changes "from earlier centuries to today as involving something like 'development', as the demise of a 'traditional' society and the rise of the 'modern'". Taylor's conceptualisation of culture could be appropriate when considering the economic dimension of collaboration in globalisation. The supply chains change the culture of consumption and production worldwide by creating a new global culture. The CSC is a means to create a more sustainable future for the planet and its inhabitants. Therefore, the CSC can be seen as a way to change the culture of consumption and production by reducing waste and increasing sustainability.

CSCs can be influenced by culture, as different cultural groups may have different preferences, expectations and behaviours regarding the products and services they consume and produce. For example, some cultures may be more willing to adopt circular practices, such as sharing, leasing, repairing or recycling products, while others may prefer to own, dispose or replace products. Therefore, CSCs require a careful understanding and adaptation of the cultural factors that affect the demand and supply of products and services.

4.4 The Legal Dimension

Laws define specific duties regarding right and wrong behaviours. If one set of laws were applied across all supply chains, the difficulties that might arise along the chain would be under the purview of a court. But that is seldom the case. Legality varies along the supply chain as borders are crossed. Take the simple notion of the United States legalising medical marijuana. It is legal in some states but not in others. In some states, it is only legal as a treatment for certain disorders.

For CSCs and a CE, some aspects of legal requirements may need to be coordinated. Transporting goods manufactured in Mongolia through China for sale in the United States must be legal. Moving remnants or reusable goods back through the supply chain must fit the laws of all partners along the supply chain. Without that coordination, CSCs and a CE become more difficult to achieve.

The legal dimension of TMSC and CSCs is an important aspect that relates to the laws, regulations, standards and norms that govern the ethical and environmental performance of the supply chain actors. The legal dimension can have various implications for the design, implementation and evaluation of TMSC and CSC practices:

- The legal dimension can provide incentives or disincentives for the adoption of TMSC and CSC practices, depending on the level of enforcement, compliance and sanctioning of the relevant laws and regulations. For example, some countries have introduced laws that require companies to report on their environmental and social impacts, such as the UK Modern Slavery Act. These laws can encourage companies to improve their transparency and accountability in their supply chain operations.
- The legal dimension can create opportunities or challenges for the collaboration and coordination among different supply chain actors, depending on the degree of alignment, harmonisation and integration of the legal frameworks across different jurisdictions and sectors. For example, some regions have developed common standards and certifications for CE practices, such as the EU Ecolabel.

These standards can facilitate the exchange of information and best practices among supply chain partners.

- The legal dimension can influence the innovation and competitiveness of the supply chain actors, depending on the extent of protection, promotion and support of the intellectual property rights, research and development, and market access of CE products and services. For example, some initiatives have created platforms or networks for the sharing and diffusion of CE innovations, such as the Circular Economy Club or the World Economic Forum. These platforms can foster the creation and adoption of new solutions and business models for a CE.

Therefore, the legal dimension of TMSC and CSCs is a crucial factor that can shape the outcomes and impacts of these approaches. The legal dimension can also vary significantly across different contexts and situations, depending on the political, economic, social and cultural factors that affect the development and implementation of laws and regulations. Hence, it is important for the supply chain actors to be aware of and adapt to the legal dimension of their operating environment.

4.5 The Ethical Dimension

For this chapter, we adopted the stance that ethics come from external sources, while morals are adopted by and cling to the individual. Ethical theories and principles can influence the behaviour of the players in TMSC. Some players even issue conscious statements of ethical principles. This is essential to communicate the ethical principles associated with a CE and CSCs.

Codes of conduct/standards are the rules or guidelines that define the ethical and social responsibilities of the organisations and individuals involved in the supply chain. CSCs can benefit from codes of conduct/standards, as they can help to ensure the quality, safety, transparency and accountability of the circular practices and processes. For example, codes of conduct/standards can specify the requirements for product design, material sourcing, production methods, waste management, customer service, etc. that support the CE principles. Codes of conduct/standards can also help to communicate and align the values and expectations of the stakeholders involved in the supply chain.

The United Nations' Universal Declaration of Human Rights directs ethical statements at every aspect of the supply chain, although, as the name suggests, it should apply beyond the confines of supply chains. The International Labour Organization (ILO) has a standards initiative that not only has an on-going development of standards but also has programmes to oversee their implementation and continued compliance with them (ILO, 2019).

Such codes of ethics are issued by many government organisations, non-governmental organisations, trade organisations and businesses. But compliance monitoring is not always available, not always sought and seldom transparent. A major advantage of these standards is that they can be broadcast for a CE and CSCs.

These litanies of ethical statements may be influenced by one or more ethical theories and principles that range from utilitarianism to Kant's Categorical Imperative. The West tends to ignore the influences of the East in many of these statements, whereas the East tends to view the West as naïve (Karimova & LeMay, 2021). Some ethical codes apply only to specific groups of people, usually social groups that developed the ethical statements in the first place. For example, the Rules of Professional Conduct for the State Bar Association of California apply to members of that bar association and no one else.

At this point, no such code of ethics exists for supply chains, nor is it likely to develop any time soon, particularly by universal consensus. The complex, modern, IT-enabled supply chain lacks transparency and crosses too many cultural, national and ethical boundaries. Meanwhile, in TMSC codes of ethics may conflict with one another theoretically but do not clash with one another in practice. They do not clash with one another because the person who adheres to one code may not be aware that the other person in the transaction adheres to a different, conflicting code. Transparency could be a double-edged blade.

4.6 The Practical Dimension

The Aristotelian idea of phronêsis, practical wisdom or "street smarts", gives us a practical view of TSMC (Small, 2004; Karimova & LeMay, 2019). Despite potential conflicts, supply chains must still work. A seller in China may differ morally from a buyer in Florida. Does it matter if the transaction still takes place? In such a simple example, there is no need for an exchange of moral views.

However, key elements must come together for CSCs to form and work. The decision-makers must agree that reusing, repurposing and recycling have value. The value can be ethical or economic. Preferably, it would be both.

We will add detail to this example since it brings up a key value for all supply chains: trust. If the people on each side of the transaction know each other, they can decide whether to trust one another. That may need direct contact but may come from learning—repeated transactions.

Here is the more detailed example. A small Florida-based manufacturer buys aluminium tubing from a small manufacturer in China. They have little chance to know one another, let alone trust one another. Thus, they must develop a method they both trust. The Florida buyer paid upfront, while the Chinese manufacturer ordered parts to make the Florida order only after payment. This system demanded trust from the buyer instead of the seller, but it worked.

These transactions seemed to hold economic content, not moral content. How would this work for circularity? As it stands, it needs help. The Florida firm sells goods to consumers and businesses. If consumers and businesses do not recycle or reuse the product, the supply chain will not be circular. Post-consumer aluminium tubing has a market, but someone must first put the tubes into that supply chain.

When larger organisations take part, transparency alongside the potential for clashes may increase. This is a result of the media and other watchdogs concentrating more on the larger corporations. The larger players abuse CSR for plausible deniability to protect their brands. CSC practices may make this more difficult since CSCs change supply chain relationships.

Circularity, if adopted, alleviates this apparent dual standard. The misappropriation of CSR points to the need for TMSC or a similar model in creating CSCs and a CE. Circularity requires involvement from the entire supply chain, from conception to consumption. It also needs to educate each of the stakeholders on the ramifications of their decisions. If we plan to move towards a circular global system, we must address the social, environmental and economic needs throughout the entire CSC. The goal of this chapter is to stimulate a conversation on this topic.

5 Conclusions

TMSC encompasses CSR and expands on the ideas related to it. It suits the ethical and moral analysis of a CE and CSCs better than CSR alone. It is naïve to think that corporations can transform supply chains based in countries where they have little influence or control. Apple may influence politics in many countries, but it does not have an invading army. It cannot tell the DRC to stop child labour in mining operations. Neither can mining firms like Freeport-McMoRan. The DRC government and business officials need to adopt the principles of circularity. Otherwise, circularity becomes more difficult to achieve in their markets.

Moral ideas will clash along supply chains, but it is difficult to predict such clashes. A transformation as broad as circularity will create ethical and practical problems. In this chapter, we applied the novel concept of TMSC to circularity. We hope it alerts managers to problems that may arise as circularity increases.

TMSC and CSR address the same major issues: labour exploitation, sustainability and corruption. TMSC broadens the scope of actors and spreads responsibility for circularity to consumers and officials along the supply chain. A CE and CSCs depend on more than corporations alone. Otherwise, CSR will remain "astroturfing", "greenwashing" or otherwise based on appearance alone. We hope that TMSC helps put the importance of circularity into a broad social context. Future research should address the concept of CE within the TMSC.

References

Amnesty International. (2016). *"This is what We Die For": Human Rights Abuses in the Democratic Republic of the Congo Power the Global Trade in Cobalt*. https://www.amnesty.org/en/documents/afr62/3183/2016/en/

Beske-Janssen, P., Johnson, M. P., & Schaltegger, S. (2015). 20 years of performance measurement in sustainable supply chain management–what has been achieved? *Supply Chain Management: An International Journal, 20*(6), 664–680. https://doi.org/10.1108/SCM-06-2015-0216

Campbell, G. A. (2020). The cobalt market revisited. *Mineral Economics, 33*(1–2), 21–28. https://doi.org/10.1007/s13563-019-00173-8

Carroll, A. B. (1999). Corporate social responsibility: Evolution of a definitional construct. *Business & Society, 38*(3), 268–295. https://doi.org/10.1177/000765039903800303

Carroll, A. B. (2008). A history of corporate social responsibility: Concepts and practices. In A. Crane (Ed.), *The Oxford handbook of corporate social responsibility* (pp. 19–46). Oxford University Press.

Dahlsrud, A. (2008). How corporate social responsibility is defined: an analysis of 37 definitions. *Corporate Social Responsibility and Environmental Management, 15*(1), 1–13. https://doi.org/10.1002/csr.132

Davids, M. (1999). Global standards, local problems. *Journal of Business Strategy, 20*(1), 38–40. https://doi.org/10.1108/eb039982

Eriksson, D. (2016). The role of moral disengagement in supply chain management research. *European Business Review, 28*(3), 274–284. https://doi.org/10.1108/EBR-05-2015-0047

Freeport-McMoRan Inc. (2020, December 14). *Freeport-McMoRan Completes Sale of Undeveloped Project in the Democratic Republic of Congo for $550 Million* [Press release]. https://investors.fcx.com/investors/news-releases/news-release-details/2020/Freeport-McMoRan-Completes-Sale-of-Undeveloped-Project-in-the-Democratic-Republic-of-Congo-for-550-Million/default.aspx

Geisendorf, S., & Pietrulla, F. (2018). The circular economy and circular economic concepts—A literature analysis and redefinition. *Thunderbird International Business Review, 60*(5), 771–782. https://doi.org/10.1002/tie.21924

Gert, B., & Gert, J. (2017). *The Definition of Morality*. The Stanford Encyclopedia of Philosophy. https://plato.stanford.edu/archives/fall2017/entries/morality-definition/

Guo, H., Gu, S., & Su, Y. (2018). Bargaining power choices with moral hazard in a supply chain. *Discrete Dynamics in Nature and Society, 2018*, Article ID 2068287. https://doi.org/10.1155/2018/2068287

Ha-Brookshire, J. (2017). Toward moral responsibility theories of corporate sustainability and sustainable supply chain. *Journal of Business Ethics, 145*(2), 227–237. https://doi.org/10.1007/s10551-015-2847-2

Hofstetter, J. S., De Marchi, V., Sarkis, J., Govindan, K., Klassen, R., Ometto, A. R., Spraul, K. S., Bocken, N., Ashton, W. S., Sharma, S., Jaeger-Erben, M., Jensen, C., Dewick, P., Schröder, P., Sinkovics, N., Ibrahim, S. E., Fiske, L., Goerzen, A., & Vazquez-Brust, D. (2021). From sustainable global value chains to circular economy—Different silos, different perspectives, but many opportunities to build bridges. *Circular Economy and Sustainability, 1*(1), 21–47. https://doi.org/10.1007/s43615-021-00015-2

International Labor Organization. (2019). *Ending Child Labour, Forced Labour and Human Trafficking in Global Supply Chains*. https://www.ilo.org/wcmsp5/groups/public/%2D%2Ded_norm/%2D%2D-ipec/documents/publication/wcms_728062.pdf

Karimova, G. S. (2020). Governance von globalen Wertschöpfungsketten - Welche Verantwortung haben Konsument*innen? In L. Heidbrink & S. Müller (Eds.), *Consumer Social Responsibility. Zur gesellschaftlichen Verantwortung von Konsumenten* (pp. 143–157). Metropolis.

Karimova, G. S., Heidbrink, L., Brinkmann, J., & LeMay, S. A. (2023). Global standards and the philosophy of consumption: Toward a consumer-driven governance of global value chains. *Business Ethics, the Environment & Responsibility*, 1–15. https://doi.org/10.1111/beer.12648

Karimova, G. S., & LeMay, S. A. (2019). The moral supply chain, Phronêsis, and management education. *Teaching Ethics, 19*(2), 255–276. https://doi.org/10.5840/tej202091481

Karimova, G. S., & LeMay, S. A. (2021). The Belt and Road Initiative, world order, and international standards: Continuity, adaptation, or discontinuity? *Journal of Global Ethics, 17*(1), 71–90. https://doi.org/10.1080/17449626.2021.1946836

Karimova, G. S., & LeMay, S. A. (2022). Consumer social responsibility (CnSR) in the circular economy of global value chains: What does it mean, and why does it matter? *International Journal of Circular Economy and Waste Management, 2*(1), 1–19. https://doi.org/10.4018/IJCEWM.302207

Kelly, A. (2021, November 8). Cobalt: DRC miners toil for £30p an hour to fuel electric cars. *The Guardian*. https://www.theguardian.com/global-development/2021/nov/08/cobalt-drc-miners-toil-for-30p-an-hour-to-fuel-electric-cars

Lee, Y., Song, S. H., & Cheong, T. (2018). The value of supply chain coordination under moral hazard: A case study of the consumer product supply chain. *PLoS ONE, 13*(3), Article e0194043. https://doi.org/10.1371/journal.pone.0194043

LeMay, S., Helms, M. M., Kimball, B., & McMahon, D. (2017). Supply chain management: The elusive concept and definition. *The International Journal of Logistics Management, 28*(4), 1425–1453. https://doi.org/10.1108/IJLM-10-2016-0232

Mason, R., & Lalwani, C. (2006). Transport integration tools for supply chain management. *International Journal of Logistics: Research and Applications, 9*(1), 57–74. https://doi.org/10.1080/13675560500534599

Milošević, I., Arsić, S., & Stojanović, A. (2023). Corporate social responsibility, circular economy and sustainable development: Business changes and implications in project-oriented companies. In V. Obradović (Ed.), *Sustainable business change* (pp. 111–143). Springer.

Mining Technology. (2020). *Freeport Sells Interests in DRC Kisanfu Project to CMOC for $550 million*. https://www.mining-technology.com/news/freeport-sells-interests-in-drcs-kisanfu-project-to-cmoc-for-550m

Montag, L. (2023). Circular economy and supply chains: Definitions, conceptualizations, and research agenda of the circular supply chain framework. *Circular Economy Sustainability, 3*, 35–75. https://doi.org/10.1007/s43615-022-00172-y

Neo, P. (2023). *Haribo pushes halal and multipack strategies to boost growth*. Food Navigator Asia. https://www.foodnavigator-asia.com/Article/2023/08/28/haribo-pushes-halal-and-multipack-strategies-to-boost-growth

Niarchos, N. (2021, May 31). The dark side of Congo's cobalt rush. *The New Yorker*. https://www.newyorker.com/magazine/2021/05/31/the-dark-side-of-congos-cobalt-rush

Pászto, V. (2020). Economic geography. In V. Pászto, C. Jürgens, P. Tominc, & J. Burian (Eds.), *Spationomy* (pp. 173–192). Springer.

Pinto, A. J., & Thomas, D. (2016). *Freeport sells prized African copper mine for $2.65 billion*. Reuters. https://www.reuters.com/article/us-freeport-mcmoran-tenke-cmoc-idUSKCN0Y015U

Poole, C. (2016). *Freeport-McMoRan Sells Congolese Copper Mine to CMOC for $2.65 Billion*. The Street. https://www.thestreet.com/markets/emerging-markets/freeport-mcmoran-sells-congolese-copper-mine-to-cmoc-for-2-65b-13562877

Preiss, J. (2019). Freedom, autonomy, and harm in global supply chains. *Journal of Business Ethics, 160*(4), 881–891. https://doi.org/10.1007/s10551-018-3837-y

Russell, H., Smith, K. T., & Smith, M. (2017). Using social audits to detect and eliminate human trafficking in corporate supply chains. *Internal Auditing, 32*(1), 31–38.

Schlegelmilch, B. B., & Öberseder, M. (2007). Ethical issues in global supply chains. *Symphonya. Emerging Issues in Management, 2007*(2), 12–23. https://doi.org/10.4468/2007.2.03schlegelmilch.oberseder

Small, M. W. (2004). Wisdom and now managerial wisdom: Do they have a place in management development programs? *Journal of Management Development, 23*(8), 751–764. https://doi.org/10.1108/02621710410549602

Taylor, C. (1999). Two theories of modernity. *Public Culture, 11*, 153–174. https://doi.org/10.1215/08992363-11-1-153

Thomsen, L., & Hess, M. (2022). Dialectics of association and dissociation: Spaces of valuation, trade, and retail in the gemstone and jewelry sector. *Economic Geography, 98*(1), 49–67. https://doi.org/10.1080/00130095.2021.1989302

World Economic Forum. (2020). *Making mining safe and fair: Artisanal cobalt extraction in the Democratic Republic of the Congo*. https://www.weforum.org/whitepapers/making-mining-safe-and-fair-artisanal-cobalt-extraction-in-the-democratic-republic-of-the-congo

Guli-Sanam Karimova is assistant professor and postdoctoral researcher at the Institute for Ethics in Management at the Vienna University of Economics and Business, Austria. In her current research, she focuses on the role of morality and ethics in organisations from a business ethics perspective.

Stephen LeMay is Full Professor of Marketing and Logistics at the University of West Florida. He has published more than 75 refereed journal articles, two textbooks, one research book and a collection of logistics education materials. He received his undergraduate degree in Magazine Journalism from Northwestern University; his MBA in Operations Management from the University of Tennessee, Knoxville; and his DBA in Transportation and Logistics at the University of Tennessee, Knoxville.

Part III
Practical Implementations of Circular Economy in Global Supply Chains

Life Cycle Sustainability Assessment as a Tool for Businesses in a Circular Economy: Business Examples from Egypt

Ola Mamdouh

1 Introduction

The adoption of sustainable supply chain management practices by companies helps to achieve the 17 Sustainable Development Goals (SDGs) launched by the United Nations in 2015. Circular economy is considered as the roadmap to realise these SDGs. The life cycle sustainability assessment (LCSA) is a tool that businesses can use to operationalise circular economy.

This chapter introduces circular economy as a derivative of the green economy versus the brown economy and the business-as-usual pattern. It then explains the LCSA tool that can be used by businesses to adopt a circular supply chain rather than a linear one. Additionally, the chapter showcases real-life examples of businesses that apply sustainability practices and under which phase of the circular supply chain these practices fall. It also demonstrates the respective Rs that each practice is related to. The circular supply chain phases and the Rs framework are both explained. Finally, the chapter illustrates which SDGs the sustainable practices of companies are related to. In particular, SDG 12—responsible consumption and production—is directly impacted. Other SDGs are also impacted and discussed for each company's sustainability practices.

O. Mamdouh (✉)
Coventry School of Business, The Knowledge Hub Universities, Cairo, Egypt
e-mail: ola.mamdouh@tkh.edu.eg

E. Fröhlich et al. (eds.), *Circular Economy in Sustainable Supply Chains*, CSR, Sustainability, Ethics & Governance,
https://doi.org/10.1007/978-3-031-70749-0_18

2 Circular Economy

The traditional economic development pattern is the business-as-usual way. It is also known as the brown economy. The economic growth of recent decades has been accomplished mainly through drawing down natural resources, without allowing stocks to regenerate, and through allowing widespread ecosystem degradation and loss. With the global population growing and the natural resource supply not being able to meet demand, the old pattern of development needs to shift (UNEP, 2011). For instance, the green economy is a development pattern that allows a switch from the brown economy. It is broadly one that results in improved human well-being and social equity, while significantly reducing environmental risks and ecological scarcities. Since 2011, the United Nations Environment Programme defined the green economy as a system where growth in income and employment should be driven by public and private investments that reduce carbon emissions and pollution, enhance energy and resource efficiency, and prevent the loss of biodiversity and ecosystem services (UNEP, 2011).

After the brown economy has proved its failure, circular economy may be considered as a derivation of the green economy. Circular economy is a system that regenerates nature and ensures that no material goes to waste. Through practices including maintenance, reuse, renovation, remanufacture, recycling and composting, goods and resources are kept in circulation in a circular economy. Even the life cycles of products are lengthened. By separating economic activity from the use of finite resources, circular economy addresses climate change as well as other global problems, including waste, pollution and biodiversity loss (Kirchherr et al., 2023).

2.1 *The Main Principles of Circular Economy*

Circular economy is intended to benefit businesses, society and the environment. Building economic, natural and social capital are the goals of a circular economy, which strives to separate economic growth from the consumption of scarce resources. It revolves around the R frameworks, such as reduce, reuse and recycle (Blomsma & Brennan, 2017; Ghisellini et al., 2016; Sihvonen & Ritola, 2015; Brennan et al., 2015; Zhu et al., 2010). Potting et al. (2017) have created a 9R framework for circular economy (Fig. 1), in which the Rs are refuse, rethink, reduce, reuse, repair, refurbish, remanufacture, repurpose, recycle and recover. This framework also categorises the Rs on a continuum between linearity and circularity.

Circular economy could be considered a way to operationalise sustainable development for businesses (Ellen MacArthur Foundation, 2014; EY, 2015). According to the Ellen Macarthur Foundation (2020), the circular economy is an economy designed to keep materials in use, eliminate waste and regenerate natural systems.

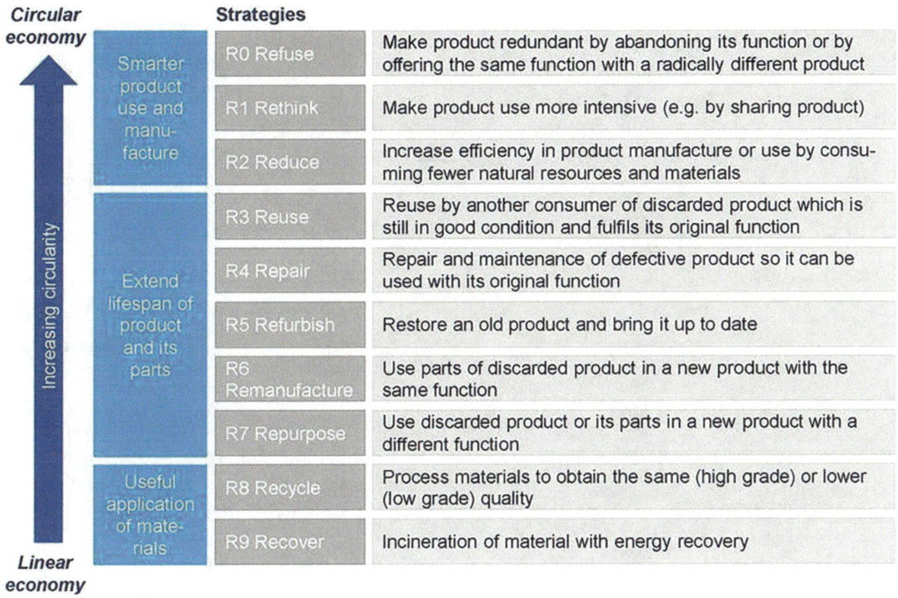

Fig. 1 The 9R framework (adapted from Potting et al., 2017, p. 5, as cited in Kirchherr et al., 2017, p. 224)

3 Life Cycle Sustainability Assessment: A Tool for Circular Economy

A procedure for evaluating the environmental impacts of a good or service over the course of its whole life is known as life cycle sustainability assessment (LCSA), commonly referred to as life cycle analysis. It is frequently used to identify the best performing product, service or other solution, at a given point in time, in terms of specific environmental impacts. It could also be considered a tool used by businesses to fit into the circular economy.

The cradle-to-grave or cradle-to-cradle LCA tries to measure the environmental effect of a product or service. Cradle-to-grave covers raw materials to disposal. Meanwhile, cradle-to-cradle represents the ideal situation where final disposal provides raw materials for the next manufacturing cycle. This is a circular supply chain rather than a linear throw away supply chain (Norton, 2012).

The LCSA concept (Fig. 2) states that collecting raw materials from natural resources (such as mined minerals, forests, fisheries, etc.) is the first step in the life cycle of a good or service. Both energy and waste are produced during this cycle. The utilisation of raw materials in a production process involves additional energy and natural resource (air, water, etc.) consumption. Produced goods need to be packaged, stored and distributed, which uses even more resources and energy. After that, the buyer uses and maintains the goods and related services, using additional energy or resources. In the final phase, consumers either dispose of or recycle

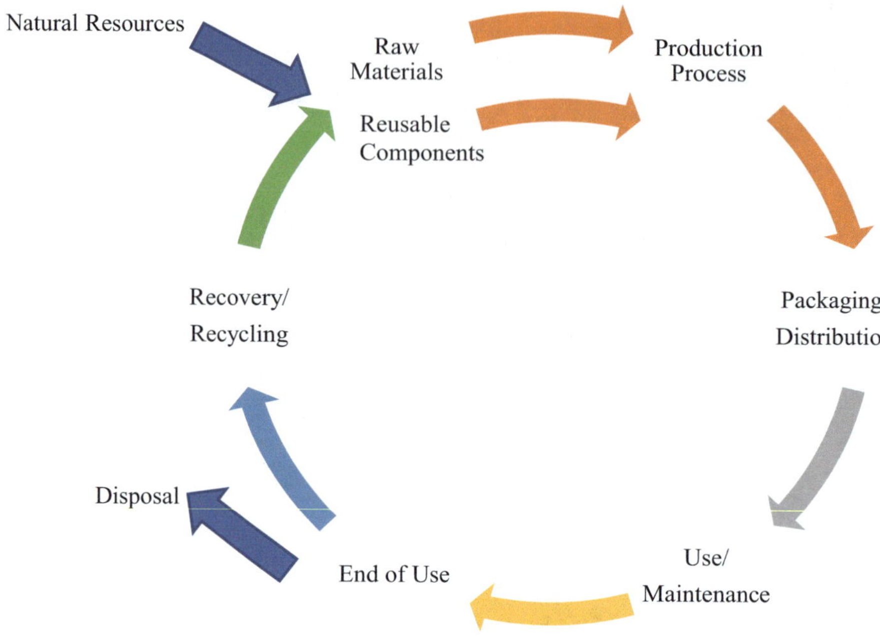

Fig. 2 Life cycle stages of a product (own illustration based on Norton, 2012)

products that are no longer needed, whether worn out, broken or undesired (Norton, 2012).

The following actions must be completed to conduct an LCSA for a product (Norton, 2012):

- **Goal and Scope Definition** Identify the product(s) or service(s) to be evaluated, select a functional basis for comparison and specify the needed level of detail.
- **Inventory Analysis of Extractions and Emissions** Each process' energy and raw material requirements, as well as its emissions to the atmosphere, water and land, are quantified before being merged in a process flow chart.
- **Impact Assessment** A small number of impact categories are used to classify the impacts of resource use and emissions, which can then be weighted for importance.
- **Interpretation** The findings are presented in the most informative manner possible, and possibilities and needs for reducing the impact of the product(s) and service(s) are identified.

When the output of a supply chain is dumped in a landfill, it often harms the environment severely. One of the major chances to improve their sustainability is for companies to create goods that are less resource-intensive, recyclable and

Fig. 3 Life cycle analysis of a Coca-Cola can (Norton, 2012)

remanufactured after being used. A closed loop supply chain is how Chopra and Meindl (2016) explained circular economy. Circular economy could also be achieved by embedding sustainability across the supply chain. This can be done by incorporating sustainability in terms of inputs or outputs through the different drivers of the supply chain: facilities, inventory, transportation, sourcing, pricing and information (Chopra & Meindl, 2016).

Figure 3 displays the assessment of the life cycle stages of a Coca-Cola can. It shows the materials, water and energy associated with each stage, as well as the carbon emissions and waste that are produced. When this is measured and quantified, it could be managed (Norton, 2012).

Research was conducted on the environmental impacts of the full life cycle of carbonated soft drinks manufactured and consumed in the UK (Amienyo et al., 2013). It was found that packaging is the main hotspot for most environmental impacts, contributing between 59 and 77%. While the ingredients accounted for 7 to 14% of the environmental impacts mainly due to the sugar, the production stage accounted for 5 to 10% mainly due to the energy for filling and packaging. The refrigeration phase of the drink at the retailer increased global warming potential by up to 33%. Lastly, the transport and distribution contributed up to 7% to the environmental impacts (Amienyo et al., 2013).

4 The Business Case for Sustainability

Sustainably advantaged products can drive sales and create tomorrow's options for growth by helping customers reduce their footprints (Chopra & Meindl, 2016). This strengthens a company's brand and reputation, giving it better access to capital and talent, and in turn resulting in a competitive advantage.

Companies try to hit two birds with one stone by reducing environmental impact alongside costs. When companies embed sustainability practices that reduce materials, water or energy (inputs) and waste or greenhouse gases (outputs), it is also creating economic benefits. This makes the business case for sustainability to companies as illustrated in Fig. 4. Additionally, it encourages companies to consider the social impact they have on the community they are working within.

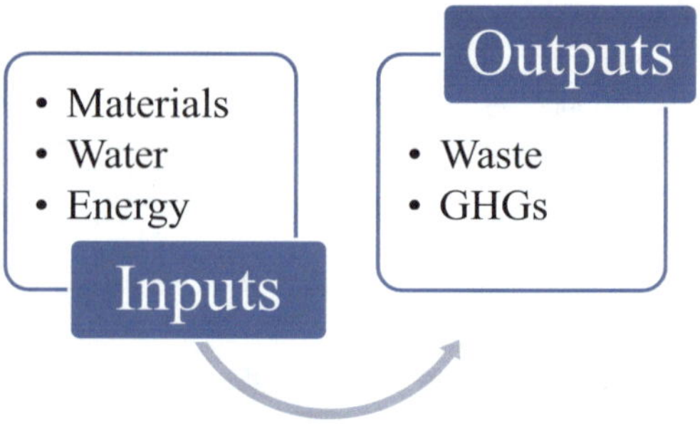

Fig. 4 Business case for sustainability (own illustration)

5 Business Case Studies from Egypt

5.1 Reform Studio

Reform Studio (https://reformstudio.net) is a young Egyptian company that started in 2012 and invented the new eco-friendly material plastex. They produce fashion products (accessories, bags and shoes) and home products (furniture and home accessories). All their products and designs are made with plastex, which is weaved from used and discarded plastic bags. This accomplishes the principle of circular economy—circulating products and materials—by using discarded plastic bags as an input material for their production process. Alternatively, it could also be said that they eliminate waste and pollution by using the plastic bag waste produced by others in their production instead of being disposed to landfills.

The sustainability efforts of Reform Studio are considered to fall under the design and production phases of the LCSA. Their business model also affects the end-of-life phase of the discarded plastic bags, as they are used as the raw material for Reform Studio's production.

Reform Studio has collaborated with other business organisations that have sustainability placed at the heart of their strategies, such as supplying Ikea with the special tote bag collection (Fig. 5). The founders of Reform Studio believe that design can solve stubborn problems. For instance, Egypt has a waste management problem, and plastic waste is one of the main sources of environmental damage in the world. Reform Studio found a way to lengthen the life cycle of plastic from minutes to years by embedding it into their design process and weaving discarded plastic bags into useful products.

Other than the environmental impacts that Reform Studio have addressed, they also have positive social impacts. Namely, they are aiming to revive the old craft of weaving, which is near extinction due to fast fashion and mass production. Reform Studio is reviving it, especially in rural areas of Egypt, by encouraging talented craftsmen and allowing them to share their skills with as many people as possible. They also empower underprivileged women with limited resources and education

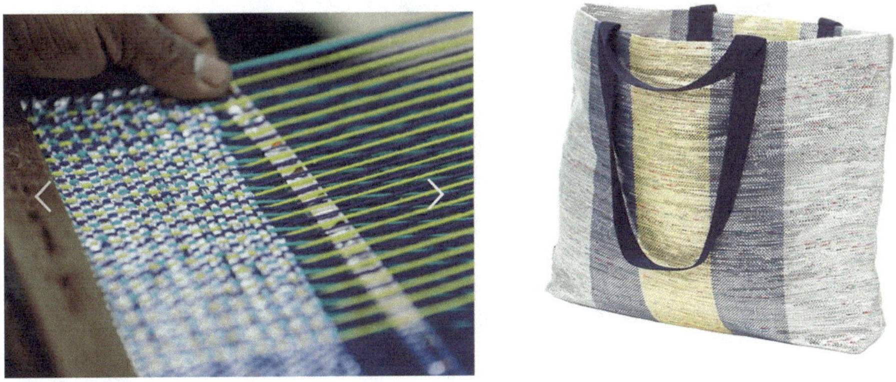

Fig. 5 The tote bag designed and produced by Reform Studio in collaboration with Ikea (https://reformstudio.net/)

Fig. 6 Fashion and furniture products produced by Reform Studio made from weaving discarded plastic bags (https://reformstudio.net/)

by creating job opportunities to a craft, weaving, that has already been engraved in their heritage and culture for generations. Figure 6 shows several examples of weaved products produced by Reform Studio.

Reform Studio also seeks to achieve some of the SDGs. Its business model mainly and directly targets SDG 12, responsible consumption and production, since they have lengthened the life cycle of the discarded plastic bags by embedding them as a raw material into their products. Hence, they require less virgin raw material, as well as less energy and water that would be needed to extract and transform this raw material from nature. Other SDGs, like SDG 1 (no poverty) and SDG 5 (gender equality), are also being achieved by providing job opportunities to underprivileged women to earn a living and improve their quality of life. Table 1 summarises the sustainability practices of Reform Studio and how they align with the SDGs and Rs.

Table 1 Aligning Reform Studio's sustainability practices with the related SDGs and [Rs]

Sustainability practice conducted by Reform Studio & its link to the [Rs]	Related SDG
Reform Studio business model that revolves around lengthening the life cycle of the discarded plastic bags by embedding them as a raw material into their products [Recycle]	Responsible Consumption and Production (SDG 12)
Providing job opportunities to underprivileged women	No Poverty (SDG 1) and Gender Equality (SDG 5)

5.2 *Dabur*

Dabur Egypt is a subsidiary of Dabur International, and Dabur International is a wholly owned subsidiary of Dabur India, which is the fourth largest fast moving consumer goods company in India. Today Dabur (https://www.dabur.com) is India's most trusted name and the world's largest Ayurvedic and natural health care company.

Dabur Egypt, established in 1996, has emerged as a prominent market leader in Egypt's personal care products industry. The company has carved a niche for itself by specialising in the manufacturing of a wide range of personal care products, including skin care, hair care and oral care products.

Dabur Egypt has incorporated sustainability practices into its production and packaging phases. This case study elaborates on how the packaging phase practices targeted the SDGs.

Packaging is an essential part of any product that serves many purposes. It mainly contains the product and must protect the contents that it is carrying, but it also facilitates transportation, shipping, handling and stocking of the products. In addition, it acts as a marketing and communication tool providing information to customers (Draskovic, 2007; Emblem, 2012). Packaging has an environmental responsibility as well that shall be thought of, planned and designed by the company's management (Molina-Besch & Pålsson, 2020). Dabur adopted several sustainable practices to take up this environmental responsibility of packaging.

For example, the design of the Vatika oil replacement tube was changed (Fig. 7). The old design had a metallised layer that gave a shiny look to the tube. However, it also made it infeasible to recycle the tube due to the combination of different layers, each of a different material, which makes it difficult to segregate before the recycling process. Removing this metallised layer from the tube while still maintaining an attractive colourful design enabled the tube to be recyclable and hence lengthening its life cycle (Mustafa, 2023).

Fig. 7 Change in package design to remove the metallised layer to facilitate the recycling process of the tube (Mustafa, 2023)

Dabur has also changed the material of its secondary package of Vatika hair henna (Fig. 8) from a metallised carton (shiny metal layer on carton) to a non-metallised mono carton (no metal layer on carton). As explained previously, the mono carton packages can be recycled more easily than the metallised carton packages with their combination of the different materials (metal and paper). Switching from metallised to non-metallised mono cartons led to 25% cost saving (Mustafa, 2023).

These practices by Dabur are considered to fall under the packaging phase of the LCSA. It also affects the end-of-life phase since the new package design allows recyclability.

Furthermore, Dabur reduced the dimensions of their carton packages (Fig. 9). This led to cutting material cost as well as cutting transportation cost by loading more cartons inside the container. Reducing the number of containers and the travel distance helps reduce carbon emissions. This practice falls under the packaging and transportation phases of the LCSA (Mustafa, 2023).

Practices adopted by Dabur aim to achieve numerous SDGs. It directly targets SDG 12 (responsible consumption and production) by reducing waste and lengthening the life cycle of the product. It also affects SDG 13 (climate action), especially

Fig. 8 Change in package design to remove the metallised layer and make the new design a mono carton to facilitate the recycling process of the carton (Mustafa, 2023)

Fig. 9 Carton dimension optimisation to reduce material cost and transportation cost (Mustafa, 2023)

Table 2 Aligning Dabur's sustainability practices with the related SDGs and [Rs]

Sustainability practice conducted by Dabur & its link to the [Rs]	Related SDG
Changing to a non-metallised tube package [Recycle]	Responsible Consumption and Production (SDG 12)
Changing to a non-metallised mono carton package [Recycle]	Responsible Consumption and Production (SDG 12)
Reducing the carton package dimensions [Reduce]	Responsible Consumption and Production (SDG 12) and Climate Action (SDG 13)

due to reducing distance travelled. Table 2 summarises how these practices relate to the SDGs and Rs.

5.3 *Nestlé Egypt*

Nestlé (https://www.nestle.com/) is a Swiss multinational food and beverage company that is one of the largest and most well-known corporations in the world. Nestlé operates in numerous countries and has a vast portfolio of brands across different categories. Some of its most popular brands in Egypt include Nescafé, KitKat, Maggi and Nesquik. The company's products are available in various sectors, such as dairy, coffee, confectionery, pet care and bottled water.

Nestlé has a great responsibility towards the environment and society, given its large size and its operations in several countries across the globe. In Egypt, Nestlé embedded a sustainability practice in its Nestlé Pure Life water bottles (Nestlé, n.d.). The plastic tear-offs that covered the bottle cap and neck (cap sleeves) have been removed from Nestlé Pure Life water bottles (Fig. 10). Instead, the pack reads "if it clicks, it's safe" as an easy way for users to identify unopened bottles (i.e. listen for a click when the bottle cap is twisted). This led to the environmental benefit of waste prevention. Nearly 240 tonnes of plastic tear-offs were eliminated in the first 18 months (since January 2019). This practice by Nestlé falls under the packaging phase of the LCSA. It also impacts the end-of-life phase by preventing waste from being generated in the first place. Therefore, it adopts the strategy of direct plastic elimination and the circular economy principle of eliminating waste. As seen in Table 3, this practice aims at SDG 12 (responsible consumption and production). Despite this sustainable practice by Nestlé, it is still running short of many other practices due to its huge impact worldwide as a multinational company.

Fig. 10 Nestlé water bottle without the plastic tear-off covering the bottle cap and neck (Cameron, 2021)

Table 3 Aligning Nestlé's sustainability practices with the related SDGs and [Rs]

Sustainability practices conducted by Nestlé Egypt & its link to the [Rs]	Related SDG
Removing the plastic tear-offs that covered the bottle cap and neck [Reduce]	Responsible Consumption and Production (SDG 12)

5.4 *Bekia*

Bekia (https://www.bekia-egypt.com/en/) is an Egyptian start-up working in the waste management field. It operates locally and encourages people to take part in the recycling cycle by allowing its users to swap their sorted trash for goods and services.

On Bekia's website, customers are encouraged to recycle inorganic waste, such as paper, cans, plastics, metals and more, in exchange for goods and services like groceries or metro tickets. Through the website, users can declare the waste they want to dispose of and exchange it for a range of products. Bekia's profits come from selling the separated waste they collect from individuals to factories and companies. Another business activity that Bekia conducts is collecting unused and non-operational electronic devices from customers. Bekia repairs these devices, then resells them.

Bekia's business model revolves around the end-of-life phase of the circular supply chain. Customers who segregate their waste before throwing it out are responsible customers. Bekia encourages this responsible behaviour, then collects the segregated waste and sells it to recycling factories. Hence, the lifetime of the

Table 4 Aligning Bekia's sustainability practices to the related SDGs and [Rs]

Sustainability practices conducted by Bekia & its link to the [Rs]	Related SDGs
Bekia's business model is built on reducing the negative environmental impact of waste by collecting segregated trash from customers and selling it to recycling factories [Recycle]	Responsible Consumption and Production (SDG 12)
Bekia collects unused and non-operational electronics to repair and resell [Repair]	Responsible Consumption and Production (SDG 12)

materials/products is lengthened, and less energy, water and resources are used for producing the recycled items (Ferreira et al., 2012). This targets both aspects of SDG 12: responsible consumption and responsible production. Moreover, Bekia reselling repaired electronics also falls under the end-of-life phase of the circular supply chain, again hitting SDG 12 (Table 4).

6 Conclusion

In general, a business needs to adopt sustainability practices. By doing so, it can hit two birds with one stone, reducing environmental impact as well as cutting costs or increasing revenues. Hence, the business case for sustainability is laid down. Looking through the different phases of the LCSA, companies can identify how to adopt sustainability practices in any or more of the phases. By reducing materials, energy or water (inputs) and waste or greenhouse gases (outputs), companies can adopt sustainability practices across the different product life cycle phases: extraction of raw materials, design, production, packaging, distribution and transportation, use, end-of-life and disposal/recover.

Table 5 summarises the sustainability practices conducted by the companies mentioned in this chapter, as well as links them to the related phase in the circular supply chain and the related SDG. It also shows how these practices relate to the different pillars of sustainability—economic, environmental and social. Additionally, it identifies which of the Rs each sustainability practice hits; they all fall under recycling, reducing and repairing.

It is important that companies avoid greenwashing when adopting sustainability practices. Future research is required in this area, identifying the sustainability practices of companies and then evaluating whether they are greenwashing or not. Further research could also be done on the sustainability practices of companies in other phases of the LCSA.

Table 5 Relating the companies' sustainability practices to the [Rs], circular supply chain phases, SDGs and sustainability pillars

Company name	Sustainability practice & its link to the [Rs]	Related phase in the circular supply chain	Related SDG	Related sustainability pillar
Reform Studio	Reform Studio business model that revolves around lengthening the life cycle of the discarded plastic bags by embedding them as a raw material into their products [Recycle] Providing job opportunities to underprivileged women	Design Production End-of-life	SDG 1 SDG 5 SDG 12	Environmental Economic Social
Dabur Egypt	Changing to a non-metallised tube package [Recycle] Changing to a non-metallised mono carton package [Recycle] Reducing the carton package dimensions [Reduce]	Packaging Distribution & Transportation End-of-life	SDG 12 SDG 13	Economic Environmental
Nestlé Egypt	Removing the plastic tear-offs that covered the bottle cap and neck [Reduce]	Packaging End-of-life	SDG 12	Economic Environmental
Bekia	Bekia's business model is built on reducing the negative environmental impact of waste by collecting segregated trash from customers and selling it to recycling factories [Recycle] Bekia collects unused and non-operational electronics to repair and resell [Repair]	End-of-life	SDG 12	Economic Environmental

References

Amienyo, D., Gujba, H., Stichnothe, H., & Azapagic, A. (2013). Life cycle environmental impacts of carbonated soft drinks. *The International Journal of Life Cycle Assessment, 18*, 77–92. https://doi.org/10.1007/s11367-012-0459-y

Blomsma, F., & Brennan, G. (2017). The emergence of circular economy: A new framing around prolonging resource productivity. *Journal of Industrial Ecology, 21*(3), 603–614. https://doi.org/10.1111/jiec.12603

Brennan, G., Tennant, M., & Blomsma, F. (2015). Business and production solutions: Closing loops and the circular economy. In H. Kopnina & S. O. Eleanor (Eds.), *Sustainability: Key issues* (pp. 215–236). Routledge.

Cameron, R. (2021, March 10). *Sustainable Packaging Nestlé* [Presentation]. ESG Bernstein Conference. https://www.nestle.com/sites/default/files/2021-03/bernstein-conference-2021-sustainable-packaging-rob-cameron.pdf

Chopra, S., & Meindl, P. (2016). *Supply chain management: Strategy, planning & operation* (6th ed.). Pearson Education.

Draskovic, N. (2007). The marketing role of packaging: A review. *International Journal of Management Cases, 9*(3–4), 315–323. https://doi.org/10.5848/APBJ.2007.00034

Ellen MacArthur Foundation. (2014). *Towards the circular economy Vol. 3: Accelerating the scale-up across global supply chains.* https://www.ellenmacarthurfoundation.org/towards-the-circular-economy-vol-3-accelerating-the-scale-up-across-global

Ellen MacArthur Foundation. (2020). *Upstream innovation: A guide to packaging solutions.* https://www.ellenmacarthurfoundation.org/upstream-innovation/overview

Emblem, A. (Ed.). (2012). *Packaging technology: Fundamentals, materials and processes.* Elsevier.

EY. (2015). *Are you ready for the circular economy? The necessity of an integrated approach.*

Ferreira, B., Monedero, J., Martí, J. L., Aliaga, C., Hortal, M., & López, A. D. (2012). The economic aspects of recycling. In E. Damanhuri (Ed.), *Post-consumer waste recycling and optimal production* (pp. 99–128). https://doi.org/10.5772/34133

Ghisellini, P., Cialani, C., & Ulgiati, S. (2016). A review on circular economy: The expected transition to a balanced interplay of environmental and economic systems. *Journal of Cleaner Production, 114*, 11–32. https://doi.org/10.1016/j.jclepro.2015.09.007

Kirchherr, J., Reike, D., & Hekkert, M. (2017). Conceptualizing the circular economy: An analysis of 114 definitions. *Resources, Conservation and Recycling, 127*, 221–232. https://doi.org/10.1016/j.resconrec.2017.09.005

Kirchherr, J., Yang, N. H. N., Schulze-Spüntrup, F., Heerink, M. J., & Hartley, K. (2023). Conceptualizing the circular economy (Revisited): An analysis of 221 definitions. *Resources, Conservation and Recycling, 194*, Article 107001. https://doi.org/10.1016/j.resconrec.2023.107001

Molina-Besch, K., & Pålsson, H. (2020). A simplified environmental evaluation tool for food packaging to support decision-making in packaging development. *Packaging Technology and Science, 33*(4–5), 141–157. https://doi.org/10.1002/pts.2484

Mustafa, M. (2023). *Sustainability in Packaging - Dabur* [PowerPoint slides]. Dabur Egypt Limited.

Nestlé. (n.d.). *What is Nestlé doing to tackle plastic packaging waste?* https://www.nestle.com/ask-nestle/environment/answers/tackling-packaging-waste-plastic-bottles

Norton, M. (2012). *Sustainability: Duty or opportunity for business?* Routledge.

Potting, J., Hekkert, M. P., Worrell, E., & Hanemaaijer, A. (2017). *Circular economy: Measuring innovation in the product chain* (2544). PBL Netherlands Environmental Assessment Agency. https://dspace.library.uu.nl/handle/1874/358310

Sihvonen, S., & Ritola, T. (2015). Conceptualizing ReX for aggregating end-of-life strategies in product development. *Procedia CIRP, 29*, 639–644. https://doi.org/10.1016/j.procir.2015.01.026

UNEP. (2011). *Towards a green economy: Pathways to sustainable development and poverty eradication - A synthesis for policy makers.* https://sustainabledevelopment.un.org/content/documents/126GER_synthesis_en.pdf

Zhu, Q., Geng, Y., & Lai, K. H. (2010). Circular economy practices among Chinese manufacturers varying in environmental-oriented supply chain cooperation and the performance implications. *Journal of Environmental Management, 91*(6), 1324–1331. https://doi.org/10.1016/j.jenvman.2010.02.013

Ola Mamdouh is an assistant professor of business administration at Coventry School of Business, the Knowledge Hub Universities, Cairo, Egypt. She earned her PhD in the field of sustainable supply chain management from Ain Shams University, Cairo, Egypt. Her main research fields are operations management, supply chain management and sustainable development. She served as the head of the sustainability and innovation committee at the Centre of Innovation Governance and Green Economy (CIGGE), the British University in Egypt and was responsible for assuring that the Principles of Responsible Management in Education (PRME) are embedded within the teaching and research of the faculty.

Structural Framework Development for Smooth Circular Supply Chain Management Adoption

Monika Vyas and Gunjan Yadav

1 Introduction

Globally, rising resource scarcity is due to the consumption of finite resources, accompanied by various forms of waste generation in vast amounts. Resource efficiency is a global concern that the G7 countries committed to in 2015 (Satpathy et al., 2016). The manufacturing sector consumes a lot of energy, and the highest energy consumption is by Indian manufacturing industries (Soni et al., 2017). Manufacturing sectors increase the circularity gap due to the adoption of linear processes and models (Franco, 2017).

1.1 Background and Motivation

A shortage of semiconductors in the technology market has led to disturbed supply chains and an increase in product wait times (Frieske & Stieler, 2022). Additionally, global epidemics have disturbed supply chains in almost all sectors to some extent. Resilient and responsive supply chains are in demand. This calls for the efficient tackling of resources to move towards sustainable development. The transformation of the supply chain into a circular one yields beneficial results, such as reducing virgin material, as secondary material can be made available by reusing and

M. Vyas
Department of Mechanical Engineering, Swarrnim Startup & Innovation University, Gandhinagar, India

G. Yadav (✉)
Department of Mechanical Engineering, Prashanti Institute of Technology & Science, Ujjain, India

E. Fröhlich et al. (eds.), *Circular Economy in Sustainable Supply Chains*, CSR, Sustainability, Ethics & Governance,
https://doi.org/10.1007/978-3-031-70749-0_19

recycling. Some industries, such as automotive, have focused on secondary materials, but end-of-life management lacks the most optimised approach in terms of economic and environmental aspects (Satpathy et al., 2016). Difficulties and barriers impede the application of circular economy (CE) practices in the supply chain (Govindan & Hasanagic, 2018).

1.2 Study Objectives

The drivers are building blocks to facilitate this process. This study focuses on the main drivers that promote circular supply chain management (CSCM). Its implementation increases material circulation and product life by adopting regenerative and restorative cycles (Homrich et al., 2018). Policies and standards for recycling and waste management are stringent in developed countries. Meanwhile, developing countries are still far behind in CSCM (Mangla et al., 2018).

A need exists to identify the most impactful drivers that influence a firm's decision-making process for CSCM adoption. The study aims to develop a relationship model between identified drivers using the interpretive structural modelling (ISM) approach. ISM is the most suitable approach to this problem, as it has the potential to develop the hierarchical structure of identified drivers by deriving their driving and dependency powers.

The second section covers the adopted research method for framework development and depicts a process for ISM analysis. Framework analysis and conclusions with implications are incorporated in sections three and four, respectively.

1.3 Literature Review

The Scopus database was explored with the search string "circular supply chain management" until 2022. Results were concentrated, and papers were read and identified to locate significant drivers that promote CSCM and applications of different multi-criteria decision-making (MCDM) techniques adopted by various researchers. For clarity, the literature review is subdivided into two parts.

1.3.1 Circular Supply Chain Management and Drivers

The term CSCM was coined to apply the principles of CE to the supply chain (Farooque et al., 2019). Drivers are also known as promoters (Govindan & Hasanagic, 2018; Masi et al., 2017). Table 1 showcases CSCM drivers that were identified from the literature and were finalised with expert input.

Table 1 CSCM drivers

Driver No.	Driver	Description	References
1	Higher authority vision	The higher authority develops a basic framework	Moktadir et al. (2020)
2	Technology adoption	It increases speed, accuracy and efficiency of operations	Oropallo et al. (2021)
3	Consumer activeness	Their role in closing the loop is necessary	Wandosell et al. (2021)
4	Financial resources	It helps in execution	Kazancoglu et al. (2021)
5	Long-term CSCM planning	It serves as a long-term gain	Govindan and Hasanagic (2018)
6	Government policy	Policy and norms result in better adoption	Hazen et al. (2020a)
7	Employee involvement	Link for execution	Sawe et al. (2021)
8	Improved overall performance	A better output can be achieved	Braun et al., (2018) and Walmsley et al. (2018)
9	Supplier support	Supplier network for process	Khalili Nasr et al. (2021)
10	Sustainable development	Govern resources and energy in an environmental, economic and societal approach	Sehnem et al. (2019) and Zhang et al. (2021a)
11	Transparency	Accessibility by each associated link	van Capelleveen et al. (2021)

1.3.2 Applications of Multi-Criteria Decision-Making Methods

MCDM methods find their application in the problem of prioritisation or selection of the best alternative. In most cases, each choice consists of multiple conflicting criteria, and hence decision-making is a complex phenomenon (Magableh & Mistarihi, 2022; Zhang et al., 2021b). Table 2 sheds light on various MCDM methods and their applications. Each method possesses its own advantages; thus, its selection is based on the type of problem. DEMATEL is adapted to locate the cause-and-effect relationship (Si et al., 2018). The analytic hierarchy process (AHP) assigns weight to the parameters, but it does not show interdependency (Liu et al., 2020). ISM shows the contextual relationship based on the driving and dependency values among variables (Mangla et al., 2018).

Table 2 Application of MCDM methods

Sr. No.	Study	Method	References
1	A decision support approach employing the PROMETHEE method and risk factors for critical supply assessment in large-scale projects	PROMETHEE	da Cunha et al. (2022)
2	Evaluating the circular supply chain implementation barriers using Pythagorean fuzzy AHP-DEMATEL approach	Fuzzy AHP-DEMATEL in CE	Lahane and Kant (2021)
3	Hybrid BWM-ELECTRE-based decision framework for effective offshore outsourcing adoption: a case study	BWM-ELECTRE for outsourcing	Yadav et al. (2018)
4	Interpretive structural modelling and MICMAC analysis for identifying and benchmarking significant factors of seismic soil liquefaction	ISM-MICMAC factors related to seismic soil liquefaction	Ahmad et al. (2019)
5	Prioritising the solutions of reverse logistics implementation to mitigate its barriers: a hybrid modified SWARA and WASPAS approach	SWARA and WASPAS in logistics	Prajapati et al. (2019)

2 Research Method

The structured research steps initiate with a data search of the Scopus database with the search string "circular supply chain management". The research flow adopted is shown in Fig. 1. To achieve the main aim of framework development for CSCM adoption, the preliminary stage consists of a literature review. It is performed in two phases to identify the drivers and MCDM methods, as discussed in section one. The identified 15 major drivers are changed to 11 important drivers with the expert suggestions to modify the two similar and exclude the two least important sub-drivers. Eight industry employees, including one managing director and others from the fields of marketing, sales, production, and manufacturing, are selected as experts. The ISM work is developed with expert consultation, and consistency checks are applied. Detailed steps are depicted in this section. In addition, framework analysis results are achieved that demonstrate the study's implications.

2.1 Framework Development by Interpretive Structural Modelling

The MCDM approach has major applications in the complex decision-making problems of real life. The five below-listed steps (Yadav & Desai, 2017) are executed to develop and analyse the ISM framework and derive the final results.

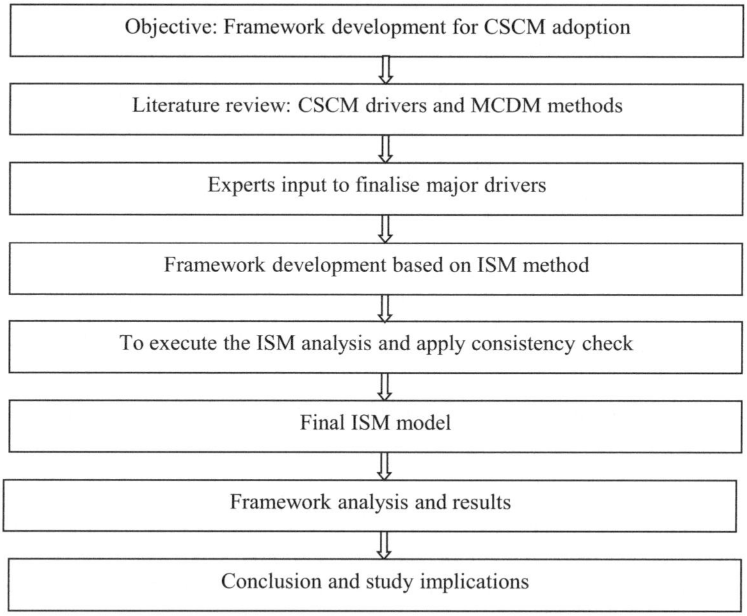

Fig. 1 Adopted research method (own illustration)

Table 3 Contextual relationship replacement

SSIM value ↓	Binary replacement	
(*i*, *j*) entry	(*i*, *j*) entry	(*j*, *i*) entry
V	1	0
A	0	1
X	1	1
O	0	0

Step I: To Identify CSCM Parameters

Firstly, important drivers are listed here, and their contextual relationship can be obtained as $(xij=)n$ variables.

Step II: To Obtain a Structural Self-Interaction Matrix

Pairwise comparisons for each identified variable are constructed. To check the relationships of the system with a selected n number of factors, suppose that 1 experts are considered to obtain the direct influence that factor f has on factor *j* in a decision group. Table 3 shows the terms applied to construct the structural self-interaction matrix (SSIM), namely V, A, X and O (Ahmad et al., 2019). To develop contextual relationships, these four terminologies are adopted depending on how they promote or help another variable. V is used when *i* helps to reach *j*, while A is applied when j helps to reach i. If both promote each other, then X is substituted; however, if no relation appears, then O is employed to construct the SSIM.

Step III: To Develop a Reachability Matrix from the SSIM

There is a need to convert the SSIM into an initial reachability matrix. It is obtained by converting the SSIM matrix into a binary matrix by substituting 0 or 1 in place of V, A, X, and O as per the substitution rules shown in Table 3.

Step IV: To Remove Transitivity Links

Further, a transitivity check is applied, and the final reachability matrix is obtained. The output for each driver is obtained as driving power and dependency power.

Step V: Inconsistency Check, Level Partition and Final ISM Model Development

In order to obtain the hierarchical structure, a series of iterations are carried out. Two sets, reachability and antecedent, are obtained from the final reachability matrix. A reachability set is formed of the driver itself and others that are reached because of it; similarly, an antecedent set is a structure formed of the driver itself and others that affect that particular enabler. An intersection set reveals the common values of both the reachability and antecedent sets. The first iteration derives the base or initial level, and the same process of iterations is continued until all levels are obtained. The highest driving power secures the bottom level in the hierarchical structure, while the lowest driving power secures the topmost level. This is obtained by counting the number of iterations and removing the previously identified levels to derive the next level. The final ISM model is constructed based on the obtained levels.

2.2 *Framework Analysis*

Initially, 15 major drivers, supported by the literature review, were handed to the expert members. The final 11drivers that support CSCM were finalised with two modifications and two exclusions. Further pairwise comparison relationships between each driver are obtained, and by putting the relationship in terms of V, A, X and O, a matrix of 11*11 known as the SSIM is obtained.

By using the rules of binary substitution, the SSIM gets converted into binary numbers. The application of transformation and transitivity rules, if any, results in the final reachability matrix. The transitivity rule states that if driver i is related to j and driver j is related to k, then it can be mentioned that driver i is directly related to j and the final reachability matrix is obtained. Here, the 11*11 matrix is converted into binary numbers 0 or 1 and is termed as the final reachability matrix.

Two sets, named reachability and antecedent, are obtained from the final reachability matrix. The reachability set is a group that consists of the self-enabler and the others that have reached it due to that self-enabler, whereas the antecedent set comprises the self-enabler and the others that can reach out to it. The intersection set results in the common values of both sets related to each driver.

Level partitioning is done by separating the drivers at each level and removing the previous levels. The process of execution continues until all levels are obtained. Therefore, level partitioning places each driver at its respective position in the hierarchical structure based on its powers and completes the ISM model.

3 Findings and Discussion

Upon achieving all the levels, the drivers and their relationships are built through a diagraph relating to each driver. The final ISM model framework is obtained after removing the links and based on the corresponding levels of the iterations. Successive levels I, II, III, IV and V are shown in Table 4. Figure 2 depicts the final ISM model based on and derived from Table 4.

To construct the final ISM model, the corresponding drivers of first-level (I) partitioning are placed at the bottom of the ISM diagram. Further, level II drivers are positioned above, and so on until the topmost level V drivers. The model shows the hierarchical structure from bottom to top. While the bottom level has the highest driving power, successive top layers have a higher dependency power. In other words, the topmost level is the output that has the most dependencies on drivers. The identified 11 drivers exist across the five different levels.

The ISM framework's output shows that the five hierarchical levels correspond to the 11 key drivers. The initial base level (I) has the highest driving power, consisting of higher authority vision, financial resources and government policy. Similar results show that the government and top management are the basis for the implementation of CE. For instance, some studies highlighted that government roles prominent at national level implications can also foster pilot projects and are critical for the change (Abuabara et al., 2019; Dalhammar, 2016). It promotes the other

Table 4 Result of all five-level partitioning

No.	Reachability set	Antecedent set	Intersection set	Level
1	1,2,3,5,7,8,9,10,11	1	1	I
2	2,7,8,9,10,11	1,2,4,5,6,7,8	2,7,8	III
3	3,8,10,11	1,3,4,5,6,8	3,8	III
4	2,3,4,5,7,8,9,10,11	4	4	I
5	2,3,5,7,8,9,10,11	1,4,5,6	5	II
6	2,3,5,6,7,8,9,10,11	6	6	I
7	2,7,8,10,11	1,2,4,5,6,7	2,7	III
8	3,8,10,	1,2,3,4,5,6,7,8,9,10,11	3,8,10	V
9	2,8,9,10,11	1,2,4,5,6,9	2,9	III
10	8,10	1,2,3,4,5,6,7,8,9,10,11	8,10	V
11	8,10,11	1,2,3,4,5,6,7,9,11	11	IV

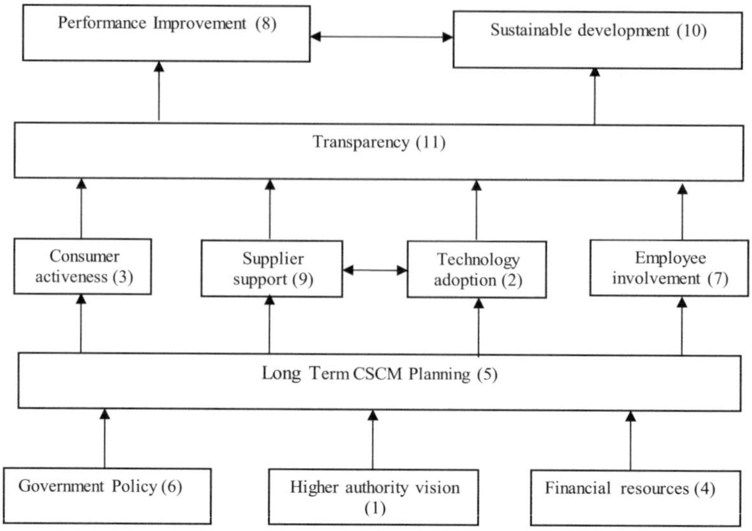

Fig. 2 ISM model (own illustration)

drivers at level II and results in long-term CSCM planning. Other research on supply chain showcases strategic decisions at each stage to deliver desired outputs. It involves waste management, consumers, green initiatives, planning models, etc. (Hazen et al., 2020b; Taghikhah et al., 2019). Level III consists of four equally significant drivers: consumer activeness, supplier support, employee involvement and technology adoption. Technologies put forth in the Industry 4.0 era result in beneficial processes in the supply chain (Abideen et al., 2021). Also, consumer awareness and adoption changes execute the laid down results in terms of remanufactured products, additionally proposed to develop incentive schemes for the employee (Wang et al., 2020). This encourages level IV, which is transparency in the supply chain. Highly dependent drivers obtained at the top level (V) are improved performance and sustainable development. Similarly, triple points and sustainable development in the food-agro-industrial supply chain were highlighted (Gutierrez-Franco et al., 2021). Performance measures for sustainability need to be addressed and social and environmental performance need to be treated as equal to economic performance. Altogether, consumers are the driving link for other stakeholders and can promote performance and sustainable development (Taghikhah et al., 2019). Strong drivers are capable of driving others and possess low levels of dependence. They are the most influential factors that need more focus and are the basis for better output. Analysis helps management with strategic decisions and the identification of the most impactful factors of CSCM implementation.

4 Conclusions

The main objective of this work was to identify and develop a framework using the ISM method. The position of drivers based on their driving and dependency powers, as obtained from the ISM method, was located. This resulting hierarchical model of 11 key drivers will provide a vision for managers and policymakers.

Additionally, it will help to focus on implementing CSCM. The high driving and low dependency drivers appeared at the bottom level and are of prime importance. They propagate the other drivers above until the highest level is achieved. For example, to achieve sustainability and performance improvement, government policy for CSCM regulations and laws have to be strict. Support and vision from higher authority with finance are basic driving parameters that can develop long-term execution. Intermediate drivers are associated with four broad spectrums related to technology adoption, supplier support, employer involvement and consumer activeness. Failure or less involvement of one of these driving parameters can create an imbalance. Their proper structure imparts transparency in the supply chain. It showcases the involvement of each stakeholder with a vision for sustainability that improves the performance of the organisation.

4.1 *Study Implications*

This study has implications for researchers and policymakers. It is also a roadmap for researchers in the domain. To implement CSCM or convert the supply chain into a circular one, the drivers play a significant role. Various drivers have different impacts on the implementation, and they facilitate other driving parameters.

Results support that high driving and less dependent drivers should be highly focused on by practitioners for the smooth adoption of CSCM. Therefore, the derived framework is a strategic tool, revealing the initial core drivers that policymakers need to focus on for implementation decisions.

Managers come across many drivers at different stages of supply chain operations. Their adoption and execution enable beneficial performance. Higher driving power leads to the achievement of desired or set outputs. The role of consumer and supplier integration is directly associated with the smooth functioning of CSCM; hence, the government and firms can adopt awareness campaigns to raise society's awareness of resource optimisation and waste management techniques. Sustainability and improved performance of the firm can be rewarding for the planet as well as all stakeholders.

4.2 Limitations and Future Scope

The involvement of experts in this study may have some limitations because a few important research outputs or drivers may have been missed due to the applied selection criteria and execution steps. Alongside a literature review, a questionnaire can be adopted in further research to remedy this.

Future studies can also incorporate sub-drivers, different types of industrial segments and various validation techniques. Additionally, MICMAC analysis, as well as other MCDM or mathematical modelling, can be adopted. Further studies can even be adjusted to examine the performance and obstacles associated with CSCM. For instance, the study of technology and social aspects in detail can be further extended with CSCM.

References

Abideen, A. Z., Pyeman, J., Sundram, V. P. K., Tseng, M. L., & Sorooshian, S. (2021). Leveraging capabilities of technology into a circular supply chain to build circular business models: A state-of-the-art systematic review. *Sustainability, 13*(16), Article 8997. https://doi.org/10.3390/su13168997

Abuabara, L., Paucar-Caceres, A., & Burrowes-Cromwell, T. (2019). Consumers' values and behaviour in the Brazilian coffee-in-capsules market: Promoting circular economy. *International Journal of Production Research, 57*(23), 7269–7288. https://doi.org/10.1080/00207543.2019.1629664

Ahmad, M., Tang, X. W., Qiu, J. N., & Ahmad, F. (2019). Interpretive Structural Modeling and MICMAC Analysis for identifying and benchmarking significant factors of seismic soil liquefaction. *Applied Sciences, 9*(2), Article 233. https://doi.org/10.3390/app9020233

Braun, A. T., Kleine-Moellhoff, P., Reichenberger, V., & Seiter, S. (2018). Case study analysing potentials to improve material efficiency in manufacturing supply chains, considering circular economy aspects. *Sustainability, 10*(3), Article 880. https://doi.org/10.3390/su10030880

da Cunha, R. A., Rangel, L. A. D., Rudolf, C. A., & dos Santos, L. (2022). A decision support approach employing the PROMETHEE method and risk factors for critical supply assessment in large-scale projects. *Operations Research Perspectives, 9*(May), Article 100238. https://doi.org/10.1016/j.orp.2022.100238

Dalhammar, C. (2016). Industry attitudes towards ecodesign standards for improved resource efficiency. *Journal of Cleaner Production, 123*, 155–166. https://doi.org/10.1016/j.jclepro.2015.12.035

Farooque, M., Zhang, A., Thürer, M., Qu, T., & Huisingh, D. (2019). Circular supply chain management: A definition and structured literature review. *Journal of Cleaner Production, 228*, 882–900. https://doi.org/10.1016/j.jclepro.2019.04.303

Franco, M. A. (2017). Circular economy at the micro level: A dynamic view of incumbents' struggles and challenges in the textile industry. *Journal of Cleaner Production, 168*, 833–845. https://doi.org/10.1016/j.jclepro.2017.09.056

Frieske, B., & Stieler, S. (2022). The "semiconductor crisis" as a result of the COVID-19 pandemic and impacts on the automotive industry and its supply chains. *World Electric Vehicle Journal, 13*(10), 1–14. https://doi.org/10.3390/wevj13100189

Govindan, K., & Hasanagic, M. (2018). A systematic review on drivers, barriers, and practices towards circular economy: A supply chain perspective. *International Journal of Production Research, 56*(1–2), 278–311. https://doi.org/10.1080/00207543.2017.1402141

Gutierrez-Franco, E., Polo, A., Clavijo-Buritica, N., & Rabelo, L. (2021). Multi-objective optimization to support the design of a sustainable supply chain for the generation of biofuels from forest waste. *Sustainability, 13*(14), Article 7774. https://doi.org/10.3390/su13147774

Hazen, B. T., Russo, I., & Confente, I. (2020a). Circular economy: Recent technology management considerations. *Johnson Matthey Technology Review, 64*(1), 69–75. https://doi.org/10.1595/205651320x15716717619537

Hazen, B. T., Russo, I., Confente, I., & Pellathy, D. (2020b). Supply chain management for circular economy: Conceptual framework and research agenda. *International Journal of Logistics Management, 32*(2), 510–537. https://doi.org/10.1108/IJLM-12-2019-0332

Homrich, A. S., Galvão, G., Abadia, L. G., & Carvalho, M. M. (2018). The circular economy umbrella: Trends and gaps on integrating pathways. *Journal of Cleaner Production, 175*, 525–543. https://doi.org/10.1016/j.jclepro.2017.11.064

Kazancoglu, Y., Ekinci, E., Mangla, S. K., Sezer, M. D., & Kayikci, Y. (2021). Performance evaluation of reverse logistics in food supply chains in a circular economy using system dynamics. *Business Strategy and the Environment, 30*(1), 71–91. https://doi.org/10.1002/bse.2610

Khalili Nasr, A., Tavana, M., Alavi, B., & Mina, H. (2021). A novel fuzzy multi-objective circular supplier selection and order allocation model for sustainable closed-loop supply chains. *Journal of Cleaner Production, 287*, Article 124994. https://doi.org/10.1016/j.jclepro.2020.124994

Lahane, S., & Kant, R. (2021). Evaluating the circular supply chain implementation barriers using Pythagorean fuzzy AHP-DEMATEL approach. *Cleaner Logistics and Supply Chain, 2*(December), Article 100014. https://doi.org/10.1016/j.clscn.2021.100014

Liu, Y., Eckert, C. M., & Earl, C. (2020). A review of fuzzy AHP methods for decision-making with subjective judgements. *Expert Systems with Applications, 161*(December), Article 113738. https://doi.org/10.1016/j.eswa.2020.113738

Magableh, G. M., & Mistarihi, M. Z. (2022). Applications of MCDM approach (ANP-TOPSIS) to evaluate supply chain solutions in the context of COVID-19. *Heliyon, 8*(3), Article e09062. https://doi.org/10.1016/j.heliyon.2022.e09062

Mangla, S. K., Luthra, S., Mishra, N., Singh, A., Rana, N. P., Dora, M., & Dwivedi, Y. (2018). Barriers to effective circular supply chain management in a developing country context. *Production Planning and Control, 29*(6), 551–569. https://doi.org/10.1080/09537287.2018.1449265

Masi, D., Day, S., & Godsell, J. (2017). Supply chain configurations in the circular economy: A systematic literature review. *Sustainability, 9*(9), Article 1602. https://doi.org/10.3390/su9091602

Moktadir, M. A., Kumar, A., Ali, S. M., Paul, S. K., Sultana, R., & Rezaei, J. (2020). Critical success factors for a circular economy: Implications for business strategy and the environment. *Business Strategy and the Environment, 29*(8), 3611–3635. https://doi.org/10.1002/bse.2600

Oropallo, E., Secundo, G., Del Vecchio, P., Centobelli, P., & Cerchione, R. (2021). Blockchain technology for bridging trust, traceability and transparency in circular supply chain. *Information and Management, 59*(7), Article 103508. https://doi.org/10.1016/j.im.2021.103508

Prajapati, H., Kant, R., & Shankar, R. (2019). Prioritizing the solutions of reverse logistics implementation to mitigate its barriers: A hybrid modified SWARA and WASPAS approach. *Journal of Cleaner Production, 240*(December), Article 118219. https://doi.org/10.1016/j.jclepro.2019.118219

Satpathy, I., Malik, J. K., Arora, N., Kapur, S., Saluja, S., & Bhattacharjya, S. (2016). Material consumption patterns in India. *Deutsche Gesellschaft Für Internationale Zusammenarbeit (GIZ) GmbH*, 1–20.

Sawe, F. B., Kumar, A., Garza-Reyes, J. A., & Agrawal, R. (2021). Assessing people-driven factors for circular economy practices in small and medium-sized enterprise supply chains: Business strategies and environmental perspectives. *Business Strategy and the Environment, 30*(7), 2951–2965. https://doi.org/10.1002/bse.2781

Sehnem, S., Vazquez-Brust, D., Pereira, S. C. F., & Campos, L. M. S. (2019). Circular economy: Benefits, impacts and overlapping. *Supply Chain Management, 24*(6), 784–804. https://doi.org/10.1108/SCM-06-2018-0213

Si, S. L., You, X. Y., Liu, H. C., & Zhang, P. (2018). DEMATEL technique: A systematic review of the state-of-the-art literature on methodologies and applications. *Mathematical Problems in Engineering, 2018*(1). https://doi.org/10.1155/2018/3696457

Soni, A., Mittal, A., & Kapshe, M. (2017). Energy intensity analysis of Indian manufacturing industries. *Resource-Efficient Technologies, 3*(3), 353–357. https://doi.org/10.1016/j.reffit.2017.04.009

Taghikhah, F., Voinov, A., & Shukla, N. (2019). Extending the supply chain to address sustainability. *Journal of Cleaner Production, 229*, 652–666. https://doi.org/10.1016/j.jclepro.2019.05.051

van Capelleveen, G., van Wieren, J., Amrit, C., Yazan, D. M., & Zijm, H. (2021). Exploring recommendations for circular supply chain management through interactive visualisation. *Decision Support Systems, 140*, Article 113431. https://doi.org/10.1016/j.dss.2020.113431

Walmsley, T. G., Varbanov, P. S., Su, R., Ong, B., & Lal, N. (2018). Frontiers in process development, integration and intensification for circular life cycles and reduced emissions. *Journal of Cleaner Production, 201*, 178–191. https://doi.org/10.1016/j.jclepro.2018.08.041

Wandosell, G., Parra-Meroño, M. C., Alcayde, A., & Baños, R. (2021). Green packaging from consumer and business perspectives. *Sustainability, 13*(3), 1–19. https://doi.org/10.3390/su13031356

Wang, Y., Zhu, Q., Krikke, H., & Hazen, B. (2020). How product and process knowledge enable consumer switching to remanufactured laptop computers in circular economy. *Technological Forecasting and Social Change, 161*(December), Article 120275. https://doi.org/10.1016/j.techfore.2020.120275

Yadav, G., & Desai, T. N. (2017). Analyzing Lean Six Sigma enablers: A hybrid ISM-fuzzy MICMAC approach. *TQM Journal, 29*(3), 488–510. https://doi.org/10.1108/TQM-04-2016-0041

Yadav, G., Mangla, S. K., Luthra, S., & Jakhar, S. (2018). Hybrid BWM-ELECTRE-based decision framework for effective offshore outsourcing adoption: A case study. *International Journal of Production Research, 56*(18), 6259–6278. https://doi.org/10.1080/00207543.2018.1472406

Zhang, A., Wang, J. X., Farooque, M., Wang, Y., & Choi, T. M. (2021a). Multi-dimensional circular supply chain management: A comparative review of the state-of-the-art practices and research. *Transportation Research Part E: Logistics and Transportation Review, 155*(October), Article 102509. https://doi.org/10.1016/j.tre.2021.102509

Zhang, X., Li, Z., Wang, Y., & Yan, W. (2021b). An integrated multicriteria decision-making approach for collection modes selection in remanufacturing reverse logistics. *Processes, 9*(4), Article 631. https://doi.org/10.3390/pr9040631

Monika Vyas is an academician as Assistant Professor of Mechanical Engineering at Government College, L. D. Engineering College, Ahmedabad, India. She is also a PhD research scholar in the stream of Mechanical Engineering at Swarrnim Startup & Innovation University, Gandhinagar, India. She has presented and published more than fifteen papers in various journals and national and international conferences. She is lifetime member of the Indian Society of Technical Education.

Gunjan Yadav is currently working as a Professor of Mechanical Engineering at Prashanti Institute of Technology & Science, Ujjain, India. He has more than 16 years of teaching, training, consultancy and research experience. He has completed his PhD in Mechanical Engineering from Sardar Vallabhbhai National Institute of Technology, Surat, India. He is on the reviewer panel of several renowned international journals. Presently, his *h*-index is 21 and *i*-10 index is 32. He has contributed 80+ papers at national/international level to various journals (SCI/Scopus indexed), seminars and conferences. He conducts training programmes and guides scholars in areas like sustainable manufacturing, Lean Six Sigma, QMS, GT, renewable energy management, etc.

Value-Based Selling of Green Products: A Radical New Approach to Commercial Growth

Alexander Nowroth

1 Why the Market Is So Hungry for Digital and Green Products

1.1 Be the First Who Gets It Right: "The Winner Takes It All"

As of mid-2023, the revenue and earnings of the transport market are in sharp decline in virtually all modes, and to some extent this also applies to transport volumes (Phaata, 2023). This trend may well continue into 2024. After peaking in 2021 and early 2022, the market could only decline, but the current negative trend is further exacerbated by the recession risks created by the war in Europe and the post-pandemic environment.

Most logistics companies have recognised this and are now focusing on volume growth. This is certainly a way to cushion, or at least partially avoid, a yo-yo effect after the significant increase in profit and turnover in 2021 and 2022. Although a one-sided focus on increasing sales volume is all too human, this is not a sustainable mitigation of the yo-yo effect, particularly in the face of declining margins.

Instead, companies should dare to try something new: use the higher profits from 2021 and 2022 to make both new green and existing sophisticated products (e.g. IT solutions) successful in terms of product management, and especially pricing and sales approach. So far no company has managed to *monetise green products seriously*, let alone *scale them.*

The sad reality for most logistics companies is that the conversion rate—the proportion of existing customers who are willing to pay more for these products than previously for a standard product—lies between 1 and 2% based on my

A. Nowroth (✉)
Lebenswerk Consulting Group, Düsseldorf, Germany
e-mail: anowroth@lebenswerkconsulting.com

E. Fröhlich et al. (eds.), *Circular Economy in Sustainable Supply Chains*, CSR, Sustainability, Ethics & Governance,
https://doi.org/10.1007/978-3-031-70749-0_20

practical experience from various industry projects. In many cases it is even less than 1% (Nowroth, 2023). Therefore, the first companies to successfully monetise and scale technically sophisticated as well as green products can expect the famous "the winner takes it all" effect. This means a rapidly increasing market share in the green product area, as well as a large lead over the competitors, through the help of a new pricing model combined with a newly generated margin increase.

1.2 Regulatory Constraints: "Fit for 55" (EU Sustainability Target)

The European Climate Act, adopted in 2021, sets a legally binding climate target to reduce emissions in the EU by at least 55% by 2030 compared to 1990 (European Commission, 2023). This climate target is known as "Fit for 55" (European Commission, 2021).

The pressure on companies to reduce CO_2 emissions will therefore increase from year to year, starting with very large companies and ending with SMEs as illustrated by Fig. 1.

Thus the bigger your client, the more likely they are to be under exactly that kind of pressure.

Additionally, maritime shipping will also be included in the Emissions Trading Scheme (ETS) from 2024 onwards (it is understood that air freight will follow a little later). Accordingly, all ships calling at European ports are to be gradually included in emissions trading from 2024. This applies to 100% of emissions from shipping within Europe with a three-year start-up phase of 40% in 2024, 75% in 2025 and 100% in 2026 (energiezukunft, 2022).

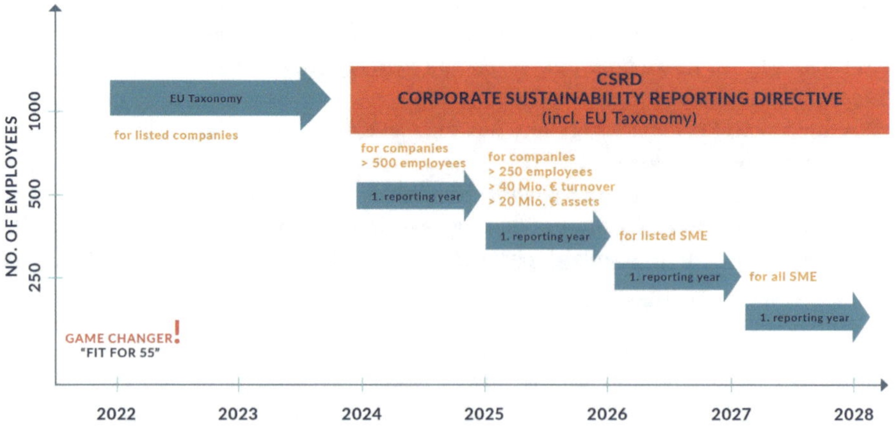

Fig. 1 Effectives of the EU-Taxonomy Fit for 55 (Waves Sustainability, 2023, p. 3)

Due to the introduction of a stricter emissions system (International Maritime Organization, 2023) and the EU sustainability reporting requirements, the market is already demanding green and net-zero products, as many companies are falling short of their CO_2 targets. Some have even set themselves the goal of becoming completely emission-free by 2035 or 2040.

Shipping and logistics companies (and certainly companies from other industries) have been developing highly interesting new green logistics or carbon zero products for years. Examples range from carbon-neutral biofuel, which is 20 to 30% more expensive than conventional fuel (International Energy Agency, 2020), to carbon-neutral container shipping, which costs 400 to 700 US dollars more per 20-foot standard container (Asia-Europe) (Pico, 2021; Rojon et al., 2021). Furthermore, there are already numerous attractive IT tools that can be used to measure CO_2 emissions along the supply chain and with the help of which, for example, CO_2 emissions can be reduced significantly by changing route planning. These products are mostly sold massively below value because they are charged per container, per ton or per shipment without considering the added value they generate.

Logistics companies are caught in another dilemma: while everyone acknowledges the responsibility for combating climate change, record high inflation and war-related uncertainties are putting decision-makers in crisis mode. In practice, this means customers do not try anything new and avoid increased spendings. For many logistics companies, this forces them to stick to the existing pricing models, which in turn destroys their margins.

1.3 Customer Groups That Demand Green Products

My experience from dozens of international consulting projects shows that initially logistics service providers make quite serious attempts to sell green products, but greatly scale down or stop the corresponding efforts altogether as soon as the first customers decide against such a purchase. Statements such as "customers simply do not want to pay more for this yet" or "it is too difficult to sell such products, especially across a certain customer range" are the usual tenor of the sales team.

This is already the first mistake because *not every customer group is suitable* for green logistics products.

In general, the formula applies, also illustrated by Fig. 2: the higher the product value, the more the industry is subject to regulatory forces, the more profitable the product and the more the customer focuses on the B2C segment, the greater the willingness to buy green logistics products.

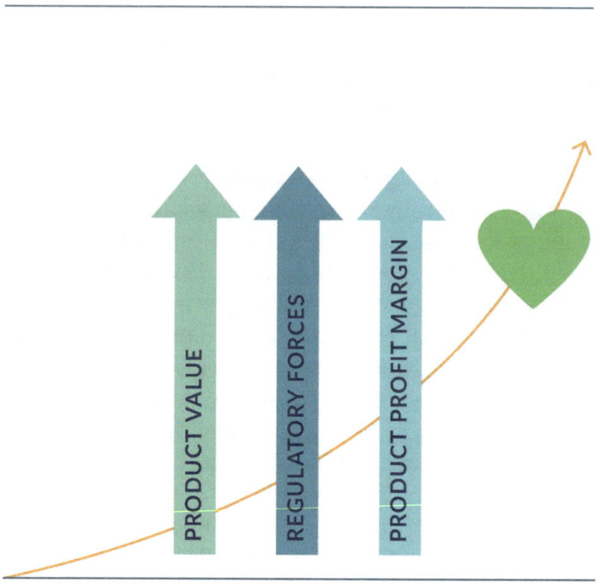

Fig. 2 Product value, regulatory forces and product profit margin along the string of desire to improve margin (own illustration)

The following industries have a high affinity for green logistics products:

- Apparel and luxury goods
- Consumer electronics
- Chemical and plastics

These sectors have a low affinity for green logistics products:

- Automotive industry
- Bulk goods

Even if your client operates in an industry with little regulatory forces, the product value combined with the proximity to a B2C client base are strong drivers.

1.4 Why Customers Choose Green Products

It is crucial to understand the motives underlying the purchase of green products. These motives can be divided into four categories:

Existing Customers. Your existing customers trust you and have done so for years or even decades. Thus it is much easier to sell a new product to such a customer than to launch a large-scale campaign for a new product aimed at new customers, most of whom have no green associations with you or your company.

Due to the established relationship of trust, existing customers are more open to new, namely green products than a potential customer who is unfamiliar with your company, which has only recently started selling this new product. Your acquisition costs will therefore be lowest with existing customers.

CO_2 Target. The more ambitious the CO_2 target, i.e. the nearer in the future it lies, the more affinity your customers have for green products. After all, the CO_2

savings must come from somewhere. Apart from production and energy emissions, the biggest CO_2 polluter in many producing industries is logistics, or more specifically the transport of goods.

Public-Listed. The more a company is in the public eye, the greater the social pressure on corporate decisions—this is especially true for public-listed companies. In the coming years this will increase even further, particularly when it comes to consumer goods companies such as P&G or Nestlé, because they are mostly active in the B2C market and thus more subject to social trends than companies operating in the B2B market.

Bonuses of the Management and the Board of Directors. The more ambitious the CO_2 target of a company, the higher the probability that the management's monetary incentive is partly linked to a CO_2 reduction. Based on information derived from the trusting relationship with your clients and from publicly available sources (such as annual reports), you can easily learn what criteria are used to define the bonuses of certain directors and management. The number of companies that link the incentivisation of their executives to the reduction of CO_2 emissions is steadily increasing. Therefore, it is in the own interest of a decision-maker to look for solutions that lead to CO_2 savings.

In the following, we will mainly deal with the value-based mechanics, according to which entrepreneurial benefits are put in relation to the costs. However, the most important thing when selling a new product in the B2B space is to uncover the personal agenda behind the corporate agenda. Bonuses are tailored to the individual, and if you can demonstrate that investing in your solution is likely to increase these bonuses, new budgets may open up.

An overlap of the purchasing motives—a CO_2 target, a public-listed company and available bonuses—results in existing customers who are most likely to buy the green products, as shown in Fig. 3.

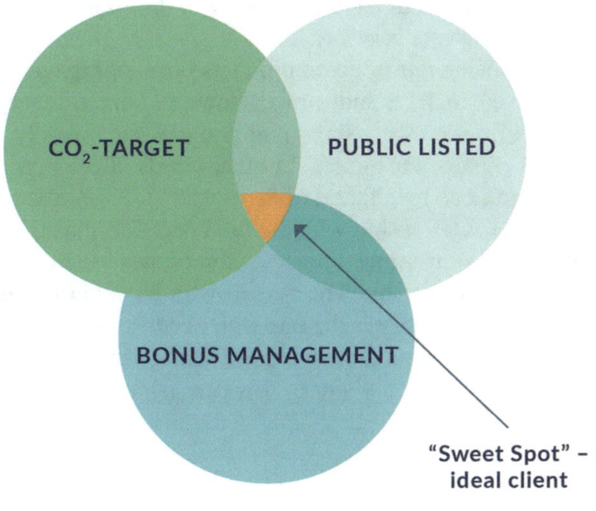

Fig. 3 CO_2 target, public-listed and bonus management intersect at ideal client "sweet spot" (own illustration)

2 How to Increase the Conversion Rate

2.1 The "1% Curse": Why 99% of Companies Massively Underperform in Conversion Rates for Digital and Green Products

A CEO recently told me, "Our customers find our new green products very exciting, but do not want to pay more for them". However, as previously noted, the end customers of shipping lines and logistics service providers are under legal pressure to reduce CO_2 emissions, for example the plastics and chemical industries (Die Bundesregierung, 2021). Therefore, there is a corresponding need.

Yet the logistics service providers hardly sell any of these green products to their customers, so that the investment in the new product either does not pay off at all, or they fall short of their massive organic growth opportunities. To give a concrete example, the German Lufthansa, like other airlines, offers the possibility to pay a surcharge on the airfare to offset CO_2 emissions. All you must do is tick the box when booking. The surcharge is between 20 and 200 US dollars, depending on the distance. Nevertheless, not even 1% of the customers choose this option (Haße, 2023).

This leads to the crucial question: how much higher could the conversion rate possibly be?

2.2 Rethinking the Customer Experience

While it is more difficult to focus on an individual customer in the B2C context, this is quite different in the B2B context. Most logistics companies operate large key account organisations that deal with thousands of major customers and thus decision-makers every day.

The closing rate of green products is rarely higher than in the Lufthansa example cited, even in B2B industries. However, my research (Nowroth, 2023) and the results of my projects show that you can successfully sell green products to up to 15% of existing customers. In other words, there is potential for an increase in the conversion rate by a factor of 5.

A close relationship with the final decision-maker is essential. But in most cases the existing customer contact—sometimes established for many years—is the *wrong* contact person. The person who can sign off an investment into this completely new category is the one who needs to be convinced.

You may assume that your good relationship with your previous decision-maker will automatically transfer to this new contact, for example because of an internal recommendation. However, this is far from the truth. Instead, you must start almost entirely from scratch.

If the new decision-maker is closely connected with your existing contact person, this is certainly an easier entry point, but still requires trust combined with real budget authority.

Real decision-makers with sole budget responsibility, unlike lower-level managers, always look at *both costs and value or benefits,* and quickly view costs as an investment once they have been guided through such a business case process. As long as they buy-in to the significant added value, they are happy to spend more. Only 2% of company employees are a so-called "economic buyer" who has their own budget and can thus say "yes" and "no" to an offer. All the others may be "non-deciders", but they are internal referral sources who can pave your way to the final decision-maker, hence called "key recommender". It is essential for the successful initiation of value-based deals that the acquisition process focuses throughout on these 2% decision-makers, visualised in Fig. 4.

Logically, this means that key account managers must approach much higher-level decision-makers to successfully sell green products than if they want to sell conventional products. This is the only way to understand the personal agenda of the new contact and be ready for the declination of the three basic components that make up a value-based offer: objectives, success criteria and entrepreneurial benefits.

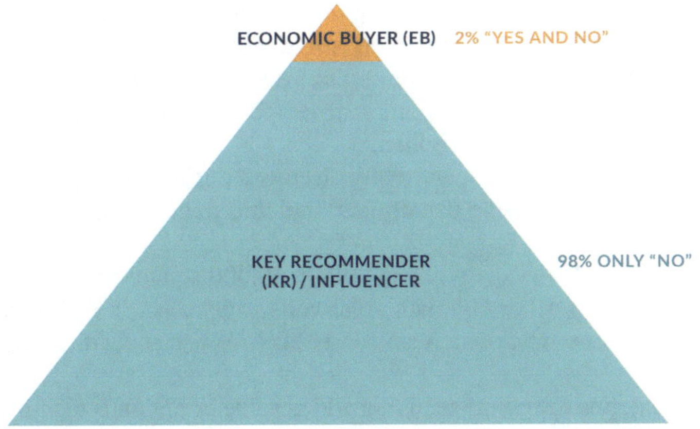

Fig. 4 2% "YES and NO" as well as 98% "only NO" (own illustration)

3 The "Magic Bullet" Value-Based Pricing Model: The New Margin Turbo

3.1 From "Per Unit" to "Value-Based" in Practice

As previously established, the basic marketing approach for existing standardised logistics products, such as container transport (TEU), road transport (FTL) or storage capacity (per pallet or square metre), is based on a price per unit. That model has been around for decades and hence is harder to change.

But some products like IT or green products (e.g. a supply chain risk management software) have only been around for a few years and are still sold on a per unit base (i.e. per user or licence per month).

Therefore, the metric is unit-based rather than value-based, which could increase profitability very significantly.

In contrast to the "per unit" model described above, "value-based" means that you base the fee you demand in return for your product on the tangible and intangible value that your product or service will trigger or create for your customer.

The fee you charge does *not depend* on how many units (e.g. software licences, CO_2 tonnage or man-hours) you sell, but on *how much benefit you will most likely create*. The following formula applies: the higher the value of your service, the higher your fee. This fee is a fixed sum paid by your client before or during your service.

Compared to conventional freight transport, green transport that saves more CO_2 offers considerable advantages, both tangible and intangible. Let us assume that you ship 10% of the 5000 containers, that you would otherwise import by conventional sea freight from the US to the UK for one of your key accounts, in a CO_2 neutral way. Table 1 below lists some examples of tangible and intangible added values from the point of view of your customer.

However, these benefits are never really discussed during the acquisition process (certainly not by the purchasing department) and thus not considered in the pricing of the final offer.

Another example: let us say you can build a 50,000 square metre warehouse in (A) the conventional way or (B) with solar cells, fully electric forklifts, etc. The price difference between variant A and B is 30%. What about the difference in added value?

If you do *not* define the concrete higher added value of option B together with the final decision-maker (e.g. in numbers), why would anyone be willing to pay a premium over an already existing known product? In fact, hardly anyone would do so (except perhaps the 1% from the previous Lufthansa example).

Table 1 Tangible and intangible advantages

Tangible advantages	Intangible advantages
CO_2 reduction to be allocated to CO_2 reimbursement through emissions trading = X US dollars (Organisation for Economic Co-operation and Development, 2023)	Better reputation of your company but also that of your customer on the market
Paying out higher bonuses to managers (incl. to the decision-maker)	Qualification for possible governmental subsidies
Better financing conditions within the framework of the EU-taxonomy, which has a strong influence on the conditions of banks and financial products.	Better reputation of the supply chain organisation within the group through additional CO_2 savings
Increase in turnover through possible servicing of the demand for green offers on the customer side.	Faster achievement of the net-zero target (net-zero emissions target)
Saving on otherwise necessary compensation to achieve the CO_2 targets	Greater attractiveness on the labour market for young talents who place a great deal of value on the social responsibility of their employer
Existing successful but traditional products can be kept on the market longer if necessary.	

3.2 *The Architecture of a Value-Based Offer*

Here is the detailed structure of a value-based offer based on the most important guiding questions, which vary depending on whether you sell green or logistics technical products.

First, you must physically meet with the decision-maker described previously. to reach a conceptual agreement on the following issues:

Objective Start with the objective. Limit yourself to one or at most two objectives.

- What is the impact and target state that your client is aiming for?
- What is different for the company and for the decision-maker personally when they have achieved the goal, such as reducing CO_2 emissions?

Success Criteria. Together you agree on KPIs, which are used to measure whether you are getting closer to achieving the objective. It is important that these are criteria that you can ensure with your service. However, they are not entrepreneurial values, otherwise a false expectation would arise. Below are some practical examples.

- The number of containers transported CO_2 neutrally.
- A successful implementation of a CO_2 IT tool that provides ideas for CO_2 saving route optimisation.
- To what extent do end-user sales increase for your customer because their customers can now generate more sales? Perhaps because the use of your products enables them to enter completely new markets or better achieve their sustainability goals.

Entrepreneurial Value/Benefits. Entrepreneurial value is about tangible (expressed in dollar figures) and intangible (not expressed in dollar figures) effects.

- What are the concrete effects on the CO_2 balance sheet of transporting a certain quantity of goods with CO_2 neutral modes of transport for your customer?
- How high is the upselling potential of the greener logistics service that helps your customer open up new markets/customers?
- What does your decision-maker personally gain from the product? Behind every corporate goal (e.g. CO_2 target) is a personal goal. The trick is to uncover this thanks to a strong mutual trust. Think here, for example, of the improved reputation within the organisation that follows the CO_2 reduction of a department.
- What are the tax and regulatory benefits of reducing CO_2 emissions for your client, e.g. cheaper insurance premiums, tax relief, increased attractiveness for investors or lower investment costs for greener companies?

You will then find a range of monetary and non-monetary benefits that very quickly add up to a 5:1 or even 10:1 return on the amount you charge for the green product. This will make a decision-maker seriously consider buying your green product.

Your remuneration is a fixed sum based not on success but the expected value that can be achieved. Ensure that the offer you draw up in the respective case is sent to the decision-maker, and only to him, since its content is primarily based on personal input and agreement on his part.

We can demonstrate proof from our projects that the probability of closing such an offer is between 50 and 70%.

3.3 Making Entrepreneurial Benefits Measurable

Of course, it is vital to provide fact-based evidence for monetary value generated through green products. This is demonstrated below with a practical example from Germany. Depending on which country your client mainly operates in, different regulations may apply and must be checked on a case-by-case basis.

The price for CO_2 emissions is regulated by authorities, i.e. in Europe by European emissions trading. Anyone who harms the planet will have to pay for it, and increasingly more in the coming years. For end consumers in Germany, this new CO_2 price will initially be 25 euros per ton from January 2021 onwards. It will then gradually rise to 55 euros in 2025 (Die Bundesregierung, 2023). For 2026, a price corridor of at least 55 and at most 65 euros will apply (Die Bundesregierung, 2019).

For the logistics sector, such as road transport, significantly higher rates will apply in Germany. Accordingly, a CO_2 differentiation of the truck toll and a CO_2 surcharge of 200 euros per ton of CO_2 are to be introduced on 1 January 2024. Emission-free trucks are to be exempt from the charge until the end of 2025, after which 25% of the regular rate will be levied (Zweites Deutsches Fernsehen, 2023).

These developments form the perfect basis for a clearly measurable monetary added value by logistics providers who manage to actively reduce CO_2 emissions for customers by intelligently linking conventional and green products. The less CO_2 emitted during transport, the lower the costs.

3.4 The Unique Selling Proposition Against Your Competition

All in all, the "value-based" pricing model offers several advantages for your company, including:

- a significant increase of the profitability of technically sophisticated and green products compared to conventional products;
- a significantly higher closing rate for proposals, as you create a customised offer together with the "real" decision-makers, which they can approve immediately; and
- an increased reduction of CO_2 emissions as you work with more customers.

It is very difficult for competitors to copy this approach, as each fee derived from the "value-based" pricing model is unique in that it is based on the value that is expected to be achieved. Therefore, the price for your service will always vary.

Figure 5 shows the relationship between a linear profit development with a per unit charging basis compared to the value charging basis. With the scaling of clients billed according to the "value-based" pricing model, a significant margin effect is established after some time. It grows exponentially with the number of customers

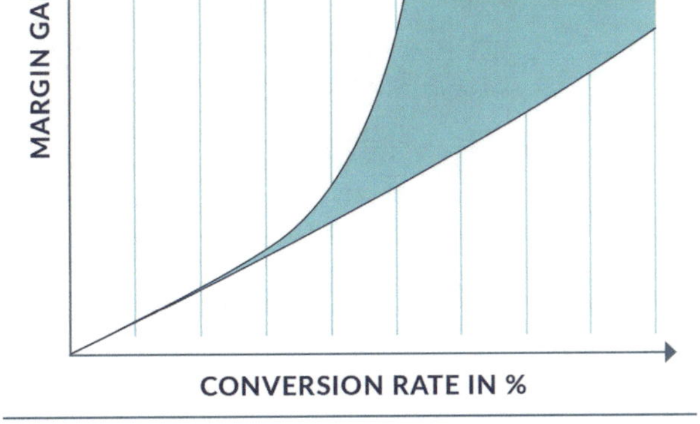

Fig. 5 Relationship between margin gain and conversion rate (own illustration)

who successfully follow the "value-based" pricing model. Additionally, another advantage is that the more your team sells value-based, the less you are dependent on additional operational staff to grow. This is a great peripheral added value, especially invaluable in times of (professional) staff shortages.

Successfully monetising green or technologically sophisticated products, i.e. selling them value-based, is difficult but highly desirable. It is even more difficult, but all the more desirable, to scale this, i.e. to embed this approach as an absolute core competence in the key account organisation.

Unfortunately, as Fig. 6 demonstrates, most companies remain in the bottom left quadrant known as the "status quo". Very few manage to create both a high conversion rate and a high value-based share of customers. But if they achieve both, they have succeeded in gaining a real competitive advantage. You too can achieve this with a consistent approach.

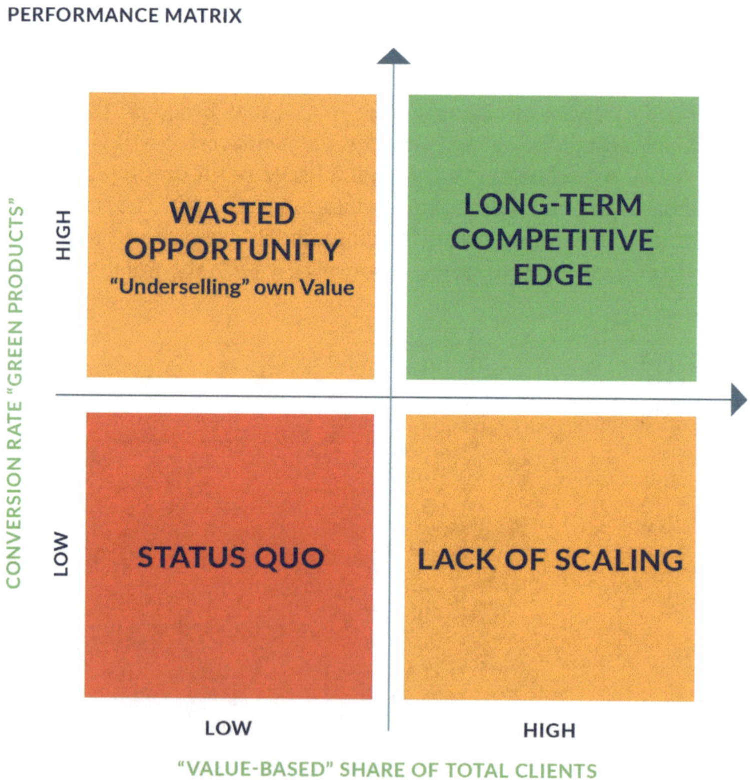

Fig. 6 Performance matrix (own illustration)

4 Implementation Guide for Commercial Organisations

Finally, the most important requirements for a successful practical implementation will be explained.

4.1 Decision-Maker Level

Difference in the Job Level of the Decision-Maker. Most companies and key account managers have been working with their procurement contacts, category managers, etc. for years. These stakeholders have existing budgets associated with previous categories. With the introduction of environmentally friendly products, such as carbon-neutral fuel, a category has emerged that they have not dealt with before. This means that there is usually no established budget for it and thus a decision in this regard is clearly beyond their area of responsibility.

To solve this issue sales must enter at a much higher level, i.e. at the level of the chief product officer (CPO) or even the chief financial officer (CFO). The question that arises is: who benefits most from a reduction in CO_2 emissions? If, for example, the CFO's bonus depends on how quickly he can bring emissions to zero (and most decision-makers in large companies have a bonus linked to demonstrable CO_2 reductions), then it is essential to involve him in the acquisition process. This involvement should occur as early as possible and preferably via a recommendation from the previous decision-maker. Another advantage is that the higher-ranking decision-maker has the authority to approve your project immediately.

Do Not Be Fooled by Sustainability Departments. In many companies today there are so-called sustainability departments. However, similar to human resources departments, they almost never have real decision-making power, i.e. as described in Sect. 2.2 they can only say "no" not "yes". While this department should be involved in the decision due to their valuable influence, it is even more important to keep this group as small as possible because the more people involved in the process, the longer the sales cycle usually takes. Thus, find out who will truly get the *most* benefit from your green product and who has the real decision-making power.

4.2 Customer Groups: Existing vs. New

If you start with existing customers, who have a strong basic trust in you (often developed over many years), you are much more likely to get through to the ultimate decision-makers. As previously indicated, it is also easier and cheaper to sell a new product to an existing customer than to sell a new product to a new customer. The latter may succeed in individual cases, but in practice the first way is the easier one.

Start with your top 100 customers and proceed according to the criteria listed previously:

1. CO_2 target
2. Public-listed
3. Bonuses of the management and the board of directors

Finally, carry out an analysis of your entire existing customer portfolio in order to identify the customers who are under the greatest pressure and/or have expressed a desire in a customer survey to be able to use more environmentally friendly services.

End-to-End Upsell. It is important to find out which new markets your clients want to enter strategically. There might be some markets that you can help your clients enter by reducing emissions in the supply chain. For example, the UN imposes very strict CO_2 targets on its suppliers. There are many industries, such as the chemical industry, that have or will have similar requirements in the near future.

4.3 What Other Products Can Be Sold on a "Value-Based" Pricing Model?

Naturally, other types of products can also be sold using the value-based approach described in Sect. 3.2. In general, the more "consultative" a product is and the higher the expected value generated, the more suitable it is for the value-based approach. To identify suitable products, create a brief portfolio analysis of all your products, mainly including software products that go far beyond the standard repertoire and require a not inconsiderable amount of implementation effort on the part of your customers. For example, a diagnostic CO_2 tool may be suitable because it measures CO_2 emissions, but it is no longer a real competitive differentiator. However, it is a different story, if you combine this product with a much more complex configuration of the tool and perhaps two or three employees to help your customers *actively and measurably reduce* CO_2 emissions.

Let us consider one last example: you are planning an implementation with two to three people from your team, introducing a digital warehouse management system. Up to now, you have been selling your man-hours with a factor of 1.3, i.e. 1.3 times your costs are paid by the customer, resulting in a gross margin of 30% for you. Alternatively, you could change this same service to the value-based business case structure introduced in 3.b. What is your client's objective in the implementation? How do you measure success? Answering these questions, you can quickly showcase an entrepreneurial value, and put the service of your two to three employees in relation to the presumably greater value they will achieve for your customer.

However, the "value-based" pricing model is not applicable to all products. The container freight price, for instance, will always be calculated on a per unit basis, not least because this has been its prevailing pricing basis for decades. In stark contrast, there are numerous "young" products that can follow their own, new approach. Furthermore, they can be combined as "value-based" with existing "per unit" pricing models.

5 Conclusion

The bottom line is that you can increase the current margins of your customers and the return on investment of your sales organisation by up to 25% if you consistently follow the value-based approach. This approach is not limited to green products, and can also be applied to other strategic products, such as logistics IT solutions.

Ultimately, it makes your organisation much more resilient to margin declines as well as disruptions from competitors. Not only because a "value-based" pricing model is more difficult to copy, but above all because a systematically implemented value-based approach provides a great additional competitive advantage.

References

Die Bundesregierung. (2019). *CO2-Bepreisung*. https://www.bundesregierung.de/breg-de/schwerpunkte/klimaschutz/co2-bepreisung-1673008

Die Bundesregierung. (2021). *Intergenerational contract for the climate: Climate Change Act 2021*. https://www.bundesregierung.de/breg-de/schwerpunkte/klimaschutz/climate-change-act-2021-1936846

Die Bundesregierung. (2023). *Ermäßigter Steuersatz für Gas, weniger Stromkosten*. https://www.bundesregierung.de/breg-de/schwerpunkte/entlastung-fuer-deutschland/entlastung-energieabgaben-2125006

energiezukunft. (2022). *Emissionshandel: Die Klimasünden der Schifffahrt bekommen einen Preis*. https://www.energiezukunft.eu/politik/die-klimasuenden-der-schifffahrt-bekommen-einen-preis/

European Commission. (2021). *'Fit for 55': Delivering the EU's 2030 Climate Target on the way to climate neutrality* (52021DC0550). European Commission. https://eur-lex.europa.eu/legal-content/EN/TXT/?uri=CELEX%3A52021DC0550

European Commission. (2023). *European Climate Law*. https://climate.ec.europa.eu/eu-action/european-green-deal/european-climate-law_en

Haße, D. (2023). *Mehr Lufthansa-Passagiere zahlen für Emissionsausgleich*. Airline.de. https://www.airliners.de/lufthansa-passagiere-zahlen-emissionsausgleich/68030

International Energy Agency. (2020). *Indicative shipping fuel cost ranges*. https://www.iea.org/data-and-statistics/charts/indicative-shipping-fuel-cost-ranges

International Maritime Organization. (2023). *2023 IMO Strategy on Reduction of GHG Emissions from Ships*. https://www.imo.org/en/OurWork/Environment/Pages/2023-IMO-Strategy-on-Reduction-of-GHG-Emissions-from-Ships.aspx

Nowroth, A. (2023). *Sales and business models in the logistics industry: Ensuring growth with innovative strategies*. Springer.

Organsiation for Economic Co-operation and Development. (2023). *Emission trading systems*. https://www.oecd.org/env/tools-evaluation/emissiontradingsystems.htm

Phaata. (2023). *Container shipping lines see a sharp decline in revenue and profit in the first quarter of 2023.* https://phaata.com/en/logistics-market/container-shipping-lines-see-a-sharp-decline-in-revenue-and-profit-in-the-first-quarter-of-2023-35803.html

Pico, S. (2021, February 17). Customers will pay higher price for Maersk's first carbon neutral ship. *Shippingwatch*. https://shippingwatch.com/carriers/Container/article12768953.ece

Rojon, I., Lazarou, N.-J., Rehmatulla, N., & Smith, T. (2021). The impacts of carbon pricing on maritime transport costs and their implications for developing economies. *Marine Policy, 132*, Article 104653. https://doi.org/10.1016/j.marpol.2021.104653

Waves Sustainability. (2023). *How to successfully master EU-Taxonomy: Requirements, approach & solutions.* https://waves-sustainability.com/wp-content/uploads/White-Paper-EU-Taxonomy-EN-Version-April-2023.pdf

Zweites Deutsches Fernsehen. (2023). *Güterverkehrsbranche warnt: Steigen Verbraucherpreise wegen Lkw-Maut?* https://www.zdf.de/nachrichten/wirtschaft/lkw-maut-inflation-verbraucherpreise-100.html

Alexander Nowroth is the CEO of Lebenswerk Management GmbH, based in Germany and the author of the book *Sales and Business Models in the Logistics Industry*, published internationally in 2022. Alexander Nowroth is an expert in implementing high-performance strategies to create high-performance organisations and is responsible for over 100 successful international projects. He is also a guest lecturer and advisory board member at renowned international universities.

Green and Collaborative Logistics in the U.S-Mexico Border: The Blue5PL Case

Olivia Hernandez-Pozas and Rodrigo Martinazioli

1 Introduction

In the last years, the region of North America has experienced various supply chain disruptions. First, the COVID-19 pandemic increased the necessity for more frequent, lower-volume shipments. To prevent the halt of a production line and to fill the shelves in a store on time, the number of shipments has risen. In other words, there is a tendency to ship less per truck but more times. Since then, truck trailers have been moving more air than goods.

Second, other global and regional events—such as the 2020–2023 chip shortage, the imminent retirement cliff of U.S. drivers, the pseudo-migration of Mexicans as B1 drivers, the new regulations in the U.S. and Mexico, and the imbalance of northbound/southbound trade—alongside a stronger demand have pushed northbound cross-border trucking rates to unprecedented levels.

Finally, the U.S.-Mexico-Canada Agreement (USMCA), which entered into force in 2020, along with the economic war between the U.S. and China has made the case for "nearshoring" in Mexico. Nearshoring consists of the relocation of production facilities of American and other foreign companies to Mexico, close to their customers, to reduce supply chain risks. Consequently, the flow of U.S.-Mexico cross-border logistics transportation is expected to keep on growing substantially in the years to come.

O. Hernandez-Pozas (✉)
Institute for the Future of Education & EGADE Business School, Tecnologico de Monterrey, Monterrey, Mexico
e-mail: olivia_hdz_p@tec.mx

R. Martinazioli
Blue5PL, Monterrey, Mexico
e-mail: rodrigo@blue5pl.com

E. Fröhlich et al. (eds.), *Circular Economy in Sustainable Supply Chains*, CSR, Sustainability, Ethics & Governance,
https://doi.org/10.1007/978-3-031-70749-0_21

Any increase in the flow of transportation typically brings positive economic effects, such as the construction of new infrastructure and, undeniably, new jobs. Nevertheless, it can also have a negative impact on the environment. The environment is adversely affected by the excessive use of fuel, traffic jams, and transportation modes that are often hampered (Raap et al., 2017).

Indeed, logistics considerably contributes to greenhouse gas (GHG) emissions (McKinnon et al., 2015; Huge-Brodin et al., 2020). Some scholars even suggest it is responsible for 13% of such emissions (Perotti et al., 2022). For that reason, addressing logistics' environmental sustainability is a big concern for practitioners and policymakers (Klymenko & Lillebrygfjeld Halse, 2021). Global transportation as part of logistics was the fourth largest source of GHG emissions in 2019. Within this sector, the movement of passengers and freight in road transport is the largest source of emissions (6.1 $GtCO_2$-eq, 69% of the sector's total), while road transport is the fastest growing (+1.7% per year; Jaramillo et al., 2022).

More than 80% of supply chain specialists in the U.S. argue that sustainability is a priority, and more than 45% said they were under pressure to adopt supply chain sustainability practices. However, less than 50% of them were able to point to actionable sustainability objectives (Bateman et al., 2021). All these pressures and new opportunities set the table for the so-called FreightTech start-ups to quickly emerge as the promising solution to manage the situation.

In this chapter, we start by going into more detail about the latest disruptions and context in North America's logistics industry. These serve as the justification for the presented innovative case. We continue by building the foundation of our learning proposal with theoretical frameworks such as green logistics and circular economy (CE). Then, we share the Blue5PL case. Finally, we close the chapter with key lessons as the takeaway.

2 Disruptions in North America's Logistics Industry

According to the Texas Department of Transportation (2021), the U.S. and Mexico share a 1954-mile common border. 1254 miles (64%) are shared between Texas and Mexico. From the Mexican side, the states of Tamaulipas, Nuevo Leon, Coahuila, and Chihuahua run alongside the state of Texas. Mexico is the largest trading partner of Texas. The Texas-Mexico border is North America's busiest trade gateway. In fact, 70% of U.S.-Mexico trade passes through the Texas-Mexico border. Therefore, this border is a key contributor to the local, regional, state, and national economies of the U.S. and Mexico. Most of the economic development in the cross-border region relates to manufacturing production, agribusiness, logistics services, and other trade-related activities.

Logistics services comprise the movement of goods and raw materials, delivery, scheduling, storage, and marketing to customer endpoints that support economic growth (Setiawan & Koestoer, 2021). In maintaining business competitiveness and sustainability, as well as in demonstrating social responsibility, logistics providers play an increasingly important role (Li & Hanafi, 2012). Their challenge in embracing sustainability practices is how to link, balance, and solve environmental performance issues with sound business practices (Lee & Kim, 2011). Especially since this industry focuses so much on optimisation, reducing costs, increasing delivery speed, and gaining maximal revenue (Tamulis et al., 2012).

The wheels of logistics spin differently across the Atlantic. The configuration of transport, logistics networks, operational decisions, and the regulatory framework are different in North America than in Europe (Rodrigue & Notteboom, 2010). In North America, vast distances favour long-haul trucking. Freight by railway is more common in North America than in Europe. Europe, with its smaller countries and dense network of borders, relies more heavily on intermodal transportation. While North America operates with a decentralised patchwork of state and federal rules, Europe boasts a unified European Union regulatory framework. Europe understood many years ago the importance of working on ecological policies to pollute less and transport cargo efficiently. In North America, the ambition for reductions, approach, and pace of environmental protection are different than in Europe (Xie & Minjares, 2023). Although environmental protection is already in the conversation, the priorities still are that the cargo arrives on time and the cost remains low. In this context, the average utilisation of partially empty truckloads in the U.S. is 47% (Driveresearch, 2022), well below the usual average in Europe. In North America, the so-called full truckloads (FTLs) are not actually full. It is relatively common to see a 53-ft truck travelling across the country with few pallets, where the urgency of the arrival of the merchandise takes precedence over the efficiency of the route and the pollution generated in the operation.

Since 2020, along with the COVID-19 pandemic, the world and, in particular, the region of North America has experienced a chip crisis. The chip crisis affected the production of new trucks and pushed used truck prices higher. Moreover, the backlog of ordered but not built trucks in 2021 nearly tripled that of the previous year (Smith, 2021). Prices became so high that even the "soon-to-be-5-year-old" trucks were knocking on the door of six figures (J. D. Power, 2021). As a result, freight demand in North America keeps growing well ahead of truck availability, propelling shipping costs through the roof.

Every day stronger demand reaches far and wide, and the transportation industry experiences not only insufficient trucks but also drivers in short supply. Before the COVID-19 pandemic hit, the driver shortage was already a headline in the U.S. In 2023, there was a shortage of 80,000 drivers, and the forecast for 2028 calculates a shortage of 160,000 drivers (American Trucking Associations, Inc., 2021). This trend is the result of the so-called retirement cliff. A large portion of current drivers are retiring, and there are not enough young drivers getting into the driving industry. The average age of current drivers is 55 years, and 60% are more than 45 years old. Mexican drivers holding a B1 visa have contributed some relief, but this is not sufficient for cross-border shipments. They can travel into the interior of the U.S. on the condition that they only move international freight and not any domestic portion. However, this pseudo-migration of B1 drivers is now causing an enormous scarcity of drivers in Mexico, especially "transfer drivers" who just cross the border from the carrier's yard in Mexico to the customs broker's warehouses in the U.S. and vice versa. CANACAR (The National Association of Truckers in Mexico), cited by González (2023) in *El Heraldo de Juarez*, reported a lack of 60,000 drivers in Mexico.

The enforcement of new regulations also impacts the driver shortage and the miles trucks run. The electronic logging device (ELD) mandate, which started in 2018, requires carriers to attach a piece of electronic hardware to a commercial motor vehicle engine to record driving hours. This stops drivers from lying on their paper logs, which considerably helps to prevent incidents but reduces the number of miles that trucks and drivers report. Another important issue to consider is the new *Complemento Carta Porte* (CPP), which complements the bill of lading and is currently going into effect in Mexico. CPP adds supplemental information (approximately 185 data fields) to the CFDI (electronic invoice), causing major delays for southbound, northbound, and intra-Mexico freight. Many customers and carriers are still trying to implement this new process into their supply chain workflow.

According to the United States Census Bureau (2023), trade between Mexico and the U.S. hit a new record high in 2023. Northbound trade was $475,606.7 million, while southbound trade was $323,227.9 million, creating a deficit of $152,378.8 million for the U.S. If we compare it to the deficit of $2 million per month in the year 2000, the imbalance of trade has grown by about six times. Thus, these indicators give us a clear picture of important changes in context.

Because of all the above reasons, shippers (especially northbound ones) are experiencing unprecedentedly high rates. Truckload rates have increased by more than 40% since pre-pandemic times, while less-than-truckload shipments (LTL) have increased by more than 20%. On top of that, diesel prices have increased, and the Russia-Ukraine war has made it more difficult for this industry.

It is well known that logistics directly affects the environmental situation (Kurbatova et al., 2020). Certainly, North America is not the exception. Fossil fuels are choking humanity, and climate change is causing dangerous and widespread disruption in nature. Therefore, doing business with logistics firms that show a real commitment to sustainability in economic and green logistics best practices cannot be underestimated (Wang et al., 2017).

3 Circular Economy and Green Logistics

Around the world, companies are in quest of alternatives that help them reduce the negative effects of their processes on the environment. Like so, CE is one of the solutions with the greatest potential for success (Kurita et al., 2023). CE's most important distinction is the closing of production cycles that transform discarded materials into new resources for organisations with the goal of achieving net zero emissions (Bocken et al., 2016).

CE has the following principles: eliminate or reduce the use of resources, design out waste and pollution, increase the useful life cycle of products and materials, and regenerate the natural environment (Oliveira et al., 2023). CE initiatives can be developed at the macro, meso, and micro levels. Often, macro initiatives relate to public policies in the region, country, or specific city. Meso projects involve the collaboration between companies, while micro practices relate to strategic internal actions of companies and non-governmental organisations (Kurita et al., 2023).

For CE to work, all players need to know their responsibilities and comply with them. Both public and private stakeholders need to articulate the best green logistics practices that conserve the ecosystem, not only within the organisation but throughout the whole supply chain. Furthermore, collaboration between government, private investors, and industries could help close cost gaps in the short and medium term (Sankaran, 2023). Implementing CE principles in logistics can lead to more efficient transportation routes, reducing the overall carbon footprint of goods and services transported between the U.S. and Mexico.

The term green logistics also involves logistics-related initiatives aiming to reduce the negative impact on the natural environment (Evangelista et al., 2017) and to develop new solutions that replace the polluting ones (Abbasi & Nilsson, 2016). Green logistics is decisive for developing the CE (Oliveira et al., 2023). Furthermore, the use of green logistics methods may become a significant competitive advantage for transportation companies, as it brings positive results for the company as well as society at large (Setiawan & Koestoer, 2021). When discussing green logistics, the providers of logistics services are always considered the key players because, even when they act on behalf of their customers, they are accountable for the emissions (Prataviera et al., 2023).

In the context of CE, the "reduce" principle emphasises the importance of being mindful of the consumption of resources and the generation of waste. The goal is to decrease the overall environmental impact of products and services by using fewer materials and resources. Companies can apply this principle by offering energy-efficient designs of products and services, as well as by reducing the energy required for transportation. This principle can also be practised by decreasing concentrations of GHGs in the atmosphere to lessen the contribution to global warming and by encouraging collaboration among businesses, governments, and other stakeholders to share best practices and drive innovation.

4 The Blue5PL Case

Case studies are recommended when the purpose of research is to gain an in-depth understanding of contemporary issues in their real-world context, as well as to answer "how" or "why" questions (Yin, 2018). Cases are selected to suit the topic under investigation, the research questions, and the study aims. In case studies, understanding the context is key (Golich, 2000). Thus, this type of research is particularly appropriate for exploring a problem concerning different contextual factors (Gammelgaard, 2017).

Our selected case is a FreightTech start-up named Blue5PL. Blue5PL plays in the attractive $80 billion U.S.-Mexico cross-border market, a subsection of the $800 billion U.S. domestic freight market and the $60 billion Mexican domestic market. Blue5PL maximises truck/trailer utilisation in terms of weight and volume. Its platform finds the best mix of LTL and FTL shipments to fill the trailer—the shared truckload (STL; Blue5PL, 2024a). Blue5PL's STLs optimise routes, running the least possible miles, with synchronised order pickups and deliveries. Blue5PL reaches efficiencies because its solution avoids passing through consolidation warehouses.

Using this case, we demonstrate how the CE principle of "reduce" is implemented in the U.S.-Mexico cross-border logistics market. We also highlight how companies can capitalise on the green logistics practices of collaboration. Additionally, we aim to motivate other companies to insert environmental sustainability practices into their supply chains. This case is based on qualitative data collected by the first author using interviews with the CEO of Blue5PL. The first author is a member of the counselling committee of the company. The case was enriched with additional information provided by the second author. He has been in the company since its creation and functions as Head of Collaboration. As recommended by Charmaz (2014), Blue5PL's representatives had relevant related information, were able to verbalise their insights and fully agreed on the company's reported description in this chapter. Data were later complemented and validated using secondary sources, such as the firm's and industry's reports, as well as the company's website. Information was collected until this arrived at the saturation point. In-depth interviews included research questions related to industry context, value added by the company, and innovative and sustainable strategies. Essential points were then summarised.

Blue5PL's approach to the notion of fifth-party logistics (5PL) refers to managing a cluster of companies that work together towards the same goals in a collaborative environment. It relates to a network of multiple supply chains harmoniously synchronised using state-of-the-art technology to achieve common goals. Blue5PL's sustainable activities fit in what some scholars, such as Centobelli et al. (2017) and Forslund et al. (2021), have categorised as green logistics practices of distribution planning and transportation execution. Specifically, those related to shipment/freight consolidation (Martins et al., 2022), route planning optimisation (Centobelli et al., 2017) as well as collaboration with customers (Chu et al., 2019; Jozef et al., 2019) and among competitors (Cousins et al., 2019; Abbasi & Nilsson, 2016).

As asset-based carriers and second-party logistics (2PL) players, consolidators and LTL carriers are probably the nearest to the idea of Blue5PL's STL. The most important difference is that consolidators and LTL carriers typically use their physical terminals to create their shared line hauls to the final destination or another terminal for further deconsolidation and consolidation. Consolidators and LTL carriers represent the closest competitors for Blue5PL in terms of price. Despite this, Blue5PL offers a significant advantage by reducing transit time and damage risks. These are major disadvantages of consolidators and LTL carriers caused by using their terminals. Nevertheless, consolidators and LTL carriers can become good Blue5PL allies as suppliers in its last mile strategies when pure STLs are not possible and longer time in transit makes sense.

Blue5PL delivers value to customers and renders competition irrelevant with its innovative shift from the traditional third-party logistics (3PL) brokerage service to its 5PL "blue ocean" collaborative environment. When companies follow a blue ocean strategy, as Blue5PL does, they simultaneously use differentiation and low cost to create a new demand and open a new market. With blue ocean strategies, companies have opportunities to grow in a profitable and rapid way because demand and rules have been recently created. In this case, the blue ocean is an analogy for an unexplored market space where companies innovate. Opposingly, a "red ocean" is where companies have a competitive approach and industry boundaries and rules are well known and accepted (Chan & Mauborgne, 2005).

Furthermore, in 2021, Blue5PL started organising round table events with potential allies to discuss supply chain disruptions. That was the starting point for the creation of a collaborative environment. Blue5PL focused first on the automobile industry in Mexico, a progressive industry with ample connections with related companies in the U.S.

Blue5PL does not match shippers with carriers like most digital platforms do. Instead, Blue5PL matches shippers with shippers to create STLs (Blue5PL, 2024a). Through its platform, Blue5PL matches cross-border shippers to move synchronised freight, finding efficiencies and tackling carbon emissions at the same time. Based on the origin, delivery points, pallet count, dimension, weight, stackability, and commodity, Blue5PL's algorithms match shippers that can ride together on the same truck. During this process, their innovative tool examines additional criteria in the Mexican freight and cross-border industry.

Overall, the Blue5PL platform unlocks various efficiencies: less cost, less time in transit, less damages, and less CO_2 emissions (Blue5PL, 2024b). Because shippers share trailer space with other shippers, they pay only a portion of the STL cost. Shippers that participate in collaboration projects doing STLs with Blue5PL experience an average of 35% cost reduction. Shipments that move from an LTL to an STL experience an average of 2 days in transit time reduction. STL shipments are touched only at pickup and delivery, minimising the probability of damages or losses. In fact, STL shipments are 99.998% damage-free, which is far better than any LTL carrier. Carriers also benefit from higher pay per load by adding a few stops on the same route. Thus, load profitability increases substantially with Blue5PL.

Blue5PL also reduces GHG emissions by an average of 38% by combining shipments and avoiding several FTLs with empty space or inefficient LTL rides. Blue5PL is committed to climate change mitigation and designed their so-called Aqua Initiative: "where blue meets green". The Aqua Initiative has three phases (Blue5PL, 2024c):

1. **Phase 1** Blue5PL calculates the total emissions the customer's supply chain is currently generating and integrates that into a report.

2. **Phase 2** As the customer starts moving STLs, Blue5PL calculates emission reductions and provides a signed letter of lbs. CO_2 reduced.
3. **Phase 3** Any remaining emissions are offset through Blue5PL's partner programmes by buying carbon certificates and sharing them with the customer.

By maximising truckload space and weight with STLs, Blue5PL eliminates the inefficient use of several trucks and/or LTL shipments. Blue5PL reduces the total miles that these trucks, full of air, run to deliver multiple shipments. Therefore, Blue5PL reduces the total CO_2 emissions. This measurable cut is worth the attention of those many supply chain professionals looking forward to adopting sustainability actions.

Blue5PL uses international standards to calculate the carbon footprint. For example, it uses the 22.509 lbs. CO_2 per gallon of diesel fuel burned, as suggested by the GHG Protocol, and the average of 5.9 miles per gallon fuel consumption of FTL and STL shipments (United States Environmental Protection Agency, 2023). Also, for LTL calculation, Blue5PL applies de Aguiar and Woolard (2014) suggestions. This takes into consideration extra miles run by LTLs, proportional ratios versus the average LTL line haul weight, and collections and delivery miles for local operations, among other things.

5 Key Lessons

The avalanche of environmental challenges has triggered responses at all levels: micro, meso, and macro (Hart & Pomponi, 2021). Our presented case, at the meso level, shows how the CE's principle of "reduce", as well as green logistics practices in the distribution planning and transportation execution, can serve to reduce CO_2 emissions.

Blue5PL visualised a business opportunity in the growing U.S.-Mexico cross-border logistics services market. The notion of STLs enhanced by technology and collaborative logistics put in practice among the right players propelled its business case. Having a good value proposition, excellent knowledge of the U.S. and Mexico logistics context, as well as creating constructive relationships based on trust with stakeholders is paying off for Blue5PL. This start-up has already been a finalist of the Heineken Green Challenge award, a finalist of the EarthX Climate Tech prize, and has won the TIW award, the Las 100 PRO2022 and the Las 100 PRO2023 by Google Cloud.

It is worth underscoring that, in addition to less carbon emissions, the Blue5PL's sustainable practices bring other important benefits to customers, such as less cost, less time in transit, and less damages. These are all imperative for companies in this industry. So, when aiming to influence others to join the green logistics agenda, an important lesson is to clearly understand concerns stakeholders might have and craft an integrative solution considering all their priorities, always adjusting strategies to contextual particularities. As shown before, the North American context differs from the European one; hence, solutions should consider this.

Centobelli et al. (2017) argued that understanding and adoption of green logistics practices to tackle the climate crisis is still insufficient. Thus, to achieve the goals of green logistics, more people and qualified personnel who exercise their responsibility to preserve the environment are needed (Kurbatova et al., 2020). Therefore, it is crucial to implement more educational and training programmes that take into consideration CE principles and green logistics practices. Also, it is essential to improve the environmental awareness of the population as a consumer of goods and services.

We close this chapter by underscoring and reminding the reader that "moving towards a zero-carbon global economy through absolute reductions in fossil fuel usage is a sure way of mitigating climate change, and a range of environmental, social, and economic benefits would follow. [...] The possible responses to environmental challenges are as numerous and varied as the problems themselves" (Hart & Pomponi, 2021, p. 127).

References

Abbasi, M., & Nilsson, F. (2016). Developing environmentally sustainable logistics. Exploring themes and challenges from a logistics service providers' perspective. *Transportation Research Part D: Transport and Environment, 46*, 273–283. https://doi.org/10.1016/j.trd.2016.04.004

American Trucking Associations, Inc. (2021). *Driver shortage update 2021*. https://www.trucking.org/sites/default/files/2021-10/ATA%20Driver%20Shortage%20Report%202021%20Executive%20Summary.FINAL_.pdf

Bateman, A., Betts, K., Pang, J., Deshpande, A., Cottrill, K., Grau, A., & McCool, D. (2021). *State of supply chain sustainability 2021*. MIT Centre for Transportation & Logistics and Council of Supply Chain Management Professionals. https://sscs.mit.edu/wp-content/uploads/2021/07/State-Sustainable-Supply-Chains-MIT-CSCMP.pdf

Blue5PL. (2024a). *About Us*. https://www.blue5pl.com/about

Blue5PL. (2024b). *Benefits*. https://www.blue5pl.com/benefits

Blue5PL. (2024c). *Sustainability*. https://www.blue5pl.com/sustainability

Bocken, N. M. P., de Pauw, I., Bakker, C., & van der Grinten, B. (2016). Product design and business model strategies for a circular economy. *Journal of Industrial Production Engineering, 33*(5), 308–320. https://doi.org/10.1080/21681015.2016.1172124

Centobelli, P., Cerchione, R., & Esposito, E. (2017). Environmental sustainability in the service industry of transportation and logistics service providers: Systematic literature review and research directions. *Transportation Research Part D: Transport and Environment, 53*, 454–470. https://doi.org/10.1016/j.trd.2017.04.032

Chan, K., & Mauborgne, R. (2005). *Blue ocean strategy*. Harvard Business Press.

Charmaz, K. (2014). *Constructing ground theory*. Sage Publications.

Chu, Z., Wang, L., & Lai, F. (2019). Customer pressure and green innovations at third party logistics providers in China. *International Journal of Logistics Management, 30*(1), 57–75. https://doi.org/10.1108/IJLM-11-2017-0294

Cousins, P., Lawson, B., Petersen, K., & Fugate, B. (2019). Investigating green supply chain management practices and performance: The moderating roles of supply chain ecocentricity and traceability. *International Journal of Operations and Production Management, 39*(5), 767–786. https://doi.org/10.1108/IJOPM-11-2018-0676

de Aguiar, G. V., & Woolard, M. A. (2014). *Estimating carbon emissions from less-than-truckload (LTL) shipments*. MIT. https://ctl.mit.edu/sites/ctl.mit.edu/files/library/public/2014fullthesis-AguiarWoolard.pdf

Driveresearch. (2022) *Solving America's half-empty truck problem in any market condition*. https://assets.website-files.com/656b56e5fc9ae6eba29a2fcf/65b1a2399f685a50b189fe62_market_research_paper_23-compressed.pdf

Evangelista, P., Colicchia, C., & Creazza, A. (2017). Is environmental sustainability a strategic priority for logistics service providers? *Journal of Environmental Management, 198*(1), 353–362. https://doi.org/10.1016/j.jenvman.2017.04.096

Forslund, H., Björklund, M., & Svensson Ülgen, V. (2021). Challenges in extending sustainability across a transport supply chain. *Supply Chain Management, 27*(7), 1–16. https://doi.org/10.1108/SCM-06-2020-0285

Gammelgaard, B. (2017). Editorial: The qualitative case study. *The International Journal of Logistics Management, 28*(4), 910–913. https://doi.org/10.1108/IJLM-09-2017-0231

Golich, V. L. (2000). The ABCs of case teaching. *International Studies Perspectives, 1*(1), 11–29. https://doi.org/10.1111/1528-3577.00002

González, R. (2023, November 2). Déficit de choferes de carga estanca las exportaciones en Juárez. *El Heraldo de Juarez*. https://www.elheraldodejuarez.com.mx/local/juarez/deficit-de-choferes-de-carga-estanca-las-exportaciones-en-juarez-10939544.html

Hart, J., & Pomponi, F. (2021). A circular economy: Where will it take us? *Circular Economy and Sustainability, 1*, 127–141. https://doi.org/10.1007/s43615-021-00013-4

Huge-Brodin, M., Sweeney, E., & Evangelista, P. (2020). Environmental alignment between logistics service providers and shippers – A supply chain perspective. *The International Journal of Logistics Management, 31*(3), 575–605. https://doi.org/10.1108/IJLM-04-2019-0101

J. D. Power. (2021). *Commercial truck guidelines: Industry review*. https://www.wallingfordct.gov/Customer-Content/www/CMS/files/JDPowerCommTruckGuidelinesDec2021Assessor020422.pdf

Jaramillo, P., Kahn Ribeiro, S., Newman, P., Dhar, S., Diemuodeke, O., Kajino, T., Lee, D., Nugroho, S., Ou, X., Hammer Strømman, A., & Whitehead, J. (2022). Transport. In Intergovernmental Panel on Climate Change (Ed.), *Climate Change 2022: Mitigation of Climate Change* (pp. 1049–1160). Cambridge University Press. https://doi.org/10.1017/9781009157926.012

Jozef, E., Kumar, K., Iranmanesh, M., & Foroughi, B. (2019). The effect of green shipping practices on multinational companies' loyalty in Malaysia. *International Journal of Logistics Management, 30*(4), 974–993. https://doi.org/10.1108/IJLM-01-2019-0005

Klymenko, O., & Lillebrygfjeld Halse, L. (2021). Sustainability practices during COVID-19: An institutional perspective. *The International Journal of Logistics Management, 33*(4), 1315–1335. https://doi.org/10.1108/IJLM-05-2021-0306

Kurbatova, S., Aisner, L., & Mazurov, V. (2020). Green logistics as an element of sustainable development. *IOP Conference Series: Earth and Environmental Science*, *548*, Article 052067. https://doi.org/10.1088/1755-1315/548/5/052067

Kurita, A. E., Espuny, M., Campos, T. L. R., Kazançoğlu, Y., Kandsamy, J., & de Oliveira, O. J. (2023). Drivers for circular economy development: Making businesses more environmentally friendly. *Environment Science and Pollution Research, 30*, 79553–79570. https://doi.org/10.1007/s11356-023-28048-0

Lee, K., & Kim, J. (2011). Integrating suppliers into green product innovation development: An empirical case study in the semiconductor industry. *Business Strategy and the Environment, 20*(8), 527–538. https://doi.org/10.1002/bse.714

Li, D., & Hanafi, Z. (2012). A study of eco-performance of logistics services in food supply chains. In Z. Zhang, R. Zhang, & J. Zhang (Eds.), *LISS 2012: Proceedings of 2nd International Conference on Logistics, Informatics and Service Science* (pp. 223–228). Springer.

Martins, V., Nunes, D., Melo, A., Brandão, R., Braga Júnior, A., & Nagata, V. (2022). Analysis of the activities that make up the reverse logistics processes and their importance for the future of

logistics networks: An exploratory study using the TOPSIS technique. *Logistics, 6*(3), Article 60. https://doi.org/10.3390/logistics6030060
McKinnon, A., Browne, M., Whiteing, A., & Piecyk, M. (2015). *Green logistics: Improving the environmental sustainability of logistics*. Kogan Page Publishers.
Oliveira, V., Martino Neto, J., Silva, C. P., Mesia, R., & Salomon, V. (2023). *Circular economy and green logistics – Systematic literature review*. Logistics Research Network, Enegep. https://www.feg.unesp.br/Home/PaginasPessoais/profsalomon/vania-oliveira-et-al-2023.pdf
Perotti, S., Prataviera, L., & Melacini, M. (2022). Assessing the environmental impact of logistics sites through CO2eq footprint computation. *Business Strategy and the Environment, 31*(4), 1679–1694. https://doi.org/10.1002/bse.2976
Prataviera, L., Creazza, A., & Perotti, S. (2023). A call to action: A stakeholder analysis of green logistics practices. *The International Journal of Logistics Management*. https://doi.org/10.1108/IJLM-09-2022-0381
Raap, M., Moll, M., & Pickl, S. (2017). Transportation network regulation for air pollution minimization. In D. Cinar, K. Gakis, & P. M. Pardalos (Eds.), *Sustainable logistics and transportation* (pp. 21–37). Springer.
Rodrigue, J.-P., & Notteboom, T. (2010). Comparative North American, and European gateway logistics: The regionalism of freight distribution. *Journal of Transport Geography, 18*(4), 497–507. https://doi.org/10.1016/j.jtrangeo.2010.03.006
Sankaran, K. (2023). Turning black to green: Circular economy of industrial carbon emissions. *Energy for Sustainable Development, 74*, 463–470. https://doi.org/10.1016/j.esd.2023.05.003
Setiawan, D., & Koestoer, H. (2021). Comparative perspectives on modern logistics transportation based on Green Logistics in Europe and Indonesia: Concept for sustainable economy. *Journal of Mechanical, Civil and Industrial Engineering, 2*(2), 44–48. https://doi.org/10.32996/jmcie.2021.2.2.7
Smith, J. (2021, September 3). Chip shortage curtails heavy-duty truck production. *The Wall Street Journal*. https://www.wsj.com/articles/chip-shortage-curtails-heavy-duty-truck-production-11630661401
Tamulis, V., Guzavičius, A., & Zalgiryte, L. (2012). Factors influencing the use of green logistics. *Economics and Management, 17*(2), 706–711. https://doi.org/10.5755/j01.em.17.2.2202
Texas Department of Transportation. (2021). *Texas-Mexico border transportation master plan 2021*. https://ftp.txdot.gov/pub/txdot/tpp/btmp/btmp-final-report.pdf
United States Census Bureau. (2023). *Trade in goods with Mexico*. https://www.census.gov/foreign-trade/balance/c2010.html
United States Environmental Protection Agency. (2023). *Emission factors for greenhouse gas inventories*. https://www.epa.gov/system/files/documents/2023-03/ghg_emission_factors_hub.pdf
Wang, C.-N., Ho, H.-X. T., Luo, S.-H., & Lin, T.-F. (2017). An integrated approach to evaluating and selecting green logistics providers for sustainable development. *Sustainability, 9*(2), Article 218. https://doi.org/10.3390/su9020218
Xie, Y., & Minjares, R. (2023). *How U.S. and EU proposals could steer the transportation to zero-emission truck and bus fleets*. The International Council on Clean Transportation. https://theicct.org/us-and-eu-proposals-zev-truck-and-bus-fleets-sept23/
Yin, R. K. (2018). *Case study research and applications: Design and methods* (6th ed.). Sage Publications.

Olivia Hernandez-Pozas , Ph.D., is a researcher at the Institute for the Future of Education and an associate professor in International Business at Tecnologico de Monterrey, Mexico. She is a member of the CONAHCYT National System of Researchers in Mexico (SNII, level 1), as well as an experienced professor and experiential trainer. She is a member of the Academy of Management (AOM) and serves as research coordinator in the Management Education and Development

Division (MED) of the AOM. She carries out research on the following topics: the future of work, skill-based education, digital skills, life-long learning, nearshoring, logistics, and Mexico.

Rodrigo Martinazioli is co-founder and head of collaboration at Blue5PL, a start-up with an innovative logistics proposal that seeks to improve efficiency and sustainability in cross-border logistics operations. He has a bachelor's degree in Business Administration, a master's degree in Logistics and education in Industrial Engineering. He worked for almost six years in the Nuevo Leon (Mexico) automobile cluster, coordinating the supply chain committee, among other responsibilities. His 30 years of experience in several countries in America and Europe and throughout the supply chain area gave him a strategic and integral vision of logistics.

Creating the First Circular Value Ecosystem in Latin America: The Experience of a Multinational Brewery

Eduardo Aguiñaga and Ana R. Leal

1 Introduction

> Only a particularly uncurious person could be unaware of the modern scourges of the natural environment, and their increasing incursions into our lived experiences. (Hart & Pomponi, 2021, p. 1)

Circular economy (CE) has become a promising concept for achieving sustainable development, especially given the recent COVID 19 pandemic and the impact it had disrupting global supply chains and causing a price increase along the supply networks around the world. This context, coupled with stringent regulations, more environmentally conscious consumers and the scarcity of natural resources, has driven not only the realisation of the limits to growth but brought an understanding on how increasing pressures on Earth need to be tackled with a more holistic approach. Consistent with the previous trends, impacts related to climate change, air pollution, soil erosion and deforestation to mention the most common are on the rise. Recently, these impacts have been followed by the adherence of terms such as "micro-plastics" and "insectageddon" to the lexicon (Hart & Pomponi, 2021). The circular value ecosystem (CVES) concept proposed in the present research encompasses an approach that mitigates the negative effects of industrial operations on the environment and an alternative to tackle the challenges of the current economic production model.

E. Aguiñaga (✉)
EGADE Business School, Tecnologico de Monterrey, San Pedro Garza García, Mexico
e-mail: eduardo.aguinaga@tec.mx

A. R. Leal
Universidad de Monterrey, San Pedro Garza García, Mexico
e-mail: anar.leal@udem.edu

E. Fröhlich et al. (eds.), *Circular Economy in Sustainable Supply Chains*, CSR, Sustainability, Ethics & Governance,
https://doi.org/10.1007/978-3-031-70749-0_22

Despite the numerous definitions available (Ellen MacArthur Foundation, 2013; Figge et al., 2023; Kirchherr et al., 2017), it is widely accepted that CE proposes an alternative to the current linear economic production model based on a "take, make, dispose" rationale. This wasteful linear paradigm is rooted in a twentieth-century mindset, where carbon-based energy, excessive exploitation of virgin materials and uncontrolled consumerism have been the rule rather than the exception. Along with this linear approach, a narrow focus on each of the different systems, namely, economic, environmental and social has hampered the accomplishment of a transition towards truly sustainable development.

Fortunately, within the so-called "umbrella concept" of CE lies the notion of industrial operations emulating the design and operation of living systems, which are interdependent with natural systems (Julien-Saint-Amand & Moenner, 2008). This idea gave birth to the concept of industrial symbiosis (IS) which resembles such systems (Chertow, 2000, 2007; Ehrenfeld, 2008).

> Industrial symbiosis engages traditionally separate industries in a collective approach to competitive advantage involving physical exchange of materials, energy, water, and/or byproducts. The keys to industrial symbiosis are collaboration and the synergistic possibilities offered by geographic proximity (Chertow, 2000, p. 314).

Therefore, IS encompasses the sharing of residues, wastes and/or by-products among geographically proximate industrial entities to reuse, remanufacture or recycle materials. The synergic combination of this residue and waste enables not only the generation of valuable products for the firm, other neighbouring firms or stakeholders but also lowers the environmental impact by reducing the need for virgin materials. Therefore, it is through this IS approach that CE provides manufacturing industries a promising alternative to overcome the environmental, social and economic constraints.

IS represents the practical subset of the concept of industrial ecology (IE), where industries collaborate in exchange for products, by-products and wastes to reduce their collective environmental footprint (Bossilkov et al., 2005). IE is not a recent concept. In fact, the field of IE has been developing for several decades. Examples of successful cases of IS can be found in places such as the city of Kalundborg in Denmark (Ehrenfeld & Gertler, 1997; Jacobsen, 2006), Jyvaskyla region in Finland (Korhonen, 2001b; Van Berkel, 2009), Kwinana in Australia (Beers et al., 2008; Bossilkov et al., 2005) and Kawasaki in Japan (Higuchi & Norton, 2016; Van Berkel, 2009).

However, these successful examples have taken place in developed countries, where there is a strong rule of law, adequate infrastructure and more environmentally conscious citizens among other factors. Yet, little is known about how to create equivalent circular value networks in developing countries and what might be the enabling factors and challenges. Also, there is room to understand the implications of IS in such contexts regarding the environmental, social and economic outcomes for the communities in which they are embedded.

To address this gap and leveraging an insider action research (IAR) methodology (Coghlan & Brannick, 2014), this chapter introduces and exemplifies the concept of a CVES. The latter concept represents a synergistic region composed of multiple circular value chains that internalise externalities by designing out waste, as a complement to the IS concept widely adopted in developed countries by incorporating the principles of CE. To exemplify the nuances of this concept, we present the case of a multinational brewery located in northern Mexico, which endeavoured to create what has been called the first CVES of Latin America. Furthermore, we also discuss the main enablers and challenges for achieving a successful implementation of CE at a regional level in a Latin American context as well as the envisioned possibilities based on success cases.

The structure of the chapter is as follows: first, a brief definition of supply chain management and the diverse paradigms used to explain it are provided. Second, the diverse literature on CE and its underlying concepts are examined. Third, by building on the previous literature the concept of a CVES is presented. Fourth, the IAR methodology is discussed and thereafter, the Meoqui Brewery case is explored in another section. Lastly, some issues are considered, and concluding remarks are made based on the study.

2 Traditional Supply Chain Management: The Linear Production Paradigm

Supply chain management can be defined as the management of flows of products, information and financial resources throughout production systems (Nasir et al., 2017; Sanders, 2020). Although there has been an increasing interest in managing greener supply chains, there are still opportunity areas for further development and implementation of more sustainable practices. Even though the concept of corporate social responsibility (CSR) was introduced in the 50s (Bowen & Johnson, 1953), most of the knowledge developed within supply chain literature has been published since the 2000s. Furthermore, most of the academic literature regarding supply chains and CSR was published between 2009 and 2019 (Modak et al., 2020). Additionally, there are authors questioning the current production paradigm and the functionality of implementing “greener” practices into supply chain management.

Sustainable or green supply chain management can be defined as the management of flows of products, information and financial resources throughout the production system, integrating environmental concerns by minimising material flows or reducing the negative impact of production and consumption processes (Nasir et al., 2017; Sarkis et al., 2011; Srivastava, 2008). Some authors (e.g. Modak et al., 2020) have questioned this perspective, as it still aligns with the linear production paradigm by having a utilitarian approach to the environment, where the operational objectives align with the environmental objectives.

Supply chain management with a CE perspective progresses the discussion by proposing a change in the production paradigm. Within this perspective, circularity questions the linear production systems and proposes the implementation of systems that re-employ materials. Also, it pursues the objective of reaching a complementary relationship between environment and economic growth, where the negative consequences of production and consumption are decreased (Francas & Minner, 2009; McDonough & Braungart, 2002; Nasir et al., 2017). This perspective, coupled with the IE and IS view, can help explain how a supply chain can achieve circularity by becoming a CVES.

3 Circular Economy, Industrial Ecology and Industrial Symbiosis: The Antecedents of Circular Value Ecosystems

As mentioned in the previous section, the interest in CSR across the supply chain has seen an increase in the last decade. Such interest has driven the conversation forward to question the disadvantages of the traditional view of green supply chain management. In this section, CE, IE and IS are discussed to present the concept of a CVES, which will be illustrated in the next sections.

3.1 Circular Economy

A CE represents an industrial system that is restorative or regenerative by intention and design by replacing the linear "end-of-life" concept with a "closed-loop" usage of resources (Ellen MacArthur Foundation, 2013). It focuses on eliminating all waste through the careful, innovative and clean design of materials, products, systems and business models. The core of CE revolves around the idea that there must be a circular (closed) flow in which materials, raw materials and energy are used through multiple phases.

Moreover, CE represents the study of the whole economy, i.e. the monetary, energetic and resource material systems (Webster, 2017). It proposes to rethink and systematically design out material leakage and disposal, so linear systems no longer prevail, but instead circular, closed-loop processes. It distinguishes eco-efficiency from eco-effectiveness by arguing that the former is rooted in the linear flow of materials paradigm seeking to minimise material use and toxicity, whereas the latter generates cyclical, cradle-to-cradle metabolisms (McDonough & Braungart, 2002).

CE operates at three levels (Geng & Doberstein, 2008; Yuan et al., 2006). The first is called the *micro* or individual firm level. It is at this first level that the principles of CE are integrated into the operational and production procedures. Activities such as reuse, recycle, reduce, redesign, repurpose, remanufacture and refurbish are among the most common. These activities create efficiencies and change internal operations. The second is called the *meso* level and is comprised of individual firms with the purpose of creating an eco-industrial network that will benefit both regional production systems and environmental protection. In other words, this level encompasses examples of cooperation between organisations to promote the exchange of waste materials to achieve the goals of cradle-to-cradle (McDonough & Braungart, 2002). This can be achieved through energy cascading, sharing of infrastructure and the creation of linkages that foster the exchange of residue, waste and by-products for recycling purposes in the same way that industrial ecosystems operate. The final and third level is called the *macro* level. It encompasses the development of regional synergies in the form of eco-villages, eco-cities and eco-provinces, where the principles of CE are introduced at a national level.

Despite the diverse definitions of CE in the literature, for the purpose of this chapter, we opted for using the following one, as it incorporates the different levels at which CE operates as well as acknowledging the role played by the supply chain:

> A circular economy describes an economic system that is based on business models which replace the 'end-of-life' concept with reducing, alternatively reusing, recycling and recovering materials in production/distribution and consumption processes, thus operating at the micro level (products, companies, consumers), meso level (eco-industrial parks) and macro level (city, region, nation and beyond), with the aim to accomplish sustainable development, which implies creating environmental quality, economic prosperity and social equity, to the benefit of current and future generations (Kirchherr et al., 2017, pp. 224–225).

Furthermore, there are three principles that guide CE: (a) designing out waste, which refers to eliminating waste (zero-waste approach); (b) circularity of biological and technical nutrients, which takes into account the origin of the resource and the possibility of returning it to the biosphere—biological ones returned to Earth and technical ones cycled through the economy (McDonough & Braungart, 2002) and (c) energy requirements aimed at reducing the dependence over non-renewable sources.

Finally, these three principles create four sources of value creation: (a) to minimise material usage; (b) longer cycling of materials, i.e. longer usage and less virgin material; (c) cascading or diversified reuse of material streams and (d) pure uncontaminated materials of high quality (Ellen MacArthur Foundation, 2013). By turning a linear, industrial economy into a circular, closed-loop economy, the importance of resource extraction and waste management (Webster, 2017) is reduced alongside the reliance on complex supply chains and resources, which might be geographically distant. As can be noted, both the principles and sources of value creation from CE have a direct impact on the supply chain as the following section describes through the concept of IS.

3.2 Industrial Ecology and Industrial Symbiosis

IE is a multidisciplinary field based on a holistic systems view that seeks to inspire industrial systems by emulating how nature designs and operates living systems (Julien-Saint-Amand & Moenner, 2008). On the one hand, the term "industrial" refers to the way in which human beings use natural resources for linear production and not only to industrial complexes. On the other hand, "ecology" refers to the metaphor of emulating these closed-loop cycles of materials and nutrients found in natural systems (Julien-Saint-Amand & Moenner, 2008).

By relying heavily on engineering and technological solutions to environmental problems (Keoleian & Garner, 1994), IE has been known as the science behind CE due to its interdisciplinary nature and the possibility of its application in the service sector (Webster, 2017). IE operates at the same three levels of CE: first within a facility of the firm, second at the inter-firm level focusing on IS and third at a global level (Chertow, 2000). It is the second level, where the IS phenomena take place, that is the main interest of this chapter. Prior to that a definition of IS is required. IS is one of the main branches of IE (Chertow, 2000). It refers to the industrial synergies generated between diverse organisations, whereby all the actors exploit each other's residues, waste or by-products, creating an ecosystem of cooperation and collaboration for mutual economic and environmental benefit (Chertow, 2007).

There are four basic principles for IE systems, namely, roundput, diversity, locality and gradual change (Korhonen, 2001b):

1. *Roundput* is the first principle and is derived as the opposite of throughput, which implies a linear approach. It refers to the utilisation of material, waste and energy in cooperation with other firms to use the waste as feedstock. The purpose is for residues and wastes to flow in a cascade-like fashion, therefore, reducing the use of virgin resources by substitution.
2. *Diversity* is related to the need of having different types of industries generating different types of residues, wastes and by-products. Furthermore, the use of the term "diversity" recognises the need to engage processes (e.g. industrial activities) of symbiotic organisations to identify opportunities. By increasing diversity, knowledge broadens as well as the resource base available to the symbiotic network, thus fostering innovation (Lombardi & Laybourn, 2012).
3. *Locality* is the third principle and suggests that, for firms to create waste-sharing synergies, they must be geographically proximate or within the region. This creates both a reduction of imported materials and, consequently, a minimisation of the carbon footprint throughout the supply chain.
4. Lastly, the fourth principle is called *gradual change*. It implies that by leveraging the principle of diversity, companies should undertake the process of adapting to both the amount of residues produced and the time when they are produced. This process generates a switch for the involved firms from a competitive to a cooperative relationship.

Overall, by coupling the principles of CE with the discipline of IE and its branch of IS, we argue that a more comprehensive conceptual framework can guide regions into becoming circular. This is done at the second level (*meso*), given that at this level of analysis, firms in cooperation face numerous sustainability problems, such as green supply chain management or eco-clusters (Nikolaou et al., 2021). To overcome these challenges, we contend that the CVES concept could provide a viable alternative.

4 Circular Value Ecosystem

As discussed in the previous section, current research has been focusing on optimising linear supply chains. Efficiency-focused efforts have been present for decades both in academic literature and in practice. However, efficiency perspectives have reached a plateau. No real gains in the form of waste reduction or elimination can be achieved if no paradigm change takes place. CE provides this alternative by closing the loop of materials.

The circularisation of supply chains has been a relatively recent priority on the agenda of governments, scholars, organisations and firms. Although first envisioned as a closed-loop economy by the British economist Kenneth Boulding in the 1960s, the idea of closing the loop of organisations and even regions has been around for decades. By circularising supply chains, firms and regions are able to create more sustainable wealth in the form of environmental resilience, economic rents and social benefits.

However, little guidance has been provided regarding the actual implementation of circularity, along with some kind of economic measurement (Lewandowski, 2016). Based on the CE principles (Ellen MacArthur Foundation, 2013; Stahel, 2016; Webster, 2017), the concept within IE and IS (Chertow, 1998; Ehrenfeld, 1997; Korhonen, 2001b), and the sustainable wealth creation based on innovation and enabling technologies (SWIT) framework (Scheel, 2016), we define a CVES as a synergistic region composed of multiple circular value chains that internalise externalities by designing out waste. An IE guided by principles of a CE is economically viable and competitive, socially responsible and inclusive, and environmentally recoverable and resilient. The proposed concept differs from others, such as a hub, cluster, industrial park or even an eco-industrial park as the following suggests.

For starters, "a hub is in essence a warehouse or central depot, which is located at the centre of a set of demand regions" (O'Kelly & Miller, 1994, p. 31). Although efficient for supply chain management, its limited functionality is based on making material or product handling more orderly. Moreover, industrial clusters are geographic concentrations of interconnected companies and institutions in a particular field (Porter, 1998). Given that a cluster is of a particular field, it does not leverage the diversity of entities. Finally, an industrial park entails an area of industrial development, and an eco-industrial park is where businesses cooperate with the local community to reduce waste and pollution, share resources and help achieve sustainable development. Although the eco-industrial park seems sufficient for achieving sustainability, according to the Environmental Protection Agency field book:

> an eco-industrial park is a community of manufacturing and service businesses seeking enhanced environmental and economic performance through collaboration in managing environmental and resource issues including energy, water, and materials. By working together, the community of businesses seeks a collective benefit that is greater than the sum of the individual benefits each company would realize if it optimized its individual performance only (Doyle et al., 1996, p. 12).

We consider that despite the collaborative nature of the eco-industrial parks, they fall short of coping with the current pressures of a more constrained supply chain and more pressing regulations on the environment. This is given that it does not explicitly focus on transforming, upcycling (McDonough & Braungart, 2013) or repurposing residues, waste and by-products within these parks. In other words, not only a more efficient linear economy, but a transformation of the industrial systems towards a more regenerative, zero-waste, circular ecosystem consisting of a group of interdependent, autonomous actors that collectively generate a circular value proposition is needed (de Gomes et al., 2023). To overcome these limitations and support the concept of a CVES, we performed IAR to provide evidence that this concept can be of great importance in the transition from a linear to a CE on a regional level.

5 Methodology

5.1 Insider Action Research: Rationale and Data Collection

Action research was first introduced by a social psychologist named Kurt Lewin as an approach that guides a cyclical and simultaneous process of research and action (Lewin, 1946). In this way, action research contributes to dealing with practical concerns while creating knowledge and igniting action as an output (Coghlan & Brannick, 2014).

We found this methodology particularly suitable, as:

> action research is not about imposing expert knowledge on stakeholders but about creating collaborative environments where research experts and local stakeholders can share their very different kinds of knowledge in the process of analyzing their problems, studying them, and collaboratively designing actions that can ameliorate the problems (Greenwood, 2002, p. 127).

However, when performing research under an organisation where the researcher is also a member, IAR represents a suitable methodology within the action research framework (Coghlan & Brannick, 2014). IAR is characterised by a collaborative, reflexive and interventionist nature (Macintosh et al., 2007). Moreover, IAR has been recently considered an important way to change organisations (Coghlan & Holian, 2007).

Furthermore, "insiders have detailed and personal knowledge of the conditions of the industry, they are capable of participating in political games at the company, and they hopefully have the required social and cultural capital to influence organisational activities and strategic decision-making" (Williander & Styhre, 2006, p. 241). In other words, IAR needs to be supported by an insider's view of activities in order to create any substantial contribution to practice and theory. Therefore, we consider it suited to the research for two main reasons. First, given the insider position of one of the researchers and, second, due to the inherent interest in analysing organisational change.

5.1.1 Data Collection

In line with IAR methodology, primary internal data were gathered from multiple sources throughout the project from May 2016 until December 2019. Data in the form of internal documents, white papers, presentations, meetings memos written by employees in the company and the insider researcher as well as emails among others. Additionally, secondary data were obtained during the COVID-19 pandemic given the nature of the project involving other stakeholders, such as industries and companies. Furthermore, meetings with the CEO, vice presidents of the company, local stakeholders and government were attended, where the insider action researcher took an active part in the discussions. All collection of primary data was performed by one insider action researcher, who worked for the company in the sustainability department and was responsible for the CVES vision and implementation.

The overall methodology followed for creating the CVES was inspired by the seven SWIT model principles (Scheel, 2016) and based on the conceptual framework for the circularisation of linear value chains (Aguiñaga, 2016). The seven principles of the SWIT model outline the guiding ideas for a successful redesign of the linear chains of dominant industrial product life cycles in a determined region. Furthermore, Aguiñaga (2016) operationalised these principles through a conceptual framework which outlines several methods for creating alternative sustainable circular loops, therefore, promoting a system of closed-loop flow of materials (see Table 1).

In the following section, we describe the case study of Meoqui Brewery, which as far as the authors are aware, is the first CVES in Latin America. This case study follows this movement, understanding how regional synergies inspired by CE principles and IS can create shared value (Porter & Kramer, 2019) for all stakeholders in a region through the concept of a CVES.

Table 1 The seven principles of the SWIT model and their operationalisation (adapted from Aguiñaga, 2016; Scheel, 2016)

Principle description	Method	Method operationalisation	Examples
1. Residues and waste from production chains remain in the region.	Extended Value System	Mapping the linear production chain as well as the collateral (i.e. residues and waste) using products' and residues' industrial classification codes, such as SCIAN, ISIC, SITC and HS.	- Agricultural waste like coffee waste remaining where it is processed. - Mining waste like limestone staying in the region where it is extracted.
2. Residues and waste inserted into circular value systems may produce higher value than linear chains.	Synergy Inventory	Operationalise and depict the metabolic route concept in the roadmap by adding primary resources, ancillary inputs, residues and products from material flow analysis, and systems definition and process chain analysis from life cycle assessment into the circularisation roadmap.	- Transforming coffee waste into T-shirts. - Converting agave leaves into biodegradable plastic bottles. - Utilising African palm tree residues to create natural fibres.
3. Proper technologies can transform residues and waste into high-value products.	Regional Synergies Based on Residues and Waste	Applying the upcycling concept to residues management, operationalised the principle of diversity from industrial symbiosis and used the Environmental Protection Agency classification on the extended value system to uncover potential synergies based on residues and wastes.	- Using the vortices process for water purification from waste. - Employing biotechnological processes to convert waste into valuable products.
4. Circular business models can provide competitive advantages.	Process and Technology Identification	Sustainable evaluation of the appropriate technologies available aiming for a high benefit–cost ratio devoted to valorising residues and wastes.	- Converting food industry waste into valuable products like transforming sawing fluids into useful items.
5. Clusters of unusual entrepreneurial ventures can be constructed within a circular zone.	Linked-Producer-Product Matrix	Create a linked-producer-product matrix adding a dynamic perspective of waste and residues flow valorisation.	- Creating multiple business opportunities from transforming tortilla residues (nejayote).
6. Assembling self-sustainable communities is possible with proper conditions and relationships.	System Dynamics Modelling	Using system dynamics to model and simulate different scenarios and measure the circularity potential of a region.	- Establishing self-regulating sustainable value systems by transforming local industrial residues into compost for urban farming.

(continued)

Table 1 (continued)

Principle description	Method	Method operationalisation	Examples
7. A regional sustainable sharing value system can be articulated with proper resources and governance.	Start-up's Clusterisation Strategies	Leveraging systemic entrepreneurs encompasses the promotion of unusual business ventures based on residues and wastes.	- Examples like self-sustainable communities in Bristol (UK), Kalundborg (Denmark) and Las Gaviotas (Colombia) demonstrate this principle.

6 Case Study: Meoqui Brewery, the First Circular Value Ecosystem of Latin America

The Meoqui Brewery is property of HEINEKEN México, which is a brewing group with a long tradition in the country. The company started operations more than 130 years ago under the name of Cervecería Cuahutémoc and it was acquired by HEINEKEN in 2010. HEINEKEN México currently owns seven breweries and a malting plant and has more than 16,000 employees in México. This company represents one of the most important operations in terms of volume and profitability for HEINEKEN, the largest brewing group in Europe and the one with the largest international presence (HEINEKEN, 2023).

6.1 HEINEKEN México Sustainability Strategy

HEINEKEN México has a long history of regional economic and social development given that the brewery became a creator of companies to meet their needs and those of their employees beyond a vertical integration strategy. Among some of these endeavours was the inauguration of the Cuauhtémoc y Famosa Society Clinic for the benefit of its workers, even before the creation of the Mexican Institute of Social Security (Cázares Puente, 2014), its involvement in the foundation of the Instituto Tecnológico de Monterrey, one of the most renowned universities in Latin America (Hibino, 1992), and the construction of the Cuauhtémoc neighbourhood, where the company delivered 1318 houses to its employees (Cázares Puente, 2014).

With this CSR heritage, HEINEKEN México followed its global sustainability strategy called "Brewing a Better World" (HEINEKEN, 2016). This strategy has three pillars, encompassing different activities. The social pillar focuses on topics such as inclusion, diversity and community among others, while the responsible pillar addresses responsible consumption. Under the environmental pillar, CE is the guiding principle of the company with a strong focus on carbon neutrality, maintaining healthy watersheds and promoting circularity across the organisation.

However, in 2015 HEINEKEN México started its journey to become the number one company for sustainability in the country and made a public commitment to transition towards a CE. This ambition has not only helped the company redefine its strategy and goals, but it has become an inspiration for many people in the company. This journey started with the launch of a CE self-assessment based on the ReSOLVE framework (Ellen MacArthur Foundation, 2013), which gave the company a measure of the circularity of its operations. The results also gave the company more clarity and helped develop a more holistic and systemic approach for its circular transition. Based on the results, the company created a five-year transition plan with the aim to reach every business unit, process, product and, more importantly, every person in the organisation through the execution of both top-down and bottom-up cross-functional initiatives.

One of the first activities was the training of employees and decision-makers. Empowered personnel have been key to achieving this circular transition. So far, the company has trained more than 2000 out of his 18,000 employees and decision-makers, thanks to the creation of the first CE e-learning in Spanish in collaboration with the Ellen MacArthur Foundation. This online course enabled trained circular thinkers across the organisation to propose and deploy high-impact projects of all sizes and at all levels. Some examples of substantive projects during this progressive transition were the recycled uniforms project that seeks to recover and recycle fibres from old uniforms, as well as the creation of granola bars made from malt solid residue to distribute among vulnerable communities in México.

In addition, since the company made its commitment to transition towards a CE public, it became a member of the Circular Economy 100 (CE100)—now called Circular Economy Network—an Ellen MacArthur Foundation's pre-competitive innovation programme (Ellen MacArthur Foundation, 2023). Following this, the company announced its latest and most ambitious circular commitment, which entailed the creation of the first CVES of Latin America, with its newest brewery at the helm of this endeavour:

> The Meoqui brewery, our seventh brewery in Mexico, will be operational at the end of 2017 to satisfy growing demand. Its design is based on the principle of the circular economy. It will positively impact our distribution footprint as it covers an area that has traditionally been difficult to reach. (HEINEKEN, 2016, p. 18).

Inspired by the principles of CE, the company started working on attracting suppliers to share waste, residues and by-products to create strong relationships with the community and local entrepreneurs and strengthen enduring cooperation with the local government. By doing so, the company sought to expand its circular ambitions to an entire region of the country.

6.2 *The Circular Value Ecosystem of Meoqui*

According to the company, the Meoqui Brewery is recognised for its large-scale production of more than 6 million hectolitres per year, supplying both the domestic and international markets (HEINEKEN, 2018). The brewery demonstrates its commitment towards the environment as 100% of the waste is treated or recycled. It also uses 100% renewable energy and has a public commitment to become carbon neutral by 2030 by maximising the use of greener energies (HEINEKEN México, 2021). One example is that the window glass installed in the plant contains photovoltaic cells that capture approximately 12% of the energy required (Ponce, 2022).

This strong sustainability orientation coupled with an implementation of a CE strategy made the operation of the Meoqui Brewery the perfect opportunity for the company to truly materialise its commitment towards a more circular approach to production at a regional scale. The CVES of Meoqui started with a vision developed by the insider action researcher, including the application of the seven SWIT model principles (Scheel, 2016) and a conceptual framework for the circularisation of linear value chains (Aguiñaga, 2016) as mentioned in previous sections.

The steps that guided the creation of this CVES can be summarised in the three following phases:

Phase 1: Internal Analysis. Inspired by the material flow analysis and the waste input–output methodologies (Guedes et al., 2017; Nakamura & Kondo, 2002), this step comprised a review of the materials that the company required for its operations as well as those that were generated as residues, wastes and by-products. Also in this step, the internal closed loops of materials were identified. Figure 1 presents a summarised schematic representation of the process.

Phase 2: Extended Value Chain Analysis. By applying the concept of the extended value system (Scheel, 2016), materials and energy flows from

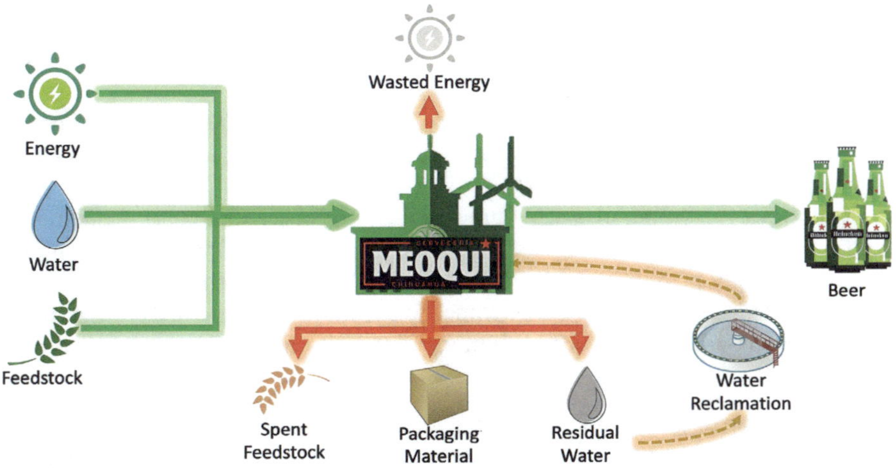

Fig. 1 First phase: Internal analysis of inputs and outputs of Meoqui Brewery (own illustration)

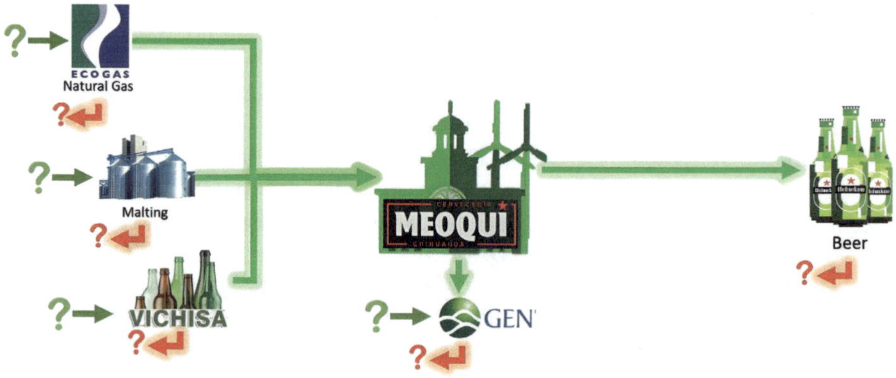

Fig. 2 Second phase: Extended value chain analysis of neighbouring companies and suppliers (own illustration)

neighbouring suppliers were identified to foster potential low-cost synergies. Within this phase, the company started to deploy an active strategy for the engagement of suppliers in CE activities. Figure 2 presents the overall process of analysing the extended value chains of the Meoqui Brewery.

Phase 3: Regional Metabolism. The third phase entailed a broader regional analysis of the most important industries, actors and organisations in the region. The final aim is to create a potential synergy map or a vision of what the CVES of Meoqui could look like with the involvement of all stakeholders. Figure 3 presents the overall interaction for the Meoqui CVES in which numerous potential synergies could occur from the involvement of the regional stakeholders to existing and potential commercial partners of the company.

In order to assess the feasibility of this initiative, this vision was analysed under the four principles of industrial ecosystems (Korhonen, 2001a) as follows:

1. **Diversity** The region of Meoqui has companies operating in different industries, such as food and beverage, agriculture and textiles, as well as companies of differing sizes.
2. **Proximity** In the region of Meoqui companies, people and materials are proximate enough to enable sharing, reuse, remanufacturing and refurbishing of wastes, residues and by-products.
3. **Roundput** Recycling of matter and cascading of energy between the actors of the system was found to be feasible. At the time of this evaluation, a neighbouring glass bottle company with excess heat available was identified.
4. **Gradual Change** Moving from competitive to more cooperative dynamic strategies increased interdependency among firms through waste and by-product sharing.

Finally, the resulting vision for the Meoqui CVES is shown in Fig. 4. At the centre of the network, with higher centrality, is the Meoqui Brewery as the anchor company. Connected by different coloured lines, which represent the sharing of

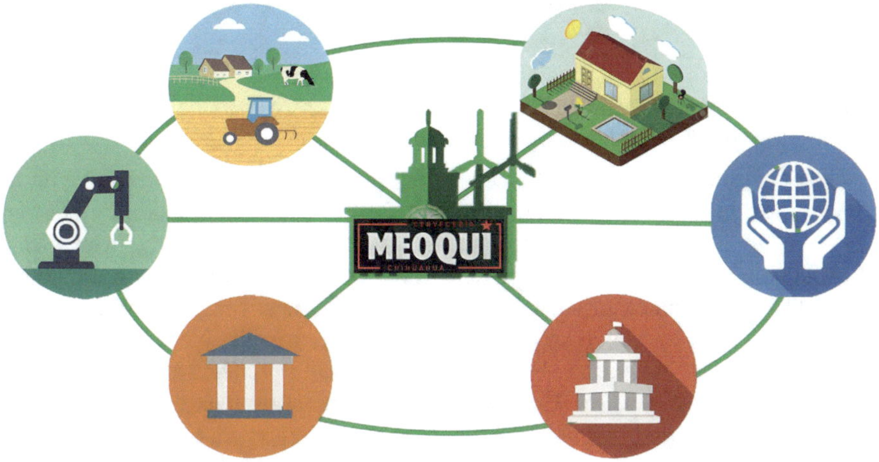

Fig. 3 The expected interaction of stakeholders for the Meoqui CVES (own illustration)

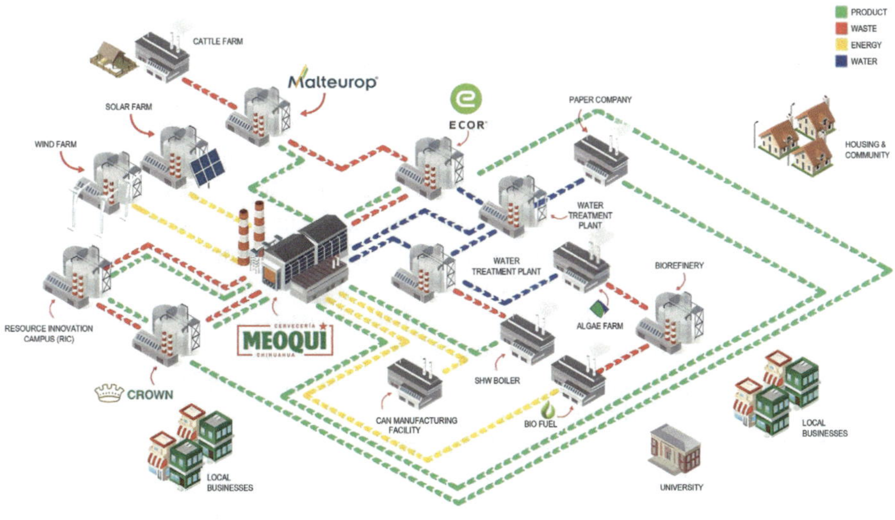

Fig. 4 The vision of the Meoqui CVES (own illustration)

either a product (green), waste (red), energy (yellow) or water (blue), these connections with the neighbouring companies would give shape to this ideal CVES.

In sum, based on the CE strategy of the company and as the region showed true potential for the creation of this CVES, the company started working on creating these synergies with the local stakeholders. In the next section, we provide the course of events involved in the conformation of this synergistic region.

6.3 Conformation of the Circular Value Ecosystem: Course of Events

> To rapidly 'create an industrial ecosystem from scratch' borders on the impossible. (Korhonen, 2001a, p. 256)

The configuration of the CVES of Meoqui has not been easy. It started with the brewery design aligned with their "Brew a Better World" sustainability strategy. The company pioneered the construction of this factory guided by the principles of CE, focusing mainly on the energy and water usage. For example, on the energy side, 100% of the electricity used in the brewery comes from renewable sources. The brewery uses photovoltaic cells that are installed in the brewery windows to generate about 12% of the total electricity needed on site. The rest is produced by solar panels and wind power. However, the real innovation regarding sustainability was not just from the state-of-the-art technology implemented, but from the disruptive way the company approached residues, wastes and by-product reuse, repurpose and sharing.

However, up until today, we contend that there is enough evidence to assert that this is the first CVES of Latin America under development, as milestones related to water usage and reclamation, use of excess heat, glass recycling and employment of by-products have been reached. These milestones are discussed in-depth in this section.

The first achievement was the circularisation of water usage. There was a growing concern and commitment of the company to hydric resources, especially in the face of scarcity. To mitigate these risks since the beginning of their operations in 2018, the company created a wastewater treatment plant whose state-of-the-art technology allows it to recover and use around 70,000 cubic metres of methane gas per month, replacing non-renewable thermal energy consumption in the brewing process (Ponce, 2022). Furthermore, sludges from this water treatment plant are granted to serve as soil improvers in neighbouring farms. Lastly, more than 60% of the water used at the Meoqui Brewery is donated after it is treated to irrigate locally cultivated areas with alfalfa or walnut.

Moreover, in 2019, the company started investing in the construction of a water reclamation plant. After its construction, the reclamation plant can recover more than two-thirds of the treated wastewater to be used again in cooling processes, general cleaning services and irrigation of green areas. These efforts have made the Meoqui Brewery the most efficient plant in water usage within the HEINEKEN company worldwide according to Guillaume Duverdier, CEO of HEINEKEN México (González, 2023). It uses less than 1.9 litres of water per litre of beer produced (Ponce, 2022). Without a doubt, closing the loop of water was a great achievement for the company, however, other equally relevant supplies remained to be analysed for circularity.

In 2018, a circularisation opportunity came from the neighbouring VICHISA glass bottle factory. Just next to the Meoqui Brewery, the most important by-product of the furnace operations of the glass company is excess heat released into the atmosphere. Realising this untapped opportunity, both companies started cooperating to build a series of pipes that transported the excess heat from the furnaces via superheated water to deliver it directly to the brewing process. By 2019, the fine-tuning of the process was achieved, which has now translated into savings of thermal energy of up to 50%. However, the synergies between these two companies do not end there, as the Meoqui Brewery sends all the leftover glass cullet back to the glass bottle factory to create new bottles out of old ones. This creates an additional closed loop of glass.

Another important raw material for beer production is malt. Malt represents 50% of the beer formula and is sourced from both local and abroad suppliers, some of them as far as the bohemian region in the Czech Republic. Aware of the potential positive impacts that could arise from substituting imports with local barley, HEINEKEN México started deploying a series of strategies focusing on supplying malt from local farmers. In parallel, and as the network of synergies kept growing, more companies started to develop an interest in becoming part of this new ecosystem. The next company to become part of this was Malteurop, a French malting farmer-maltsters company founded in 1961.

By the year 2019, Malteurop announced an investment of 120 million dollars to create a malting plant within the Meoqui CVES. This plant, which will be able to process 35,000 hectares of barley and produce up to 120,000 tons of malt per year, aims to develop a local supply chain of 100% local barley in partnership with Mexican farmers (Malteurop, 2022). Beyond the regional positive impact for farmers, the construction of this facility will provide 100% locally produced and high-quality barley to the brewery and could potentially source an abundant by-product such as barley husk. This by-product has been used by HEINEKEN México for the creation of their barley cups. These cups have replaced more than 1,000,000 old plastic cups in all the company's sports and music events (HEINEKEN México, 2019). Furthermore, another potential upcycling possibility for the malting plant's residues and by-products is the creation of packaging materials. HEINEKEN México, in collaboration with Noble Environmental Technologies, has been testing the possibility of creating secondary packaging out of cellulose-based residues, such as solid malt residue. Unfortunately, the prototypes tested so far have not been successful for scalability and the investment required outweighs the benefits at this point of time.

Finally, the most recent project announced within the Meoqui CVES was announced in June 2022 with the construction of a Green-by-design aluminium can manufacturing facility. As Mónica Bichara, Vice President of Legal Affairs pointed out, with an investment of 90 million dollars, which will be fully functional by 2024, "this plant, in a way, marks a change in terms of the integration of our supply chain" (Maza, 2022).

As it is possible to identify, the implementation of projects aligned with a CE perspective involves understanding the relationships across and between several stakeholders. This could be the incorporation of local suppliers to the supply chain (e.g. the inclusion of local farmers), identifying the opportunity areas for waste to input relationships (e.g. Meoqui Brewery using the excess energy from VICHISA)

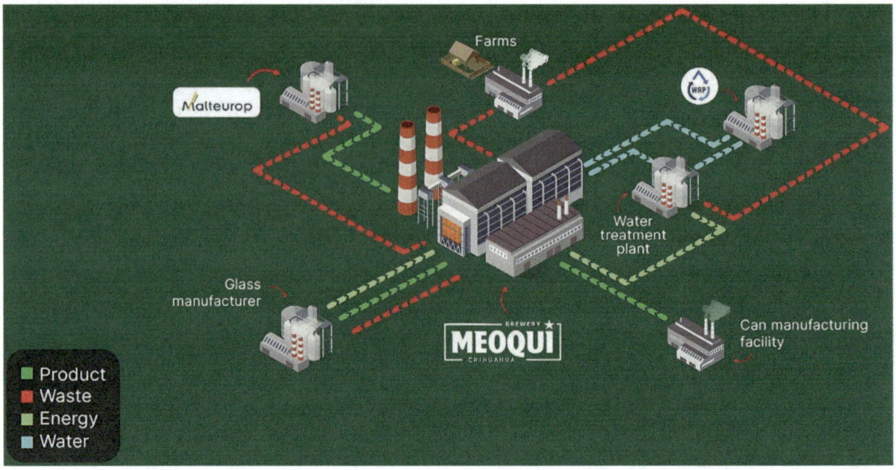

Fig. 5 Current state of the CVES of Meoqui, Chihuahua, México (own illustration)

or incorporating new firms to tackle the revalorisation opportunities for residues (e.g. generating new packaging materials with Noble Environmental Technologies). In Fig. 5, the current relationships between stakeholders in this CVES are shown.

In order to continue consolidating Meoqui Brewery as the heart of the first CVES in the region (Malteurop, 2022) it is expected that more stakeholders, such as companies, organisations, entrepreneurs and local governments, continue to get involved in this thriving ecosystem.

7 Discussion and Conclusions

The synergies of a CVES generate several sets of advantages. These are related to process efficiencies across the supply chain, the revalorisation of residues and an increasing resilience within the CVES. Despite the CE paradigm focusing on closing loops regarding resource usage, it also enables economic growth and competitive advantage for the firms involved within the ecosystem. Even though there are many upsides of implementing a CVES, forming this ecosystem does not come without its challenges. The main challenges arise from governance and conflicting goals from stakeholders, the degree of governmental involvement, access to capital, specialised knowledge and trust among stakeholders.

In the case of the process efficiencies, it is possible to see an improved productivity for the firms. Productivity increases due to the cost reductions generated by sharing the waste as an input for other processes or firms. Another related upside of allocating waste to use it as an input is that the quality of materials is uniform across the CVES.

It is possible to identify advantages that develop from the revalorisation of residues and resilience of firms. These occur because by being in proximity with a set of diverse firms, it is possible to identify a higher amount of waste valorisation opportunities than in a cluster that only has a concentration of firms from the same industry. Additionally, the coexistence of such firms might allow for greater resilience because of both a greater independence from external firms (protecting firms from external shocks) and the arising opportunities of changing suppliers or customers due to the existence of different opportunities of revalorisation.

Despite the great efforts made by all the actors in this ecosystem, there are still several challenges to overcome. To operate efficiently and create the enabling conditions for more actors to become part of this ecosystem, we argue that there are mainly five barriers.

The first is governance. When operating an ecosystem of multiple stakeholders conflicting goals may arise. One of the biggest concerns is related to negotiating a compensation for the exchanges of materials, energy and resources. Also, the implication of one of the parties not meeting the expected supply of resources. This could be caused by factors such as lowering production and changes in their processes such as the acquisition of new technology, creating fewer materials to exchange or with a different quality. These situations could disrupt the whole ecosystem operation. However, to overcome this, redundancy of connections should be fostered. This would translate into more companies with similar potential for supplying equivalent resources.

The second barrier is the potential lack of governmental involvement. We identified that political involvement in projects of such magnitude is essential. Creating a sound policy system and enabling conditions to attract companies is key to the achievement of vibrant ecosystems. A strong rule of law is crucial, especially in developing countries.

The third barrier is access to capital. There is an increasing need for funding of sustainable entrepreneurial activities as well as SMEs. These types of ecosystems require a diversity of industries, but also companies of different sizes. SMEs are of vital importance for achieving a closed-loop economy. An example of this is the IS in Kalundborg, Denmark, where global private companies work with local businesses and authorities across sectors to create successful symbiotic linkages.

The fourth challenge is advisory, training and support networks for local companies to make the transition. This entails different activities, such as promotion, formal training on the concept and methodologies of CE, and creation of support networks to foster collaboration and communication across companies, governments and organisations. At the time of writing, HEINEKEN México has launched an open-access CE e-Learning in Spanish. This has been done in collaboration with the Conscious Enterprise Center of the Tecnológico de Monterrey. This initiative was part of the efforts of the UN Global Compact Working Group on Circular Economy in Mexico to promote CE best practices and collaboration across industries.

Lastly, the fifth and probably the most challenging barrier in the Latin American context is to develop trust and transparency among stakeholders. We argue that the creation of a successful CVES requires trust and transparency among all participants, beyond materials and energy flow sharing among companies. The absence of these makes the creation of meaningful symbiotic relationships very unlikely. However, we found that the Meoqui Ecosystem's successful waste-sharing linkages among the entities were facilitated, on the one hand, by the previous long commercial relationship in the case of Malteurop and, on the other hand, by the alignment of corporate values and culture, as VICHISA—the glass bottle company—was a wholly owned subsidiary of the HEINEKEN Group.

Since the recognition of the advantages generated by approaches such as IE and CE in the sustainability field (Shrivastava, 1995), there has been a growing interest to generate a wider impact beyond the limits of a company. Instead creating ecosystems where residues and wastes are either shared or valorised. Efforts to design, construct and operate these ecosystems based on symbiotic relationships and circularity have been numerous but limited to the developed world.

However, up until the time of writing and to the understanding of the authors, there is no other successful case for Latin America. The Meoqui CVES portrays an inspirational example that could guide potential adopters in creating a CVES based on IE and CE principles.

The emergent inquiry process involved in the action research, where researchers and the organisational members partner to address a situation, is of great pragmatic value. It is important that the research results lead to better practices in supply chains and sustainability through a CE perspective amongst all the involved stakeholders, in this case in the food and beverage industry.

References

Aguiñaga, E. (2016). *Creating circular value ecosystems: A regenerative conceptual framework moving from products to systems based on internalizing externalities*. EGADE Business School.

Beers, D., Bossilkov, A., Corder, G., & Berkel, R. (2008). Industrial symbiosis in the Australian Minerals Industry: The cases of Kwinana and Gladstone. *Journal of Industrial Ecology, 11*(1), 55–72. https://doi.org/10.1162/jiec.2007.1161

Bossilkov, A., van Beers, D., & van Berkel, R. (2005). *Industrial symbiosis as an integrative business practice in the Kwinana industrial area: Lessons learnt and ways forward.* Proc. 11th International Sustainable Development Research Conference.

Bowen, R., & Johnson, F. E. (1953). *Social responsibility of businessman*. Harper and Row.

Cázares Puente, E. (2014). La Cervecería Cuauhtémoc y la industrialización del noreste mexicano. *Ciencia UANL, 17*(69), 16–21.

Chertow, M. R. (1998). The eco-industrial park model reconsidered. *Journal of Industrial Ecology, 2*(3), 8–10. https://doi.org/10.1162/jiec.1998.2.3.8

Chertow, M. R. (2000). Industrial symbiosis: Literature and taxonomy. *Annual Review of Energy and the Environment, 25*(1), 313–337. https://doi.org/10.1146/annurev.energy.25.1.313

Chertow, M. R. (2007). "Uncovering" industrial symbiosis. *Journal of Industrial Ecology, 11*(1), 11–30. https://doi.org/10.1162/jiec.2007.1110

Coghlan, D., & Brannick, T. (2014). *Doing action research in your own organization* (4th ed.). Sage.

Coghlan, D., & Holian, R. (2007). Editorial: Insider action research. *Action Research, 5*(1), 5–10. https://doi.org/10.1177/1476750307072872

de Gomes, L. A. V., de Faria, A. M., Braz, A. C., de Mello, A. M., Borini, F. M., & Ometto, A. R. (2023). Circular ecosystem management: Orchestrating ecosystem value proposition and configuration. *International Journal of Production Economics, 256*, Article 108725. https://doi.org/10.1016/j.ijpe.2022.108725

Doyle, B., Lowe, E. A., Moran, S. R., & Holmes, D. B. (1996). *Fieldbook for the development of eco-Industrial parks'*. Citeseer.

Ehrenfeld, J., & Gertler, N. (1997). Industrial ecology in practice: The evolution of interdependence at Kalundborg. *Journal of Industrial Ecology, 1*(1), 67–79. https://doi.org/10.1162/jiec.1997.1.1.67

Ehrenfeld, J. R. (1997). Industrial ecology: A framework for product and process design. *Journal of Cleaner Production, 5*(1–2), 87–95. https://doi.org/10.1016/S0959-6526(97)00015-2

Ehrenfeld, J. R. (2008). Can industrial ecology be the "science of sustainability"? *Journal of Industrial Ecology, 8*(1–2), 1–3. https://doi.org/10.1162/1088198041269364

Ellen MacArthur Foundation. (2013). *Towards the circular economy, economic and business rationale for an accelerated transition.* Ellen MacArthur Foundation Publishing. https://www.ellenmacarthurfoundation.org/publications/towards-a-circular-economy-business-rationale-for-an-accelerated-transition

Ellen MacArthur Foundation. (2023). *Who's in our Network.* https://ellenmacarthurfoundation.org/network/who-is-in-the-network

Figge, F., Thorpe, A. S., & Gutberlet, M. (2023). Definitions of the circular economy: Circularity matters. *Ecological Economics, 208*, Article 107823. https://doi.org/10.1016/j.ecolecon.2023.107823

Francas, D., & Minner, S. (2009). Manufacturing network configuration in supply chains with product recovery. *Omega, 37*(4), 757–769.

Geng, Y., & Doberstein, B. (2008). Developing the circular economy in China: Challenges and opportunities for achieving 'leapfrog development'. *The International Journal of Sustainable Development & World Ecology, 15*(3), 231–239. https://doi.org/10.3843/SusDev.15.3:6

González, I. (2023, March 2). Es Heineken Meoqui la más eficiente en uso de agua. *El Diario de Chihuahua.* https://www.eldiariodechihuahua.mx/economia/es-heineken-meoqui-la-mas-eficiente-en-uso-de-agua-20230302-2029972.html

Greenwood, D. J. (2002). Action research: Unfulfilled promises and unmet challenges. *Concepts and Transformation, 7*(2), 117–139. https://doi.org/10.1075/cat.7.2.02gre

Guedes, G. B., Paganin, L. B. Z., & Borsato, M. (2017). Material flow mapping and industrial ecosystems: A literature structured review. *Advances in Transdisciplinary Engineering, 5*, 774–781. https://doi.org/10.3233/978-1-61499-779-5-774

Hart, J., & Pomponi, F. (2021). A circular economy: Where will it take us? *Circular Economy and Sustainability, 1*(1), 127–141. https://doi.org/10.1007/s43615-021-00013-4

HEINEKEN. (2016). *HEINEKEN Annual Report.* https://www.theheinekencompany.com/sites/theheinekencompany/files/Investors/AGM/hnv-pdf/2017/heineken-nv-2016-annual-report.pdf

HEINEKEN. (2018, February 28). *HEINEKEN opens new brewery in Meoqui, Mexico. HEINEKEN Opens New Brewery in Meoqui, Mexico* [Press release]. https://www.theheinekencompany.com/newsroom/heineken-opens-new-brewery-in-meoqui-mexico/

HEINEKEN. (2023). *Our global presence. The HEINEKEN Company.* https://www.theheinekencompany.com/our-global-presence

HEINEKEN México. (2019, October 17). *¡Go green con nuestros vasos de cascarilla de cebada!* [Press release]. https://heinekenmexico.com/noticia/go-green-con-nuestros-vasos-de-cascarilla-de-cebada

HEINEKEN México. (2021). *HEINEKEN México: Informe Sustentabilidad 2021.* https://heinekenmexico.com/sustentabilidad/informe

Hibino, B. (1992). Cervecería Cuauhtémoc: A case study of technological and industrial development in Mexico. *Mexican Studies/Estudios Mexicanos*, 23–43.

Higuchi, K., & Norton, M. G. (2016). Japan's eco-towns and innovation clusters—Synergy towards sustainability. *Global Environment, 1*(1). https://doi.org/10.3197/ge.2008.010110

Jacobsen, N. B. (2006). Industrial symbiosis in Kalundborg, Denmark: A quantitative assessment of economic and environmental aspects. *Journal of Industrial Ecology, 10*(1–2), 239–255. https://doi.org/10.1162/108819806775545411

Julien-Saint-Amand, F., & Moenner, P. L. (2008). *Industrial Ecology, an Innovative Approach Serving Spatial Planning: The Example of the Tool PRESTEO (A Program to Research Synergies on a Territory).* 6th International Conference of Territorial Intelligence "Tools and methods of Territorial Intelligence", Besançon, France.

Keoleian, G. A., & Garner, W. A. (1994). *Industrial ecology: An introduction.* National Pollution Prevention Center, University of Michigan.

Kirchherr, J., Reike, D., & Hekkert, M. (2017). Conceptualizing the circular economy: An analysis of 114 definitions. *Resources, Conservation and Recycling, 127*, 221–232. https://doi.org/10.1016/j.resconrec.2017.09.005

Korhonen, J. (2001a). Four ecosystem principles for an industrial ecosystem. *Journal of Cleaner Production, 9*(3), 253–259. https://doi.org/10.1016/S0959-6526(00)00058-5

Korhonen, J. (2001b). Regional industrial ecology: Examples from regional economic systems of forest industry and energy supply in Finland. *Journal of Environmental Management, 63*(4), 367–375. https://doi.org/10.1006/jema.2001.0477

Lewandowski, M. (2016). Designing the business models for circular economy—Towards the conceptual framework. *Sustainability, 8*(1), Article 43. https://doi.org/10.3390/su8010043

Lewin, K. (1946). Action research and minority problems. *Journal of Social Issues, 2*(4), 34–46. https://doi.org/10.1111/j.1540-4560.1946.tb02295.x

Lombardi, D. R., & Laybourn, P. (2012). Redefining industrial symbiosis: Crossing academic–practitioner boundaries. *Journal of Industrial Ecology, 16*(1), 28–37. https://doi.org/10.1111/j.1530-9290.2011.00444.x

Macintosh, R., Bonnet, M., & Coghlan, D. (2007). Insider action research: Opportunities and challenges. *Management Research News, 30*(5), 335–343. https://doi.org/10.1108/01409170710746337

Malteurop. (2022, July 27). Malteurop invierte 120 MDD en Chihuahua para proveer a Heineken. *Cluster Industrial.* https://clusterindustrial.com.mx/noticia/5103/malteurop-invierte-120-mdd-en-chihuahua-para-proveer-a-heineken

Maza, A. (2022, June 27). Heineken anuncia creación de nueva fábrica de latas en Meoqui, Chihuahua. *El Heraldo de Chihuahua.* https://www.elheraldodechihuahua.com.mx/local/delicias/heineken-anuncia-creacion-de-nueva-fabrica-de-latas-en-meoqui-chihuahua-8511054.html

McDonough, W., & Braungart, M. (2002). *Remaking the way we make things: Cradle to cradle.* North Point Press.

McDonough, W., & Braungart, M. (2013). *The upcycle: Beyond sustainability—Designing for abundance.* Macmillan.

Modak, N. M., Sinha, S., Raj, A., Panda, S., Merigó, J. M., & de Sousa Jabbour, A. B. L. (2020). Corporate social responsibility and supply chain management: Framing and pushing forward the debate. *Journal of Cleaner Production, 273*, Article 122981. https://doi.org/10.1016/j.jclepro.2020.122981

Nakamura, S., & Kondo, Y. (2002). Input-output analysis of waste management. *Journal of Industrial Ecology, 6*(1), 39–63. https://doi.org/10.1162/108819802320971632

Nasir, M. H. A., Genovese, A., Acquaye, A. A., Koh, S. C. L., & Yamoah, F. (2017). Comparing linear and circular supply chains: A case study from the construction industry. *International Journal of Production Economics, 183*, 443–457. https://doi.org/10.1016/j.ijpe.2016.06.008

Nikolaou, I. E., Jones, N., & Stefanakis, A. (2021). Circular economy and sustainability: The past, the present and the future directions. *Circular Economy and Sustainability, 1*(1), 1–20. https://doi.org/10.1007/s43615-021-00030-3

O'Kelly, M. E., & Miller, H. J. (1994). The hub network design problem: A review and synthesis. *Journal of Transport Geography, 2*(1), 31–40. https://doi.org/10.1016/0966-6923(94)90032-9

Ponce, S. (2022, June 27). Trabaja Heineken en módulo de economía circular. *El Heraldo de Chihuahua*. https://www.elheraldodechihuahua.com.mx/local/delicias/trabaja-heineken-en-modulo-de-economia-circular-8512319.html
Porter, M. E. (1998, Nov/Dec). Clusters and the new economics of competition. *Harvard Business Review*. https://hbr.org/1998/11/clusters-and-the-new-economics-of-competition
Porter, M. E., & Kramer, M. R. (2019). Creating shared value. In *Managing sustainable business* (pp. 323–346). Springer.
Sanders, N. R. (2020). *Supply chain management: A global perspective*. John Wiley & Sons.
Sarkis, J., Zhu, Q., & Lai, K. (2011). An organizational theoretic review of green supply chain management literature. *International Journal of Production Economics, 130*(1), 1–15. https://doi.org/10.1016/j.ijpe.2010.11.010
Scheel, C. (2016). Beyond sustainability. Transforming industrial zero-valued residues into increasing economic returns. *Journal of Cleaner Production, 131*, 376–386. https://doi.org/10.1016/j.jclepro.2016.05.018
Shrivastava, P. (1995). Ecocentric management for a risk society. *Academy of Management Review, 20*(1), 118–137. https://doi.org/10.2307/258889
Srivastava, S. K. (2008). Network design for reverse logistics. *Omega, 36*(4), 535–548. https://doi.org/10.1016/j.omega.2006.11.012
Stahel, W. R. (2016). The circular economy. *Nature, 531*(7595), 435–438. https://doi.org/10.1038/531435a
Van Berkel, R. (2009). Comparability of industrial symbioses. *Journal of Industrial Ecology, 13*(4), 483–486. https://doi.org/10.1111/j.1530-9290.2009.00140.x
Webster, K. (2017). *The circular economy: A wealth of flows*. Ellen MacArthur Foundation Publishing.
Williander, M., & Styhre, A. (2006). Going green from the inside: Insider action research at the Volvo Car Corporation. *Systemic Practice and Action Research, 19*(3), 239–252. https://doi.org/10.1007/s11213-006-9015-7
Yuan, Z., Bi, J., & Moriguichi, Y. (2006). The circular economy: A new development strategy in China. *Journal of Industrial Ecology, 10*(1–2), 4–8. https://doi.org/10.1162/108819806775545321

Eduardo Aguiñaga is the National Director of Full-Time Programs at the EGADE Business School and a circular economy researcher, speaker and consultant. He holds a Doctorate in Business Administration. With experience as a professor and researcher, Eduardo specialises in topics like corporate sustainability and the circular economy. He has published scientific articles, contributed to book chapters and presented his research at international conferences. Additionally, Eduardo has been a speaker at significant circular economy events. He is also a consultant for circular economy projects and is currently involved with the Mexican Circular Economy Working Group of the United Nations Global Compact.

Ana R. Leal is an Assistant Professor of Management at the Business School, Universidad de Monterrey, México. Her research focuses on non-market strategies, including corporate social responsibility and entrepreneurship. Her work has been featured in the *Journal of Management Studies*, *Journal of Business Research* and *Sustainability*. Additionally, she has authored chapters on research methods and sustainable business. She is a member of the Mexican National System of Researchers.

Supporting Regional Transformation Through an Organizational Educational Network Consulting Approach: An Empirical Example Regarding the Implementation of a Regional Circular Economy

Marc-André Heidelmann and Tobias Klös

1 Introduction

A circular economy aims to reduce waste and maximise regional resource use through recycling, reuse, and "energy loops" (Geissdoerfer et al., 2017, p. 759). The idea of a circular economy (Stahel & Reday-Mulvey, 1976) has been around since the 1970s, but it has recently gained renewed attention due to its potential to promote sustainable development (Reike et al., 2018). It is an alternative to the linear approach of "take, make, and waste" that characterises traditional economic models (Weetman, 2021, p. 4). Circular economies have been recognised as a means of supporting the achievement of the Sustainable Development Goals (United Nations, 2015) and are promoted at the national and international levels through corresponding strategies.

It is also essential to consider circular economies at the regional level, especially in rural regions, which are facing significant economic, environmental, and social challenges in securing local livelihoods and supplies (Heidelmann & Klös, 2023). In such regions, the question often arises of how to shape a future-oriented reorientation of traditional economic sectors and the development of new technologies, combined with the goal of infrastructural self-sufficiency and interregional networking.

M.-A. Heidelmann (✉)
IU International University of Applied Sciences, Department of Social Science, Bremen, Germany
e-mail: marc-andre.heidelmann@iu.org

T. Klös
Heidelberg University of Education, Centre for Sustainable Education, Heidelberg, Germany
e-mail: kloes@ph-heidelberg.de

E. Fröhlich et al. (eds.), *Circular Economy in Sustainable Supply Chains*, CSR, Sustainability, Ethics & Governance,
https://doi.org/10.1007/978-3-031-70749-0_23

Regional networks and cooperation are essential aspects of cyclical economic structures at the local level. However, the problem-solving and management capabilities of networks at the regional level pose specific challenges, which can lead to "network failure" (Messner, 2012, p. 174), and have only been partially addressed by research to date (Diller et al., 2014; Heidelmann & Klös, 2023).

Additionally, network consulting plays a unique role in fostering and building long-term collaborative networks as an element of governance practices within inter-organizational network structures (Lerch et al., 2011). Regional transformation processes and the development of circular economies are often approached from the perspective of economics, geography and sometimes law and the social sciences. However, education rarely plays a role in regional transformation (Weber & Heidelmann, 2019). This applies particularly to organizational education, which can be understood as learning within, by and between organizations (Göhlich et al., 2018, p. 208). Organizational education also has its own distinct theory, methodology and intervention methods (Heidelmann, 2023).

This paper seeks to promote organizational educational network consulting (Heidelmann et al., 2023) as a strategy for transitioning towards a regional circular economy (Heidelmann & Klös, 2023). It is assumed that this could help develop suitable business models and overcome the limitations that make it difficult to build such circular value chains. Furthermore, this article discusses how these networks can link economic interests with environmental goals and promote sustainability-driven entrepreneurship.

The programme "Greening the University Canteen—With Sustainability Labs Towards a Regional Circular Economy" (Weber & Heidelmann, 2019) will be presented as an example of a trans-disciplinary and trans-organizational educational approach to organizational and network consultancy. The paper will also discuss the potential benefits and challenges of implementing this approach (Klös & Heidelmann, 2023).

Against this background, Sect. 2 first examines the fundamental economic and social problems facing rural areas, which manifest themselves in particular as a lack of public infrastructure, an ageing population, and depopulation. Section 3 addresses these structural issues and explores the potential of organizational educational network consulting to promote sustainable development in rural areas and overcome their weaknesses. On this basis, Sect. 4 presents a concrete pilot project of organizational educational network consulting, before the research design and the concrete case analysis are presented in Sect. 5. Furthermore, in Sect. 6, the potential of organizational educational knowledge for network development in the context of circular economy in the regional space is examined from the perspective of a regional stakeholder. Finally, Sect. 7 summarises the chapter and conclusions from previous sections.

2 Networks as a Potential for Regional Circular Economy

Rural regions often face the challenge of balancing the preservation of rural culture with sustainable reorientation, maintaining traditional economic sectors while embracing new technologies, and promoting infrastructure self-sufficiency alongside interregional networking. This is of vital importance for rural areas, which are particularly affected by ageing and depopulation, as well as a weak economic infrastructure (Heidelmann & Klös, 2023). Regional networks are essential for the successful and effective implementation of local governance. As a strategic process and instrument for implementing socio-ecological transformation and sustainable development in regional structures, regional governance depends largely on the involvement of relevant actors. Thus, systematic stakeholder cooperation (WBGU, 2011) shows the need to accelerate the establishment of communication processes as a constituent element of regional development processes to enable innovation from within regional structures in the medium term. Accordingly, forming and recognising new cooperation, actor constellations and networks are essential to establishing new institutions and tools in regional structures and changing existing ones (WBGU, 2011).

The circular economy approach offers new development perspectives, where economic and ecological goals can be combined, particularly for rural areas and underdeveloped, structurally weak regions. Indeed, circular economies are viewed as a regenerative system that helps to integrate economic performance, social inclusion and ecological resilience in a balanced way (Geissdoerfer et al., 2017). Circular economy models of shared value creation require the cooperation of many heterogeneous actors and thus the formation of networks. This complexity, however, represents the most significant challenge for network development, which can be addressed by organizational education.

3 Network Guidance from an Organizational Education Perspective

There are numerous network forms and types, just as there are consulting interventions and initiatives. The consulting approaches range from exclusively but clearly structured, content-oriented technical consulting to structure-open, often participatory process consulting (Sydow, 2006, p. 58). Subsequently, the organizational educational understanding of networks and their consulting is presented, followed by an outline of an organizational educational approach to typical network management problems.

3.1 Network Facilitation as a Process-Based Approach to Collaboration

Organizational education generally does not view networks as mere contexts of education and governance but examines them as "learning social entities" (Göhlich et al., 2018, p. 210). In contrast to organizational psychology, organizational sociology, or economic and administrative sciences, learning in, from and between network associations as a particular form of organization (Sydow, 1992) is the focus of research and interventions using theoretical, analytical, empirical and conceptual approaches.

From an organizational educational perspective, networks are not viewed in a narrowly disciplinary way. Rather, it applies the so-called "processual cooperation approach", which considers how experienced persons coordinate their different experiences, knowledge bases and social practices (Endres & Wehner, 1999, pp. 219–220). Network consulting is therefore not conceived as expert consulting. Instead, it is aimed at the cooperative and dialogical involvement of all decision-makers and functionaries relevant to the sustainable development of a regional circular economy (Heidelmann & Weber, 2018). Moreover, trust (Sydow & Windeler, 2003) and reflexivity are crucial for the long-term existence of regional network alliances (Weber, 2001, p. 18).

Further to the organizational educational perspective, network consulting should entail classical organizational development on the one hand and scientific accompaniment on the other hand. This type of network consulting differs from advisory network support (Weber, 2001, p. 43) in the form of network coordination, in which communication and action processes are structured and, if necessary, delegated based on the intended goals of the network. Organizational educational network consulting does not see itself as an external network consultancy based on a political mandate. Instead, it acts in an advisory capacity for network-related programmes and initiatives.

A theory-based answer to the question of how active network consultants see themselves and the world is essential for successful network consulting. Likewise, a reflexive and critical view of the consulting task and the actors involved in the process is required. Especially for the complex task of establishing circular economies, a deep understanding of intra-organizational processes within network structures is essential. Against this background, a consultancy foundation is proposed that thinks in terms of power and knowledge, and thus does not understand the contexts of consultancy as entities of an objective reality but instead as power-shaped constitutions of knowledge orders (Foucault, 1994). In the following section, Messner's (2012) conception of typical network control problems will be outlined briefly. The network management problems described by Messner will be raised once more when discussing the case analysis results.

3.2 *Addressing Problems of Network Governance from an Organizational Educational Perspective*

Numerous challenges must be overcome in network management, especially in rural areas. Messner (2012) lists seven such challenges for network control:

The Problem of (Large) Numbers. The first problem within network governance concerns the large number of actors involved, which is often trivialised. Messner (2012, p. 153) explains that the "problems bound up with negotiated coordination in networks grow as the number of actors [...] increases". This means that the often very heterogeneous actors and decision-makers who interact with each other in network alliances have different interdependencies with each other, which can enormously delay and complicate communication, coordination and, ultimately, decision-making processes (Messner, 1994, p. 568).

This problem of numbers can be made manageable, or reduced, by an action orientation of the involved actors that is directed towards common problem-solving (Messner, 2012, p. 155). From an organizational and educational point of view, finding common perspectives is made possible by imagining common visions. These need to be methodically initiated and supported by trained network consultants through appropriate consultation interventions. This is the only way to achieve institutional consolidation in network alliances, where stable cooperative relationships emerge that are generated by a specific pattern of mutually accepted organizational identities, competencies and spheres of interest.

Social Cohesion. Messner (2012) argues that there is a drawback to stabilised and cooperative networks, namely that they may lead to cognitive constrictions due to the vigorous focus on promoting unified views. This can be problematic between individuals within the same network who have formed symbiotic relationships with other network actors. In addition, new ways of development, alternatives to the existing ones and strategies beyond established options are often not sufficiently recognised, so thinking and acting in the network often threatens to become path-dependent, structurally conservative and narrow-minded (Messner, 1994, p. 573). This results in a trend towards agreement on the "lowest common denominator" (Messner, 2012, p. 172).

Organizational educational network consultation is committed to a discursive (Foucault, 1973) understanding of consulting. It seeks to enable sustainable cooperation and network building by acknowledging the dialogue of many people, especially marginalised ones. Yet, according to Messner (1994, p. 572), the common rules, conventions and routines that emerge in networks, as well as the trust, reciprocity and ability to anticipate the behaviour of other actors in this regard, can lead to an exclusionary logic and closed network structures. From the perspective of organizational education, it is precisely these tendencies that need to be addressed by network consultants (Weber & Heidelmann, 2021).

Institutional Consolidation. Messner (2012, p. 157) suggests that a mutual dependency of actors often arises within network cooperations, which suggests a tendency to continue the relationship. Thus, it also presupposes and evokes the idea of institutional consolidation, which, however, lies below the threshold of a constitution of a formal organization. According to Messner (2012, p. 157), such an organizational formation would mean abandoning the network structure. Moreover, such stable cooperative relationships make it easier to find compromises to resolve conflicts of interest between network actors (Messner, 2012, p. 157).

This institutional consolidation may lead to cumulative exit costs, making it increasingly unattractive to abandon network cooperation; since much investment has been made, it would be a high loss to leave the network again. However, this can also lead to a situation in which, in the interest of maintaining the cooperation, decision-makers may prioritise what is best for the project or community regarding stakeholders. This can lead to missed opportunities for innovation and progress and perpetuate inequalities and marginalisation. Therefore, to overcome the trap of split decision-making, it is essential to address power imbalances and prioritise transparency, inclusivity and collaboration in decision-making processes—a communication process that organizational education consultants can support. Long-term relationships and shared orientations in networks can also lead to cognitive blocks that make identifying alternatives and new solutions challenging when established development paths have to be abandoned. Organizational educational network consultation programmes offer the potential to solve cognitive blockades through a methodical expansion of perspectives.

Coordination Problem. The coordination problem, which Messner (2012) calls the 'Pareto optimum', means that no more advantageous changes are conceivable within the network in favour of at least one network participant without another participant having to be placed in a worse position as a result. This is addressed by organizational educational consulting through image-based methods with the goal of new "imaginations" (Castoriadis, 1998).

In Innovation Labs, participants are encouraged to develop shared visions of the future. The aim here is to ensure that the intended consensus-building is not based solely on argumentative weighing up, but that deeper needs and desires are brought to the fore. To evoke these in the participants, the labs increasingly use image-based methods. In this sense, the consultants do not take up a neutral role but a normative one that is committed to democratic procedures and the common good (Messner, 2012, p. 161), in line with the self-concept of organizational education (Göhlich et al., 2018).

Bargaining Dilemmas. The control problem described by Messner (2012, p. 165) as the "bargaining dilemma" means that open communication, trustful cooperation and a negotiation style aimed at joint problem-solving is necessary for the overall result of the negotiation within a network. Nevertheless, it is "not unlikely" that actors willing to cooperate and do what is necessary to attain the "'Kaldor Optimum' by means of cooperative strategies will run the risk of being cheated in matter of distribution" (Messner, 2012, p. 165).[1] The possibility that the negotiation

[1] "Kaldor Optimum" means that the benefit for those favoured is large enough to provide full compensation for all those disadvantaged by the measure (Scharpf, 1991, p. 624).

process may contain strategic action, manipulation, bluffing or threats (Messner, 2012) is reflected by a power-sensitive organizational education. Trained organizational educational network consultants always assume that exclusions are produced in the discourse space, leading to what Foucault (1994) calls 'unspeakabilities'.

Power. Against this background, the question of power and knowledge is significant for network consulting, especially for leadership behaviour (Heidelmann et al., 2023; Klös & Heidelmann, 2023). In networks, control and power resources are distributed among different actors. Unlike hierarchical decision-making structures, decisions in networks are often difficult or impossible to enforce upon reluctant network participants. Messner (2012) emphasises that imbalances exist in network structures insofar as the actors' options for action are more significant or smaller, i.e., some actors have more power than others. Such power relations are addressed within and through the dialogic arrangement of the Innovation Labs. The methods support the democratic exchange of ideas within the Innovation Labs. The moderating students as consultants have been trained to be power- and discourse-sensitive (Weber & Heidelmann, 2023).

Tensions Between Conflict and Cooperation. The tension between cooperation and conflict highlighted by Messner (2012), which manifests especially in network structures, ultimately becomes the starting point for organizational educational interventions. Instead of striving for a neutralisation of conflicts and the aim for complete social cohesion, Messner (2012, p. 171) sees the potential of networks primarily within conflicts and making productive use of them: "the chance to link productively together the logics of 'conflict and cooperation'".

Organizational education network consulting and leadership go beyond this perspective. The beginning of collective transformational practice is the transgression of existing positions (Heidelmann, 2020) and thus the transgression of conflict. Unquestioned normalities are addressed before they can inscribe themselves into the emerging cooperation structures. Organizational educational network consulting tries to ensure this by creating Innovation Labs as a place of collective boundary transgression (Heidelmann et al., 2023).

In the following section, an organizational educational consulting approach will be presented, which was conducted in 2018–2019 at the Institution of Education (Philipps-University of Marburg) and led by Prof. Dr. Susanne Maria Weber.

4 The Network Consulting Programme "Greening the University Canteen—With Sustainability Labs to the Regional Food Cycle"

The design-research project "Greening the University Canteen by Regional Innovation Labs—With Sustainability Labs to the Regional Food Cycle" was funded by the German National Council for Sustainable Development. The aim of this one-year network consultation project was the realisation of a regional economic cycle, involving all (potentially) relevant regional stakeholders. Within the three successive Innovation Labs, 60 regional stakeholders participated and were systematically brought together (Weber & Heidelmann, 2019).[2] Thus, in the 'agora' (Nowotny et al., 2001) of Innovation Labs, the 'whole system' of potential network actors in the region was brought 'into one room' (Weisbord & Janoff, 2010). Regional stakeholders worked together on different topics and problems related to the question of how the University Canteen can become more sustainable.

Numerous sub-topics and problem areas were jointly addressed in the Innovation Labs, ranging from the packaging sizes of individual products and the shortening of supply chains to the establishment of regional community structures in the producer sector. In addition to several other approaches and concrete projects, the Marburg Food Council materialised as an idea within the first Innovation Labs and was established shortly afterwards. The Food Council promotes the process towards a regional economic- and food cycle.

The joint work of the stakeholders in the Innovation Labs was organized and methodically accompanied by the university-trained student moderators (Weber et al., 2019). These students were trained in the theory and practice over the course of a year. The practical training involved the preparation, implementation and follow-up documentation of the Innovation Labs. These experiences enabled the students to navigate the complexities of regional transformation processes; accordingly, they professionalised themselves in the field of network consulting. The regional actors involved were equally enabled to form networks (Weber et al., 2019) and to work on problems in dialogue with a view to later diffusion (Weber & Heidelmann, 2021). As a participation space, the Innovation Labs aimed at transcending the mental boundaries of the participants. This entailed placing the transformation of the inner attitude by means of imaginative and image-based processes at the centre of the consulting intervention. Network consulting within the programme is not seen as an intervention brought to the process from the outside but as a scientifically-based process designed from the inside that is carried out together with the actors involved.

As a prospective social consultation service, organizational educational network consulting aims to strengthen the interaction of structurally different actors on the way to a network alliance. Science, politics, plural publics and the media are conceived as a relational network, and corresponding co-creative arrangements are created to offer space for the democratisation of expertise, mutual influence and the joint "imagining" (Castoriadis, 1998) of alternative futures. Different discursive-institutional rationalities of action become accessible through dialogue and openness for collective transformation processes. The aim of network counselling in such an organizational educational understanding is the expansion of spaces of the speakable and visible, as well as decentring certain speech positions.

[2] The acquisition of regional actors followed the path of the regional economic cycle, as well as the areas that can impact this path. Not only the actors of this direct path from producer to consumer to recycling were invited to the Innovation Labs but also the indirectly relevant actors, such as those from civil society, politics, network innovation and science (Fig. 1).

As explained in the previous section, network control problems often result from the unquestioned rationalities of the individual and collective (potential) network participants. From the perspective of organizational education, fundamental problems of network control can be levelled out by creating plurality and acceptance among network participants. The participant's acceptance of the respective consulting approaches is essential for the sustainability and permanence of circular economy structures.

5 How Regional Network Actors Experience Organizational Educational Network Consulting: The Metaphor-Analytical Research Design

In order to assess the potential effectiveness of an organizational educational network consulting approach, an exemplary metaphor-based analysis is presented here. The following illustrates the influence such an analysis can have on business cycles, including its approach on resolving network problems described by Messner (1994, 2012).

The empirical research design is based on a qualitative approach and allows for the reconstruction of the experiences of the regional actors involved in the process in a longitudinal (Brake, 2018) and triangulated (Brake, 2011) manner. Regional actors were interviewed at four points—before and after the Innovation Labs—using guided and image-based interviews as visualised in Fig. 1.

In each interview, the interviewee selected images from 60–70 images based on impulse questions[3] and unfolded their narrative using these images beyond their "elicitive" (Harper, 2002) function. However, the images do not merely have a stimulative function but are open to further interpretation and can be utilised as a gateway for metaphors. The use of metaphors to describe the actors' experiences was useful to reconstruct how potential regional stakeholders would perceive the Innovation Labs. This follows the idea of Lakoff and Johnson (2008, p. 11), who assume, in contrast to a representational model of language, that "our ordinary conceptual system, in terms of which we both think and act, is fundamentally metaphorical in nature". Forceville (2008) points out that Lakoff and Johnson's metaphor concept is not limited to linguistic metaphors but refers to all modalities of human experience. According to Junge (2014, p. 12), an image is a statement, but the

Fig. 1 The interview points in time (own illustration)

[3] Impulse questions are small initial impulses to get into the interview, for example: how would you describe your experience of the first Innovation Lab?

statement may be an insufficient description to describe the experience. Accordingly, metaphors are used to fill this gap.

In contrast to a representative model of language, Lakoff and Johnson (2008, p. 3) assume that metaphorical expressions and concepts contain reality-constituting power; they describe "our conceptual system, according to which we both think and act […]" as "fundamentally metaphorical in Nature". This makes them amenable to discourse-analytic considerations and thus to the analytical reconstruction of patterns of experience. The research design combines the metaphors as verbalised images with materialised images. Each of our interviews initially began with the choice of images, the moments of imagination evoked by these image choices becoming the focal point of transformational educational processes. Images are fundamentally open-ended communications that are visually encoded and then ultimately decoded in their meaning through their reception and impact by the viewers (Geise, 2017, p. 4).

Viewers do not simply decipher the logic of the image according to fixed rules of composition, such as text, but rather visually assemble the partial image components, thus allowing very specific contexts of meaning to emerge first. Paivio (1979, p. 33) calls this process the emergence of a "spatial grammar". In the present analysis approach, the different data types of the transcribed interview, as well as the selected images, were contrasted and analysed in an interlocking way of a triangulating analysis. Only through this interlocking analysis can the potential of the verbal (and materialised) image metaphors be fully exploited.

The following case analysis aims to reconstruct the perspective of a regional actor and his view on how the organizational education network consulting approach can support local network development.

6 Potentials of Organizational Educational Network Consulting for the Development of a Circular Economy: A Case Analysis

As previously explained, the development of networks is essential for the long-term establishment of economic cycles. The case study presented here shows how regional stakeholders—particularly farmers, who as producers of regional food are important actors for the establishment of a regional circular economy—experience such a network consultation process. Scaling potentials (Weber & Heidelmann, 2019) for other regions and cases can then be derived from these findings.

Heinz Rüdiger (H.R.) is a local stakeholder, who has been active in local organic sustainable farming for several decades.[4] He knows the challenges and opportuni-

[4]The name is a pseudonym he has chosen himself. The interview was originally transcribed in German and was translated by the authors. All quotes without a reference in the following are quotes by H.R.

ties of working with other stakeholders, particularly fellow farmers and providers. This is why he is an interesting case: he is genuinely committed to creating a better sustainable future but struggles to cooperate with other stakeholders, which he addresses as an on-going problem.

6.1 *"[B]ringing People Together" Is a Job "That Neither the Region nor the Farmers Can Fulfill": The Potential of an Organizational-Education Network Consulting Approach*

In the first interview (T1), before the first Innovation Lab, H.R. explained the main problem regarding regional network building and stakeholder cooperation. He claimed that "great idealistic goals" must meet economic requirements. In other words, projects for sustainability must "pay off economically", otherwise it would have business repercussions. His solution for this problem entails a "cooperation at eye level", in which regional stakeholders must make a joint effort ("everyone must be committed to sustainability"). This wish for a cooperation and collective effort is also metaphorically expressed within the image of a rowing boat (Fig. 2). In this respect, H.R. initially describes that cooperation can be hindered by economic risks and blockages. The group within the boat represents the regional actors, while the person at the end of the boat can be viewed as a leader, "as someone who is giving commandos", similar to the image of a bridge (Fig. 3) where someone "takes the initiative". Because all previous attempts of self-organizing have failed, the "job [...] [of] bringing people together" is something "neither the region nor the farmers can fulfil".

H.R. regards the university as the leader in Fig. 2. He argues that the university is an institution capable of organizing the regional actors. Thus, the university is placed in a pioneering role, as illustrated in the metaphorical image (Fig. 3), where someone "is leading the way" and "taking new paths". Through the university's actions, it is his hope that discussions and cooperation between the regional stakeholders would become more successful.

The university is a place where "intelligence" and "momentum" are combined. H.R. distinguishes himself from those at university, whom he describes as "young, curious people who are dedicated to research and teaching". These people "think outside the box" and "try wild new things", something a "normal farmer" like him, who is engaged in his daily work routines, would not consider. In this sense, the university is imagined as a personified boundary crossing and a place for innovative people, as illustrated in the figurative metaphor of the bridge.

In T2, after the experience of the first Innovation Lab, H.R. picks the image of the rowing boat once more (Fig. 2). This time it stands for the necessary "willingness on all sides" to cooperate and make an effort for the development process. In T3, this willingness turns into action, giving H.R. the impression of "working

Fig. 2 "Cooperation at eye level" (source unknown)

together" on a bigger goal and "rowing together in the same boat" to "move the whole thing forward". In T4, he states that there have been "many people who have been interested in" working together for sustainable development but also wanted "different things". There was also a difference in knowledge, which led to the situation that "a lot of time was spent" by him "to explain things".

In this sense, the boat metaphor demonstrates that everyone must make their contribution and take on their role in the boat to move forward ("everyone must be committed", "willingness on all sides"). At the same time, the metaphor emphasises the importance of precise goal setting and a coordinated effort to achieve the goal ("commandos", "initiative", "rowing together"). The act of rowing also symbolises

Fig. 3 "Pioneer function" (source unknown)

progress towards a shared destination, which can only be accomplished through dialogue, coordination and collective action.

In T3, the support of this process is then no longer attributed only to the university in general but to the Innovation Labs too, as represented in the metaphor of a person "watering a plant" (Fig. 4). Support is based on ensuring a plant's growth

Fig. 4 "Watering a fragile plant" (source unknown)

process: it may grow independently, but it requires sustenance from others to continue its growth. Within this metaphor, the support of the Innovation Labs for the regional nutrition cycle is imagined as the entity watering the plant, which supports the plant's growth, representing the beginning of the regional circular economy.

The metaphor of watering plants highlights a shift in focus from the pioneer metaphor, as it no longer solely centres on the university's leadership, and instead emphasises the necessity of supporting organic growth and life. Furthermore, H.R. uses the "fragile plant" metaphor to illustrate that while growth occurs naturally, nurturing is required to flourish. These metaphors not only address issues of growth and becoming independent but also draw attention to fundamental educational questions. In addition, the gardener can be viewed as an expression of a leadership practice that reflects the professional-pedagogical tradition's belief in the

Fig. 5 "Vision" of a "richly prepared table" (source unknown)

inseparable dialectical relationship between leading and letting grow (Schröer, 2018, p. 480).

In addition to this supportive role, H.R. also believes that a vision for the regional food cycle is necessary. The vision of the regional food cycle, here symbolised by a "richly prepared table" (Fig. 5), is described by H.R. as "happy people, good food, of course, ecological and from the region". The vision of a "richly prepared table" resurfaced in T4, both narratively and figuratively. In this context, he emphasises the importance of a shared vision for the future, which is the goal of the Innovation Lab as described earlier.

However, H.R. noted in T4 that this vision has not yet "been realised" and a "network has not emerged", but some positive developments have occurred, such as

Fig. 6 "Long-distance run" (source unknown)

new contacts and understanding between stakeholders, which would not have been possible otherwise. As a result, he now sees the regional stakeholders on the path towards sustainable development within the region, which is imagined as a "long-distance run" (Fig. 6), namely "perseverance". Overall, it can be said that H.R. describes the process of regional network development starting with the need for network building and consultation. He also addresses the university as the actor who should initiate this development and then provide (pedagogical) support throughout the process. The path metaphor of the bridge as boundary crossing then transforms into a path metaphor of the long-distance run, which is focused on the jointly developed vision. The following section will clarify how H.R. perceives this organizational-education process.

6.2 From "Somewhat Sceptical" to "Pleasantly Surprised", "Because the Department of Education Organized It": Perspective on the Organizational Educational Network Consulting

In T2, H.R. describes his initial impression of the organizational education advising students as "somewhat sceptical". Initially, he wondered: "What do they actually want here now?" and "Why are *they* doing this for us?". He was then "pleasantly

Fig. 7 "A refreshing experience" (source unknown)

surprised" during the process to find that the students' goal was to support the networking process among regional stakeholders. The encounter with the student counsellors is described as a "refreshing experience", as symbolised by a person diving into "very cold water" (Fig. 7). In T4, he explains that the "very important refreshing newness" is that the students go down "completely new paths" and break the "old, outdated habits of acting" (Fig. 8). This is also already present in T2 when he says that "usual thought patterns were [...] disrupted" and he wishes that the

Fig. 8 "Refreshing newness" (source unknown)

students would continue to be "ready with fresh energy" to think "completely differently; to think outside the box".

H.R. uses metaphors of freshness ("fresh energy", "refreshing newness") to describe the encounters with the students in both T2 and T4. These metaphors depict the students as a source of renewal and revitalisation. Moreover, it refers to the energetic aspect of a moment of trans-epistemic boundary crossing that is perceived as overcoming the limitations of internalised normalities and traditional ways of acting. It points to a process of reflecting on outdated ways of acting and therefore the transgression of existing positions. The consultation is also not described as expert consultation but rather as impulse and guidance. This becomes quite clear from the image in T2, in which H.R. describes the students as the "younger ones" who "should take the older ones" by the hand and "pull them along" (Fig. 9). He also contrasts the image in T2, where the senior expert advises young people, and "the young people listen enthusiastically" (Fig. 10). They are "working together for a goal" instead of one person lecturing the other. He sees the students as catalysts who bring people together and make them think. His initial scepticism was replaced by approval of the organizational educational approach in T4. Practitioners like him in farming are knowledgeable about their field, including "soil, plants, machine and animals", although they lack leadership ("leading people") and have issues

Fig. 9 "Younger people should take the older ones by the hand" (source unknown)

"organizing discussion groups". He realised within the process that the students have learned "such things: organizational education". This is why he believes that "organizational education [...] needs to be incorporated significantly into more professions" and that the event was successful "because the Department of Education organized it".

In this sense, the acknowledgement of the organizational educational approach is not only limited to the students but to the format of the Innovation Labs itself, which is described as a "completely different format" in T4. Here, he compares the Innovation Lab to the farmers' meeting he had attended before, which he describes as "stodgy". "Not a lot happens" there because "farmers are not very successful in tasks like organizing and moderating a meeting", and they also "often work alone or even in opposition to one another". The experience of the Innovation Labs was

Fig. 10 "The young people listen enthusiastically" (source unknown)

"enriching in any case", because it "opened his eyes" and showed him a "completely different perspective". Additionally, it was a "completely different way of working". For example, he highlights the work with the images in the last Innovation Lab in a particularly positive way. H.R. believes working with images is valuable, as they allow for the "visualisation of thoughts" and support a "consciousness raising".

It is clear that H.R., an experienced practitioner, is starting to accept the organizational educational consultancy approaches initially unknown to him. This also shows that the ascribed acceptance of the consulting approach not only refers to the genuine organizational educational consulting intervention in the sense of applied process consulting to achieve a predefined goal. Moreover, organizational

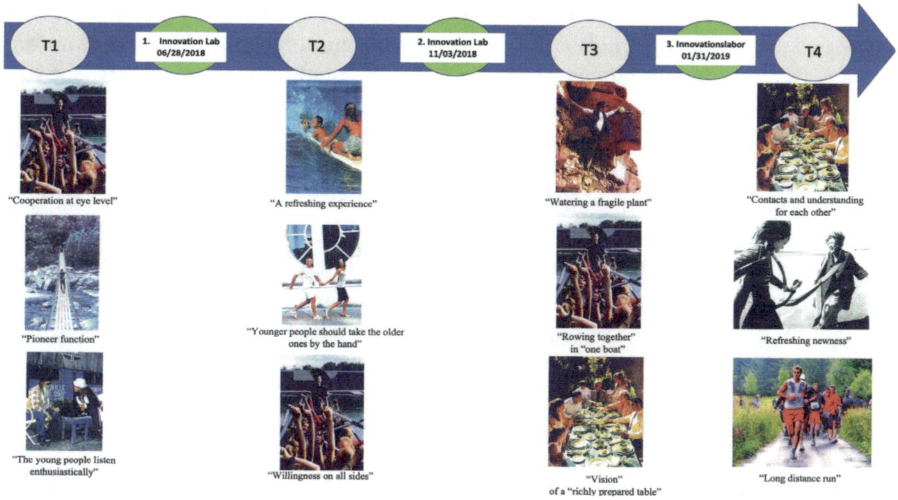

Fig. 11 The selected images by H.R. over the course of the four interview time points (own illustration)

educational knowledge is seen as suitable for providing orientation and support for the transformation process. Overall, it can be said that the Innovation Labs positively impacted the network development process in the region (Fig. 11).[5]

6.3 Network Problems and Solutions Through Organizational Education Network Consulting

After presenting the case and inductively extracting the essential metaphors, these results will be theoretically related to Messner's (1994, 2012) network problems described earlier. Reference is made only to those problems that can be addressed using the case.

6.3.1 From a Large Number of Actors Towards a "Shared Vision"

As previously described, the "large number of actors" (Messner, 2012) in a network with different dependencies can delay and complicate communication, coordination and, ultimately, decision-making processes. This problem is also present within the Innovation Labs, as H.R. describes that with the increasing numbers of actors involved ("many people who are interested in this"), there are an increasing number

[5] For example, the successful formation of a food policy council for the region of Marburg (Ernährungsrat Marburg, 2020) followed an idea within the first Innovation Lab.

of different interests ("different things wanted") and knowledge, which leads to a delay and an increasing amount of time spent on communication ("a lot of time was spent with me explaining the connections"). However, the cooperation between several stakeholders becomes possible within the Innovation Labs. This collective dimension, i.e. the successful network cooperation, is applied across the process using the metaphors that express both a collective effort ("rowing boat", "a group of runners") and a path ("long distance") towards the collective goal ("a set table"). Moreover, the development of a shared vision is viewed as an essential element between actors aiming to join forces, as was illustrated by the image of the rowing boat. The organizational educational image-based method aims to facilitate this through the "visualisation of thoughts" and a "formation of consciousness".

6.3.2 From Endless Disagreement to "People With Visions"

H.R. emphasises the importance of this shared vision as an integral part of the project in the form of the vision of the "set table", which stands for the food cycle. From an organizational educational point of view, finding a shared vision helps to overcome the problem described by Messner (2012) as the tension of disintegration, adaption and endless disagreement. The emphasis on image-based methods as a way of reflecting on one's own and others' way of thinking indicates the potential for image-based methods to find shared perspectives.

6.3.3 From Social Cohesion to "Thinking from Outside of the Box" and "Going New Ways"

H.R. recounts a situation of "social cohesion" (Messner, 2012) at the farmers' meetings, which he describes as "stodged". The refreshing experience of the Innovation Labs offers the potential for reflection from an outside-the-box perspective, which is made possible through contact with other stakeholders and students. He sees the organizational-educational advisory team as able to organize collective processes, whereas the farmers and agricultural groups cannot. In the Innovation Labs a transgression of those normalities becomes possible in the sense of visibilities ("opened my eyes, completely different perspective"), mind ("thinking outside of the box") and a ceasing of acting together ("completely different way of working") under the guidance of organizational educational guidance ("going new ways").

6.3.4 From "Working Against Each Other" to "Completely Different Perspective": Conflict and Cooperation

The beginning of collective transformational practice is the transgression of existing positions (Heidelmann, 2020, p. 270) and thus the transgression of conflict. H.R. describes this process to other stakeholders and students in the first Innovation

Lab. He states the conflict first, namely that farmers tend to work alone or even against each other. As a result of the project, which has inspired him, he has become more open to further reflection. When he and others involved in the project realised that they are facing similar problems, they exchanged contact details and began to work together more closely.

7 Conclusion

This paper aimed to demonstrate the potential for organizational educational network consulting to develop a regional economic cycle focused on network development. It was shown that rural areas face major challenges, which require cooperation and network development between regional actors to transition to circular economies (Geissdoerfer et al., 2017). In addition, the network problems, according to Messner (1994, 2012), were introduced, and the need for organizational educational consulting was highlighted. Organizational education does not consider organizations and networks as contexts of education but as learning collectives. Network learning in this sense means that only individual actors enter educational processes by means of methodical arrangements. However, the targeted network, in this case the regional food cycle, requires collective educational processes. In the way of problem-oriented educational arrangements, collectives can find solutions and development possibilities against the background of a specific goal—in this case that of a regional circular economy. The entire process is conceived as an educational process in terms of organizational education and its methods and didactics are structured accordingly. Based on the results of a metaphor- and image-based case analysis of a participating regional stakeholder, it was shown that this organizational educational knowledge could successfully address network management problems and therefore effectively facilitate regional network development. Further potential exists, particularly in the increasing professionalisation (Schwarz, 2018) and institutionalisation of organizational educational network consulting as a genuine field of knowledge application.

References

Brake, A. (2011). Kombinieren, mixen, verbinden? Integration als konstitutives Element methodentriangulierender Zugänge. In J. Ecarius & I. Miethe (Eds.), *Methodentriangulation in der qualitativen Bildungsforschung* (pp. 41–64). Budrich.

Brake, A. (2018). Prozessorientierung und Längsschnittdesign als Forschungsstrategie der Organisationspädagogik. In M. Göhlich, A. Schröer, & S. M. Weber (Eds.), *Handbuch Organisationspädagogik* (pp. 307–318). Springer.

Castoriadis, C. (1998). *The imaginary institution of society*. MIT Press.

Diller, C., Nischwitz, G., & Kreutz, B. (2014). Förderung von Regionalen Netzwerken: Messbare Effekte für die Regionalentwicklung? *Raumforschung und Raumordnung | Spatial Research and Planning, 72*(5), 415–426. https://doi.org/10.1007/s13147-014-0308-5

Endres, E., & Wehner, T. (1999). Störungen zwischenbetrieblicher Kooperation – eine Fallstudie zum Grenzstellenmanagement in der Automobilindustrie. In J. Sydow (Ed.), *Management von Netzwerkorganisationen* (pp. 215–259). Gabler.

Ernährungsrat Marburg. (2020). *Gründung*. https://ernaehrungsrat-marburg.de/gruendung/

Forceville, C. (2008). Metaphors in pictures and multimodal representations. In W. R. Gibbs (Ed.), *The Cambridge handbook of metaphor and thought* (pp. 462–482). Cambridge University Press.

Foucault, M. (1973). *The archaeology of knowledge and the discourse on language*. Pantheon.

Foucault, M. (1994). *The order of things: An archaeology of the human sciences*. Vintage Books.

Geise, S. (2017). Methoden der Bildrezeptions- und Bildwirkungsforschung. In K. Lobinger (Ed.), *Handbuch Visuelle Kommunikationsforschung* (pp. 1–26). Springer.

Geissdoerfer, M., Savaget, P., Bocken, N. M. P., & Hultink, E. J. (2017). The circular economy – a new sustainability paradigm? *Journal of Cleaner Production, 143*, 757–768. https://doi.org/10.1016/j.jclepro.2016.12.048

Göhlich, M., Novotny, P., Revsbæk, L., Schröer, A., Weber, S. M., & Yi, B. J. (2018). Research memorandum organizational education. *Studia Paedagogica, 23*(2), 205–215.

Harper, D. (2002). Talking about pictures: A case for photo elicitation. *Visual Studies, 17*(1), 13–26. https://doi.org/10.1080/14725860220137345

Heidelmann, M.-A. (2020). Verantwortung (in) der Organisationsberatung – Organisationspädagogisches Proprium im Plural theoretischer Perspektiven. In C. Fahrenwald, N. Engel, & A. Schröer (Eds.), *Organisation und Verantwortung* (pp. 259–273). Springer VS.

Heidelmann, M.-A. (2023). *Organisationen und Netzwerke beraten lernen: Eine Analyse organisationspädagogischer Professionalisierung*. Springer.

Heidelmann, M.-A., & Klös, T. (2023). Optimierung des regionalen Wirtschaftskreislaufs—Die Anerkennung organisationspädagogischen Wissens im Theorie-Praxis-Transfer ländlicher Räume. In S. M. Weber, C. Fahrenwald, & A. Schröer (Eds.), *Organisationen optimieren?* (pp. 309–325). Springer.

Heidelmann, M.-A., & Weber, S. M. (2018). Organisationspädagogisches Wissen im Praxisfeld der Organisationsentwicklung und -beratung. In M. Göhlich, A. Schröer, & S. M. Weber (Eds.), *Handbuch Organisationspädagogik* (pp. 865–877). Springer.

Heidelmann, M.-A., Weber, S. M., & Klös, T. (2023). Collective leadership toward sustainability innovations: An organizational education professionalization approach. In W. Leal Filho, A. Lange Salvia, E. Pallant, B. Choate, & K. Pearce (Eds.), *Educating the sustainability leaders of the future* (pp. 603–624). Springer Nature.

Junge, M. (Ed.). (2014). *Methoden der Metaphernforschung und -analyse*. Springer.

Klös, T., & Heidelmann, M.-A. (2023). Sustainability leaders' perspectives on the potential of innovation labs: Toward collective regional leadership. In W. Leal Filho, A. Lange Salvia, E. Pallant, B. Choate, & K. Pearce (Eds.), *Educating the sustainability leaders of the future* (pp. 659–679). Springer Nature.

Lakoff, G., & Johnson, M. (2008). *Metaphors we live by: With a new afterword*. University of Chicago Press.

Lerch, F., Sydow, J., & Duschek, S. (2011). Beyond the organizational focus: Network consulting in regional clusters. In A. Buono (Ed.), *The changing paradigm of consulting: Adjusting to the fast-paced world* (pp. 185–209). Information Age Publishing.

Messner, D. (1994). Fallstricke und Grenzen der Netzwerksteuerung. *PROKLA. Zeitschrift für kritische Sozialwissenschaft, 24*(97), 563–596. https://doi.org/10.32387/prokla.v24i97.980

Messner, D. (2012). *The network society*. Routledge.

Nowotny, H., Scott, P., & Gibbons, M. (2001). *Re-thinking science: Knowledge and the public in an age of uncertainty*. Polity Press.

Paivio, A. (1979). *Imagery and verbal processes*. Lawrence Erlbaum.

Reike, D., Vermeulen, W. J. V., & Witjes, S. (2018). The circular economy: New or Refurbished as CE 3.0? — Exploring controversies in the conceptualization of the circular economy through a focus on history and resource value retention options. *Resources, Conservation and Recycling, 135*, 246–264. https://doi.org/10.1016/j.resconrec.2017.08.027

Scharpf, F. W. (1991). Games real actors could play: The challenge of complexity. *Journal of Theoretical Politics, 3*(3), 277–304. https://doi.org/10.1177/0951692891003003003
Schröer, A. (2018). Führung als Gegenstand der Organisationspädagogik. In M. Göhlich, A. Schröer, & S. M. Weber (Eds.), *Handbuch Organisationspädagogik* (pp. 479–490). Springer.
Schwarz, J. (2018). Professionalisierung der Organisationspädagogik. In M. Göhlich, A. Schröer, & S. M. Weber (Eds.), *Handbuch Organisationspädagogik* (pp. 831–838). Springer.
Stahel, W. R., & Reday-Mulvey, G. (1976). *The potential for substituting manpower for energy (Research Contract No 76/13-V/343/78-EN)*. Commission of the European Communities.
Sydow, J. (1992). *Strategische Netzwerke*. Gabler.
Sydow, J. (2006). Netzwerkberatung—Aufgaben, Ansätze, Instrumente. In J. Sydow & S. Manning (Eds.), *Netzwerke beraten. Über Netzwerkberatung und Beratungsnetzwerke* (pp. 58–84). Gabler.
Sydow, J., & Windeler, A. (2003). Knowledge, trust, and control: Managing tensions and contradictions in a regional network of service firms. *International Studies of Management & Organization, 33*(2), 69–99.
United Nations. (2015). *Transforming our world: The 2030 Agenda for Sustainable Development.* https://sustainabledevelopment.un.org/content/documents/21252030%20Agenda%20for%20Sustainable%20Development%20web.pdf
WBGU - Wissenschaftlicher Beirat der Bundesregierung Globale Umweltveränderungen. (2011). *World in transition: A social contract for sustainability*. https://www.wbgu.de/en/publications/publication/world-in-transition-a-social-contract-for-sustainability#section-downloads
Weber, S. M. (Ed.). (2001). *Netzwerkentwicklung in der Jugendberufshilfe: Erfahrungen mit institutioneller Vernetzung im ländlichen Raum*. Leske + Budrich.
Weber, S. M., & Heidelmann, M.-A. (2019). Towards regional circular economies. 'Greening the University Canteen' by Sustainability Innovation Labs. In W. Leal Filho & U. Bardi (Eds.), *Sustainability on university campuses: Learning, skills building and best practices* (pp. 415–435). Springer.
Weber, S. M., & Heidelmann, M.-A. (2021). Im Dispositiv 'Diskursiver Gestaltung': Grenzüberschreitung als Telos organisationspädagogischer Professionalisierung im Feld der Organisations- und Netzwerkberatung. In A. Schröer, S. Köngeter, S. Manhart, C. Schröder, & T. Wendt (Eds.), *Organisation über Grenzen* (pp. 77–91). Springer VS.
Weber, S. M., Heidelmann, M.-A., & Klös, T. (2019). Zukunfts-Wissen im Diskurs: Eine Didaktik Wissenschaftlicher Weiterbildung für organisationspädagogische Professionalisierung. *Zeitschrift für Hochschulentwicklung, 14*(4), 17–36. https://doi.org/10.3217/zfhe-14-04/02
Weetman, C. (2021). *A circular economy handbook: How to build a more resilient, competitive and sustainable business*. Kogan Page.
Weisbord, M. R., & Janoff, S. (2010). *Future search: Getting the whole system in the room for vision, commitment, and action*. Berrett-Koehler.

Prof. Dr. Marc-André Heidelmann is Professor for Social Work at the IU International University of Applied Sciences since 2022. After studying politics and economics, ethics, German studies and philosophy, he received his doctorate in educational science in 2021. Before his appointment as a professor at IU, Heidelmann initially researched and taught for several years at the Institute of Educational Science at the Philipps-University of Marburg before moving to the Chair of School Pedagogy with a focus on educational governance and educational change at Friedrich-Alexander-University Erlangen-Nürnberg. His research focuses in particular on qualitative-reconstructive learning and educational research. In more recent research, he also deals with questions of learning under conditions of artificial intelligence.

Tobias Klös (M.A.) is presently a member of the academic staff at the Heidelberg University of Education, where he teaches in the programme "Digital Education for Sustainable Development". Prior to this, he spent several years working at Philipps University of Marburg, in university didactics, at the Centre for Teacher Training and at the Institute of Educational Sciences. He is currently in the final stages of his dissertation in the field of organizational development for sustainable development. His research interests are organizational education, digital education for sustainable development and game studies.

Replacing Cardboard-Based Boxes with a Circular Economy Solution in Austria: A Framework for Stakeholder Assessment and Impact Evaluation of Social, Ecological and Economic Factors

Nils Kruse, Melanie Rainer, Gerald Schneikart, and Walter Mayrhofer

1 Introduction

The recently proposed European Packaging Directive confronts the courier, express and package (CEP) industry with new requirements that envision the shift from one-way packaging to a circular economy approach, such as implementing returnable transport items (RTI; European Commission, 2022). Implementing a circular economy solution requires the identification of relevant stakeholders with regard to how they are impacted and how they can impact the company proposing the new solution. By example of the CEP industry, we demonstrate how a stakeholder management approach can be applied from the perspective of a delivery service provider. We argue that such an approach—namely, a critical number of stakeholders agreeing on the same RTI solution and routing problems from reverse logistics processes—is necessary to overcome the obstacles to implementing an RTI solution in the CEP industry.

This chapter first gives an overview about the vision of a circular economy for the CEP industry in Austria followed by an overview of the stakeholder approach. Subsequently, the triple bottom line approach is presented as a tool to assess

N. Kruse (✉)
Technische Universität Dresden, Zittau, Germany
e-mail: nils.kruse@mailbox.tu-dresden.de

M. Rainer
University of Sustainability, Charlotte Fresenius Privatuniversität, Vienna, Austria
e-mail: melanie.rainer@uni-sustainability.at

G. Schneikart · W. Mayrhofer
FHWien der WKW, Vienna, Austria
e-mail: gerald.schneikart@fh-wien.ac.at; walter.mayrhofer@fh-wien.ac.at

E. Fröhlich et al. (eds.), *Circular Economy in Sustainable Supply Chains*, CSR, Sustainability, Ethics & Governance,
https://doi.org/10.1007/978-3-031-70749-0_24

stakeholder impact and carry out a stakeholder analysis for the CEP industry, and the chapter is closed by a discussion.

2 The European Courier, Express and Package Industry

2.1 Circular Economy in Logistics

In 2022, the European Union published a proposal, as part of the European Green Deal, that will move industry towards the adoption of new sustainable packaging solutions once the act becomes law in the member states. The aim is to expedite the achievement of the global sustainable development goals (SDGs) by providing the necessary legislative structures, thus enabling novel business innovations based on different circular economy concepts, such as reuse or recycling of packaging material, or the implementation of RTIs. The proposal is designed for the realisation of the following sustainability objectives (European Commission, 2022):

- Reduction of net greenhouse gas emissions by at least 55% until 2030 (compared to 1990)
- Reduction of packaging waste by 15% per capita in every member state by 2040 (compared to 2018)
- Mandatory reuse quotas for various packaging materials by 2030
- Mandatory proscription of shrink-wrap and plastic strap usage in conjunction with pallet logistics
- Mandatory quotas for usage of RTIs (instead of cardboard-based boxes)

A key element to reach those goals is using circular logistics systems that enable the increased use of RTIs by shipping companies and CEP service providers.

2.2 Returnable Transport Items

In logistics, the use of circular packaging solutions is still low (Coelho et al., 2020). The most common packing material in parcel services is cardboard-based, which is recyclable. However, reuse of fibres is limited to only a few cycles and requires reconditioning. Even though the intention of such complex recycling processes is the reduction of environmental impacts, they entail considerable carbon footprints due to the use of one-way packaging material that requires collection, backhaul and reconditioning. As RTIs can be used hundreds or thousands of times in circular logistics processes, they represent a promising alternative (de Brito et al., 2005; Koskela et al., 2014; Soysal et al., 2023). As depicted in Fig. 1, such a circular solution for RTIs can be implemented as a closed loop system between the sender (e.g.

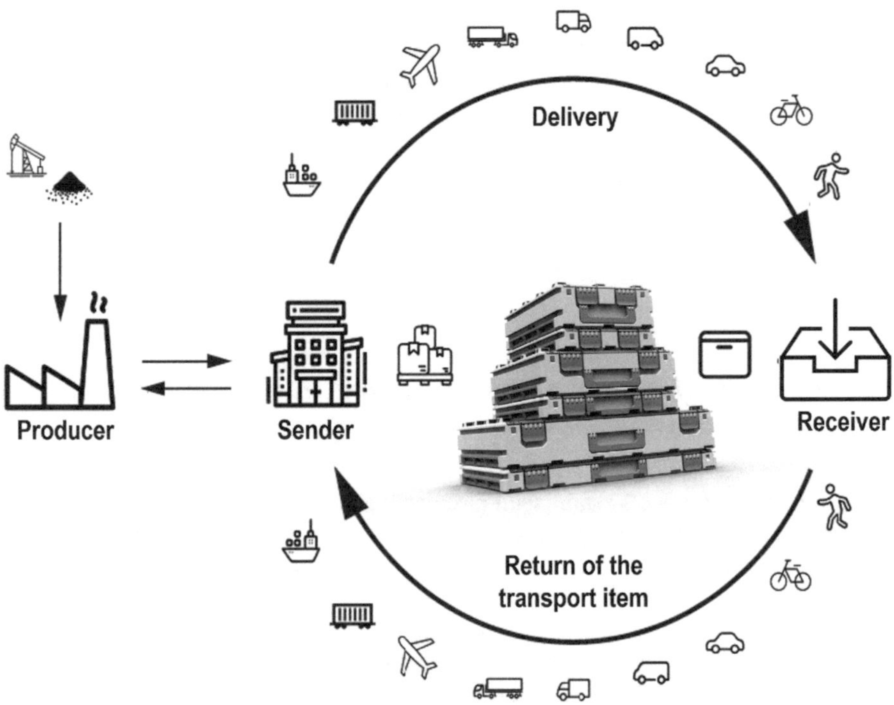

Fig. 1 Returnable transport item life cycle (own illustration)

a delivery service provider) and the receiver, encompassing all different modes of transport and taking the production and recycling of the RTI into account.

Although the current material of RTIs is commonly plastic-based, which at first glimpse does not make them appear as sustainable enablers of circular economies, they are considerably more durable than cardboard-based boxes and promise extended reuse cycles of individual RTIs, hence avoiding complex reconditioning of the cardboard-based packaging material (Coelho et al., 2020; Koskela et al., 2014). Further, RTIs require only a fraction of the resources (Table 4 in Coelho et al., 2020) and energy compared to cardboard-based packaging. However, RTIs are only considered profitable if return quotas exceed 80%. Taking current market demands and outcomes of pilot runs of RTIs into account, adoption rates need to be elevated in order to reach the required quotas (Gersmann, 2021). Two underlying challenges are to be expected as the main barriers to adoption of RTI solutions:

- The high number of stakeholders having to agree on the same enabling RTI solution to take advantage of a single system solution.
- The increased complexity of the resulting routing problem as a consequence of reverse logistics on top of the traditional forward logistics.

While the former depends on the involved stakeholders' strategies, the latter might be tackled by integrating RTIs with Industry 4.0 technology, e.g. radio frequency identification (RFID; Ilic et al., 2009), 3D-printing (Pladerer, 2009) or blockchain solutions (Samadhiya et al., 2023). Industry 4.0 solutions are the key to reaching the global SDGs, as new technologies foster sustainable innovation (Ghobakhloo et al., 2021). The resulting reverse logistics 4.0 align with the human-centric and sustainability-inducing principles of Industry 5.0 (Xu et al., 2021). Finally, circular economy concepts based on RTIs are faced with the problem of returning empty containers. However, reverse logistics also creates new business opportunities, such as using the backhaul of RTIs to return packaging material, bottles or otherwise discarded material to the recycling centre (Coelho et al., 2020).

2.3 Existing Circular Logistics Solutions in Europe and Austria

In Europe, there are well-established circular logistics solutions in certain areas—e.g. beverages (Lemos Junior et al., 2019) and groceries (Svenska Retursystem, 2023)—with considerable innovation of those multi-purpose containers. Their success often depends on the establishment of "third-party" entities (Landschützer et al., 2015), so-called "poolers", that provide service and distribute RTIs. Examples of successful pooling arrangements are the RTI systems by IFCO (IFCO SYSTEMS, 2023) or EUROBOX (EuroBox Logistiksysteme, 2023).

Austria has well-established circular RTI systems in the business areas: groceries, books, textiles, construction, electronics, automotive and pharma. The distribution of pharmaceuticals from wholesalers to apothecaries and hospitals is already employing RTIs at a rate close to 100% (Bundesministerium für Nachhaltigkeit und Tourismus, 2019). Due to high demands regarding transport conditions and logistics requirements, the pharmaceutical industry in Austria is piloting the automation and digitisation of distribution processes. The project *DigiPharmaLogNet* uses an innovative, digitally enhanced RTI by BOOXit to realise the vision of the "Physical Internet" in pharma logistics (Schneikart et al., 2023). Other such concepts are *livingpackets* (LivingPackets, 2023) and *GS1-Smartbox* (GS1 Germany, 2023) by container producers UTZ, BEKO and SCHÖLLER. A recent German study evaluated the feasibility and potential impact of RTIs and found that 86% of inquired consumers would use cardboard-based RTIs, and 52% would even pay premiums. As the reason for their readiness to accept RTIs, 70% stated the large amounts of waste that pose a burden (WKO, 2022).

RTIs often fail because they are conceived for specific purposes. The project "Mobility Modular Load Carrier" conducted by the Austrian Institute for Ecology aims at developing flexible RTI designs for multiple logistics applications. Such designs would require RTIs with multi-purpose folding and insertion mechanisms. 3D-printing serves as an enabling technology for the quick supply of RTI spare parts (Pladerer, 2009). This specific example supports the view that the success of reverse logistics processes might depend on the integration of Industry 4.0 technologies.

Irrespective of the added value of any RTI solution, the logistics industry seems hesitant in using RTIs. In a different project, "Mehrweg hat Zukunft!" ("Reusable has a future!"), concerning RTI usage in the beverage supply chain, the Austrian Institute for Ecology suggests five models for increased willingness of RTI usage (Pladerer, 2009):

- Claiming an agio per unit of ecological damage
- Distribution of limited packaging licences
- Imposing reuse quotas
- Imposing an ecological pillars model consisting of taxes for packaging material, disclosure of production and recycling costs, and a deposit system for assuring reverse flow
- A bonus system favouring ecologically advantageous packages and RTIs

The project, which was conducted in 2009, concluded that none of the models would be successful without legal frameworks forcing logistics organisations to rethink RTI-based reverse logistics (Pladerer, 2009). Interestingly, more than ten years later, the European Green Deal could eventually lay the necessary basis for these RTI success models and derivations thereof to become reality.

3 The Stakeholder Approach as Assessment Tool for Circular Economy Solutions

The concept of "stakeholders" in management theory gained traction with Freeman's (2010) seminal work *Strategic Management: A Stakeholder Approach* that was published in 1984. Since its inception, the theory has been criticised and engaged with by several thinkers as well as integrated into different approaches, such as Williamson's contractual approach to transaction theory (Freeman, 2010). Originally, Freeman's stakeholder approach was developed as an instrument for managers to deal with an increasingly turbulent environment that emerged from changing internal and external environments (Freeman, 2010). Thus, managers had to change their perspective from a production view and managerial view of the firm to a broader conception. This conception not only comprised internal stakeholders (e.g. employees and owners) or stakeholders along the value chain (e.g. suppliers and customers), but it also had to consider previously neglected stakeholders (e.g. governments and NGOs). Freeman (2010) offered the broad definition of stakeholders constituting "those groups who can affect or are affected by the achievement of an organization's purpose" (p. 46). Although stakeholder theory considers external players that are an integral part of the wider society, at its centre is the corporation and not society (Wieland & Schmiedeknecht, 2010).

One of the main challenges in applying stakeholder approaches to practical management problems has been a clear definition of what constitutes a stakeholder and how stakeholders can be prioritised. In this chapter, we first introduce a claimants-based approach to stakeholder management that looks primarily into how stakeholders are affected by a management decision. Subsequently, we employ a paradigm that conceptualises the company as a nexus of stakeholders who invest their specific resources for corporation rent. According to this approach, stakeholders are prioritised based on the resources they can invest into the company (Wieland, 2011).

The stakeholder approach involves the identification and concurrent prioritisation of relevant stakeholders. This usually involves stakeholder mapping depicting relations between the corporation and its key stakeholders (see Fig. 2). The standard stakeholder map's level of abstraction represents stakeholders as homogeneous groups, whereas, in reality, stakeholders might be heterogeneous and thus require boxes of different sizes to indicate the diverging importance of the respective stakeholders (Fassin, 2008). Therefore, such a stakeholder mapping runs the danger of underemphasising heterogeneity within respective stakeholder groups. Stakeholders are not a homogeneous and monolithic group, as their interests and power vary widely (Winn, 2001).

In contrast to all the above, Freeman (2010) argues that trade-offs must be avoided and—if this is not possible—must be remediated promptly. Acknowledging those complexities, in this chapter we will rely on a less granular version of the stakeholder map, as it satisfies the purpose of our argument. However, when implementing such a stakeholder mapping in practice, it might be useful to zoom in and consider diverging interests within stakeholder groups.

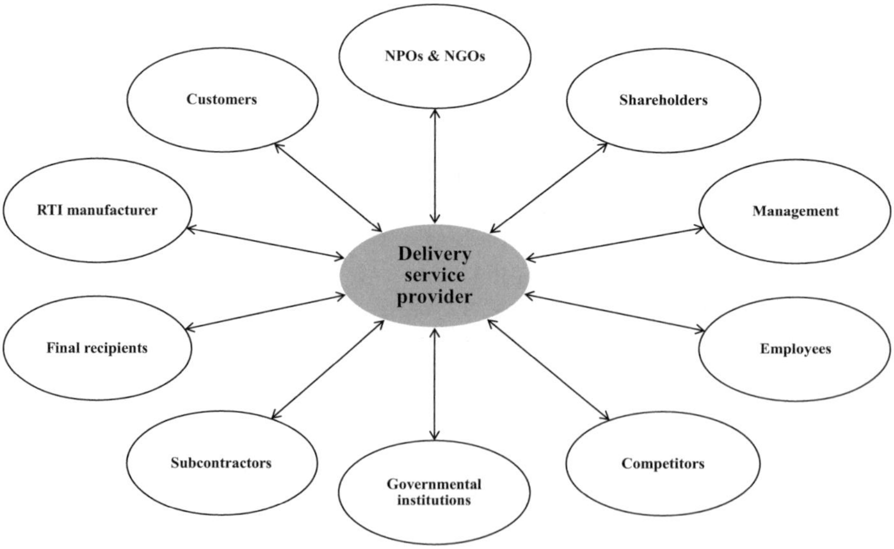

Fig. 2 Stakeholder map in the courier, express and package industry (own illustration)

3.1 *The Triple Bottom Line as Basis for a Claimants-Based Approach to Stakeholder Management*

In 1996, the term "triple bottom line" (TBL) emerged by delineating sustainability based on the three pillars "people, planet, profit" (Elkington, 1998). The concept of TBL comprises the social and environmental dimensions of sustainability as well as economic efficiency (Bansal, 2005). While the social aspect involves diversity, human rights or philanthropy, the environmental pillar focuses on the firm's impact on the natural environment (e.g. emissions). Additionally, the economic dimension is mainly related to the financial benefits for stakeholders, such as employees and shareholders (Nikolaou et al., 2013).

One way to measure TBL impact is a life cycle sustainability assessment (LCSA), which was established as a widely used environmental management tool (Arena et al., 2003). The "life cycle" includes all product-related processes, from resource inputs (extraction of raw materials, resources, energy) to outputs (transportation, usage, waste disposal). LCSA can be analysed with a focus on one TBL dimension, which results in the specification of the environmental life cycle (ELCA), the social life cycle (SLCA) as well as life cycle costing (LCC) (Hahn, 2022; Zamagni et al., 2013).

Firstly, the social dimension refers to the firm's responsibility to provide education, social justice, fair income distribution and employment within the company (Elkington, 2004) and for society at large (Brickson, 2007; Pullman et al., 2009). In this context, companies act in socially responsible ways if they (1) do not knowingly harm their stakeholders and (2) rectify whenever harm is caused (Campbell, 2007). The social aspects can be conceptualised and measured by social impacts. The latter can be assessed within the SLCA methodology, which focuses on the life cycle activities of business products and services that affect people. In this respect, all positive and negative impacts of business activities as well as potential countermeasures require evaluation (Dreyer et al., 2006). Social impact can be grouped into four different approaches: multi-sector and single-sector activities and multi-sector and single-sector outcomes (Rawhouser et al., 2019). As research primarily investigated the social impact from a multi-sector perspective, predicting impacts of firm activities, e.g. production, remanufacturing and product distribution, mainly relies on data which allows cross-sector comparison.

Secondly, the environmental dimension encompasses a company's effect related to natural capital (Elkington, 1998). The concept of environmental impact refers to bio-geophysical, socio-economic and cultural aspects (Harvey, 1998). Impacts comprise direct effects on the natural and built environment, and management is bound to assess, minimise, reduce and avoid adverse effects (Abdallah, 2017). Environmental impact assessment emerged as a governmental tool to evaluate potential environmental impacts of proposed policies and actions for governmental institutions (Harvey, 1998). ELCA assesses the environmental impact of products and services at each stage and throughout their life cycle. Based on material and energy balances, within the assessment process, all in- and outputs of each product stage are calculated and aggregated over the corresponding life cycle (Hahn, 2022).

Lastly, the TBL concept considers the economic dimension as a financial parameter for economic development, which associates the firm's performance with corporate social responsibility (CSR; Elkington, 2004). This viewpoint can be traced back to Adam Smith, who links the private interest of market actors to the general welfare improvement (Windsor, 2006) and thus connects the economic business dimension with society at large. The definitions of "economic" reach from disruptive innovation for economic development (Schumpeter, 1934) to social responsibility through financial performance (Campbell, 2007). Economic sustainability is mainly linked to economic growth, namely, the organisational creation of economic values by generating wealth for future generations and facilitating survival (Alhaddi, 2015). By means of LCC, all internal and external costs arising from the entire life cycle of a product are aggregated and help evaluate trade-offs (Hahn, 2022).

In a nutshell, the LCSA with its subcategories can serve as a suitable basis to evaluate the impact on stakeholders based on their claims. When implementing stakeholder mapping, other factors, such as the legitimacy of stakeholders, can be included in order to prioritise conflicting claims (Mitchell et al., 1997).

3.2 A Resource-Based Approach to Stakeholder Management

Proponents of a resource-based view take a different approach to stakeholder management. Wieland (2014) defines a stakeholder as "a resource owner in a cooperative team constituted by explicit and implicit contracts, whose aim is to generate a cooperation rent by realizing a specific transaction by means of an appropriate and enduring governance form" (Wieland, 2014, p. 111). The underlying concept of the firm is that of a "nexus of formal and informal contracts by means of which the owners of resources or competencies come together to form a team" (Wieland, 2014, p. 110). The idea is that resource owners can yield a cooperation rent by combining their resources compared to individually offering them on the market. This view indicates a shift from positive and negative externalities to stakeholders as resource owners who together constitute a corporation. According to Wieland (2020), stakeholders must fulfil three criteria: (1) possess resources relevant for the firm, (2) accept formal or informal contracts to contribute these resources to the team and (3) contribute to long-term cooperation rent (Wieland, 2014).

Wieland (2014) suggests a list of criteria to prioritise stakeholders, namely, in which timeframe as well as how and by which means they can contribute to a team's success. If a stakeholder group scores high on many categories, it should be prioritised over a group that scores comparatively low. The proposed criteria are (1) *contract relevance*, i.e. team members who are interested in a long-term cooperation must be given priority over groups that are only interested in specific transactions; (2) *resource relevance*, i.e. how relevant a stakeholder's resource, such as technical knowledge or moral legitimacy, is to a certain transaction; (3) *cooperation relevance*, i.e. the cooperative quality of a stakeholder, such as the willingness to accept responsibility in a cooperation project; and (4) *investment relevance*, i.e. the willingness of a stakeholder to build either team or transaction-specific resources, which indicates the durability and quality of a stakeholder's commitment to the cooperation project (Wieland, 2014). There are other transactional approaches that add an exchange relationship (from informal or formal contracts) or a power-dependence relationship as prioritisation criteria (Mitchell et al., 1997).

3.3 Integrating Both Approaches

The concept that stakeholders can be categorised as claimants and influencers is not novel but has been criticised as lacking means to prioritise stakeholders. Mitchell et al. (1997) argue that considering a sense of urgency can help managers to solve this issue. While the notion of urgency is helpful in prioritising stakeholders, it does not suffice to solve the conundrum of how to balance claims versus resources. Thus, the respective theory and resulting stakeholder definitions potentially play a role in the identification of stakeholders and their prioritisation.

Although the stakeholder approach is about concrete names and faces (McVea & Freeman, 2005), its application to any individual company is specific to that particular situation. Since a model resembles a reduction of reality, its goal is not to capture the whole complexity of the real world. This is especially true for stakeholder mapping in the abstract realm without a specific application of a company. The following chapter demonstrates applying both stakeholder approaches to the Austrian CEP industry to evaluate the benefits and potential drawbacks of implementing a circular economy solution that is sensitive to its social and ecological implications. However, this chapter can provide only a blueprint for an actual company and must be adjusted to other industries. Regardless, we believe that this blueprint can be a helpful starting point for practitioners across a variety of industries.

4 The Case of the Austrian CEP Industry

This section compares the implications of the two different stakeholder paradigms, namely, the claimant-based and the resource-based perspective. It provides the fictional—but realistic—case of an international and well-established delivery service provider. Building on CEP industry material, we take the scenario of a delivery service provider switching from cardboard-based packaging to a more sustainable circular solution. By implementing an RTI system and evaluating and comparing the social, ecological and economic impact of different stakeholders from the claimant-based (Table 1) and the resource-based perspectives (Table 2), the

Table 1 Stakeholder impact assessment

Stakeholder group	Triple bottom line impact		
	Social	Ecological	Economic
Shareholder	Medium	Medium	High
Management	Medium	Medium	High
Employees	High	Low	High
RTI manufacturer	High	High	High
Customers	Low	Medium	High
Final recipients	Low	Medium	Medium
Subcontractors	High	Medium	High
Competitors	Low	Medium	High
Governmental institutions	Medium	High	Low
NGOs & NPOs	Medium	High	Low
Communities	High/Medium	High	Low

Table 2 Relevance of stakeholder input contributed to reach critical mass in the circular economy solution

Relevance	High	Medium	Low
Contract	Management, RTI manufacturers, employees, shareholders	Final recipients, customers, subcontractors	NGOs & NPOs, communities, governmental institutions
Resources	Management, shareholders, RTI manufacturers, NPOs & NGOs	Employees, governmental institutions	Final recipients, subcontractors, customers
Cooperation	Management, employees, RTI manufacturers	Customers, RTI manufacturers, NGOs & NPOs, communities, shareholders, governmental institutions, subcontractors	Final recipients
Investment	RTI manufacturers, management	Employees, RTI manufacturers, subcontractors	NGOs & NPOs, shareholders, communities, governmental institutions, customers, final recipients

relevance of a stakeholder's contribution to the cooperation project in terms of contract, resources, cooperation and investment is considered. The exemplary stakeholders, which are mapped in Fig. 2, were identified by analysing the value chain depicted in Fig. 1. While we suggest that this approach is a suitable basis for stakeholder identification in other industries, the following provides an exemplary prioritisation of different stakeholders according to the respective paradigms.

4.1 Stakeholder Prioritisation from a Claimant-Based Perspective

Shareholders. Different types of shareholders can have diverging interests when it comes to maximising financial returns and prioritising social impact (DesJardine et al., 2023). While transactional shareholders are short-term oriented due to quick stock trading, durable investors are concerned with long-term financial returns but pay close attention to delayed returns because of CSR initiatives. In contrast, sustainable shareholders primarily focus on environment, social, governance (ESG) or socially responsible investing (DesJardine et al., 2023). Additionally, transactional shareholders are merely and highly impacted by the economic dimension rather than social and ecological aspects. For our case study, we assume the main investor to be durable, remaining positive towards CSR projects as long as their long-term financial returns are not in danger. Durable investors are highly impacted by the economic dimension and moderately affected by the social and ecological aspect.

Management. Managers are important stakeholders as they directly experience the effects of company performance. The management incorporates the role of an "ethical strategist" who is responsible for stakeholder engagement and management (Noland & Phillips, 2010). Assuming that social and environmental factors are relevant to the management's remuneration, we estimate that the two dimensions are of medium importance, whereas the company's effect on the economic dimension is high.

Employees. Employees within the CEP industry can be grouped into loaders, delivery service personnel (drivers) and administrative personnel. While the loaders, especially in the high-volume intensive parcel sector, need to have physical fitness, the delivery service employees are required to be punctual and reliable. Although all employees share the same social (e.g. fair compensation) and economic impact (e.g. economic wellbeing), they do differ in their ecological attitude (e.g. responsible production and emissions). On an aggregated basis, we assume that the impact is medium.

RTI Manufacturers. RTI manufacturers are responsible for developing, producing and delivering the RTIs to the delivery service provider. They also have a crucial role in developing the delivery system and the entire supply chain by means of RTIs (Ilic et al., 2009). RTI manufacturers are highly impacted by the distribution of the boxes on a social level—the use of RTIs has an influence on job tasks and positions (e.g. transport requirements and automatisation)—as well as at the ecological (e.g. responsible production and emissions) and economic (e.g. rotation rate and sales) level.

Customers. Customers can be grouped into business-to-business (B2B) and business- to-customer (B2C) clients. While B2C customers want fast, flexible and preferably free deliveries, B2B clients prefer automation throughout the delivery process as well as punctual deliveries. Both client groups share the same social (e.g. fair prices) and ecological (e.g. low emissions) orientation and value individualisation in size, space and the RTI's adaptability to the shipped product. B2C customers are more likely to appreciate customer experience through packaging. B2B and B2C clients are equally affected by the TBL dimensions. Specifically, they are most affected by the economic (e.g. fair order prices), moderately affected by the ecological (e.g. delivery emissions) and least affected by the social dimension.

Final Recipients. Final recipients are product users receiving and returning packages (Karcz & Ślusarczyk, 2021). Their buying decisions are affected by product (e.g. fulfilment of customer expectations) as well as transport service quality. While they value economic factors, such as a functional delivery system, including transport availability, flexibility, frequency and on-time delivery (Karcz & Ślusarczyk, 2021), they likely also care about social aspects, such as the creation of employment throughout the ordering and delivery process. Their TBL impact is low in the social but medium in the ecological (e.g. waste reduction) and economic (e.g. RTI price) dimensions.

Subcontractors. Subcontractors are often transport carriers that assume a special position in the CEP industry, as they take a similar position to employees but frequently work under precarious conditions. Transport carriers execute logistical services for their customers, namely, as third-party logistics providers (Cochran & Ramanujam, 2006). Regarding the social dimension, transport carriers pursue fair compensation (high impact). From an environmental perspective, subcontractors are merely affected to a medium degree by the transport routes and, specifically, the returnability aspect of RTIs, as environmental effects only become apparent indirectly. The stakeholder group is highly concerned with the economic dimension, as financial wellbeing to secure a long-term working relationship and adequate compensation are important parameters.

Competitors. According to the claimant-based perspective, competitors are stakeholders of the CEP industry. Based on the TBL approach, competitors are primarily and highly affected at the economic level, as a successful implementation of a product impacts the industry as well as product prices for all market actors. Regarding the social dimension, competitors value knowledge generation and transfer to foster progress but are impacted at the lowest degree. In the environmental context, competitors are moderately impacted by emissions.

Governmental Institutions. Considering the packaging and packaging waste directive, governmental institutions aim at harmonising standards in the context of the European Green Deal. Hence, governmental institutions are key stakeholders in the CEP industry as they collect taxes and pass regulations. Nevertheless, they shape regulations in collaboration with industry stakeholders and raise companies' awareness of sustainability (Ji et al., 2014). With respect to the TBL, governmental institutions are highly affected by the social (e.g. societal wellbeing) and ecological (e.g. low emissions) but less affected by the economic dimension.

Non-Profit Organisations (NPOs) and Non-Governmental Organisations (NGOs). NPOs and NGOs can have a collaborative role as stakeholders (Xu et al., 2022). In the CEP sector, the collaboration is mainly influenced by contextual factors, inter alia governmental regulations, publicness and the socio-economic setting (Xu et al., 2022). The main non-profit actors are situated in the logistic sector (e.g. BVL) but also in the area of environmental protection (e.g. WWF and Der Grüne

Punkt). Consequently, non-profit actors are moderately impacted by the social (e.g. human rights), highly impacted by the ecological (e.g. low emissions) and less impacted by the economic dimension.

Communities. Communities include local residents as directly concerned actors. Both are impacted by the TBL, as they have a high social and ecological but minimal economic effect. Regarding the social dimension, communities have a profound interest in knowledge generation to advance technological innovation. Due to traffic and noise pollution, local residents are highly affected by ecological aspects because they prefer quiet residential areas. As a consequence, implementing a circular economy solution does not only have a positive environmental impact (Greenwood et al., 2021) but could further foster community growth and knowledge transfer.

4.2 Stakeholder Prioritisation from a Resource-Based Perspective

The resource-based approach allows not only the analysis of which stakeholders are relevant for a long-term cooperation project, but also which stakeholder inputs are relevant for a particular transaction. This allows the differentiation between transactions at hand. We identified (1) the acceptance of an RTI solution by a critical mass of stakeholders and (2) the reverse logistics process of the RTI as the most serious obstacles to implementing a circular economy solution.

4.2.1 Resource-Based View on Critical Mass for RTI Solutions

When implementing an RTI-based circular economy solution, it is crucial that a significant number of stakeholders agree to the same technology to realise a single system solution (Schneikart et al., 2023), i.e. the positive effects of establishing a system of compatible transport items, like shipping containers, bottle deposits or EURO-pallets. However, not necessarily all patrons have to agree unanimously.

As suggested by Wieland (2014), the first criterion to identify relevant stakeholders is an actor's long-term interest in the cooperation. A proxy for a stakeholder group's interest can be found in the kind of contract with the firm. Although the concrete analysis might differ slightly depending on the context, we suggest that management, employees, RTI manufacturers, long-term shareholders and governmental institutions should be prioritised over NGOs and communities. Due to the critical mass for this cooperation, final recipients, customers of the delivery company and subcontractors have to get on board, although they can be regarded as lower priority.

The analysis then moves to the relevant resources, the readiness and ability to cooperate in this particular transaction, and the willingness to build up transaction-specific and team-specific resources (Wieland, 2014). Arguably, NPOs and NGOs can contribute resources like moral legitimacy to the project, whereas shareholders are needed for financial resources. We assume that internal stakeholders, such as a company's employees, its management and RTI manufacturers, have a high degree of interest in cooperation. Especially the RTI manufacturers and parts of management might be willing to develop transaction-specific skills and technologies. As mentioned, such an analysis becomes increasingly subjective when conducted in a case with more data at hand.

4.2.2 Resource-Based View on Reverse Logistics Processes

Considering the narrower challenge of managing a reverse logistics process, the most important stakeholders for this specific case are likely the RTI manufacturers, managers, shareholders, customers and final recipients, as they are directly relevant to its success. Except for the final recipients and customers, we assume that there is an intrinsic long-term willingness to cooperate based on profit motives, i.e. if the circular economy solution pays off or is cost-neutral in the long term, stakeholders are likely to cooperate. Moreover, the resource-based view allows looking beyond the time frame of cooperation to the specific resources that can be invested.

The RTI manufacturers should be able to design and produce the RTI in a manner that satisfies the needs of the relevant parties involved. Managers are needed to champion the circular economy solution within the delivery company and contribute their innovation skills and support from the top for a successful transaction. In particular, they need to organise the reverse logistics processes and successfully maintain the RTI system. Long-term shareholders must be willing to contribute their resources to the circular economy system, as—at least in the short term—the development of new technologies comes with the trade-off of deferred payoffs. Customers, namely, the companies that choose to ship their goods with the delivery companies, must be willing to accept the new system. They may either invest resources or pass on cost increases to the final recipients in order to recoup the initial investment costs. Lastly, the final recipients must cooperate to return the RTIs to the delivery company without discarding them prematurely or using them for different purposes.

5 Discussion

5.1 Different Perspectives on Stakeholder Management: More Than Merely Semantics

Comparing the claimant-based and resource-based view on stakeholder management above shows that managerial paradigms have implications for managers. When applying the claimant-based perspective, actors who are more strongly affected by the introduction of a circular economy solution—in our case, an RTI system—deserve attention as stakeholders. Utilising established instruments, such as LCSA and its various sub-methods that are based on the TBL, allows managers to measure and assess the consequences of corporate actions on their respective stakeholders. This can serve as a basis for prioritising stakeholder groups according to their significance to each TBL category.

In both cases, the RTI manufacturer plays a pivotal role as a stakeholder. However, other stakeholders, such as competitors, play a minor role because they are not recognised as stakeholders from a resource-based view. Although the final recipients are only moderately affected from a claimant-based view, they are major stakeholders from a resource-based view. This shows a significant difference in the prioritisation of stakeholders based on the applied paradigm.

5.2 Theoretical Contributions and Managerial Implications

As discussed above, the perspectives on stakeholder management—namely, a claimant-based and a resource-based approach—matter in practice because both perspectives lead to differences in the identification and prioritisation of stakeholders. It follows that the appropriateness of the approach depends on the problem at hand. On the one hand, when reaching a critical mass of stakeholders to implement a single RTI solution, we suggest that the claimant-based approach might be more suitable, as it is more inclusive and can thus garner support from a larger number of stakeholders. On the other hand, problems that require a comparatively small group of stakeholders to invest their skills and efforts, such as the issue of reverse logistics processes of RTIs, might be more suitable to a resource-based view.

Closely related is the argument that certain analytic tools are more suitable to one of the distinct theoretical perspectives. For instance, carbon accounting arguably focuses on the claimant-based view, as it shows how the company affects its environment with scope one to three emissions. Hence, it is possible to measure a company's negative externalities on society. However, calculating the overall impact on society is difficult, since inferring how a single company that is part of a complex system affects individual stakeholders is almost impossible. The company's negative externalities, such as air pollution and global warming, often only make up a small portion of the observable macro effects. For example, while it can be estimated how much carbon emissions could be reduced by introducing a circular economy solution, estimating the effect actually felt by the general public or the loaders as specific stakeholders is far more uncertain.

The early stakeholder definition of Freeman's (2010) stakeholder approach is inclusive, as it comprises any actor who can affect or is affected by a company's action. In contrast, Wieland and Schmiedeknecht (2010) define stakeholders as groups that have a formal or informal contract, own resources that are relevant to the company and want to cooperate in the long term. This begs the question of a differentiation between power (i.e. the ability to affect the company) and resources.

For reverse logistics processes, one can argue that the relevance and definition of stakeholders should not merely depend on the resources they contribute to the team to yield cooperation rents. In addition, the ability to veto or block the introduction of a certain solution might be equally relevant, although they would not be considered stakeholders in the resource-based perspective. Exploring the difference and implications between resources and power would provide a promising field for future research.

As an immediate solution, we propose that practitioners follow our framework to analyse stakeholder interests:

1. Identify and define relevant stakeholders.
2. Prioritise stakeholders who have the most significant influence and impact on your business by balancing a:
    (a) *claimant-based perspective*, reflecting on the stakeholders' TBL impact at the social, economic and environmental dimension using an LCSA; and
    (b) *resource-based perspective*, evaluating the stakeholders' impact on business relationship variables, namely, duration of the work contract, resources involved, cooperation and partnership, and investment and resource alignment.
3. Rank the stakeholders according to their impact (high, medium or low).
4. Include the highest ranked stakeholders in your business decisions.

5.3 Limitations

The proposed framework is not without its limitations. One caveat is that the proposed framework can only serve as a rough blueprint for an organisation that aims to engage in a stakeholder approach for introducing a circular economy solution. For instance, the exact impact of the respective stakeholder groups on the TBL can be calculated by means of carbon accounting. However, the necessary calculations require actual data input and specification of the underlying models, such as the average distance from the warehouse to the final receiver. Not to mention that the application in different industries or geopolitical contexts likely requires even more adjustment. Despite the level of analysis, we believe that the framework can offer a useful blueprint for practitioners to serve as a starting point.

A second limitation is that stakeholders might—contrary to the position of Freeman et al. (2010)—have diverging interests. Based on the stakeholder impact assessment and the analysis of stakeholder resources, it becomes clear that stakeholders might be unevenly affected by a company's actions. Although stakeholders might have a long-term interest in the success of the company, the risks and payoffs are often unevenly distributed. Whilst we did not offer a specific way to mediate between stakeholders, our framework can serve as a basis to identify conflicts of interest. Specific measures then need to be taken on a case-by-case basis.

6 Conclusion

This chapter demonstrated the application of different stakeholder models when introducing a circular economy solution in the form of RTIs to the Austrian CEP industry. A major contribution is the demonstration that different theoretical stakeholder paradigms might lead to differing stakeholder identifications as well as their prioritisation. The respective paradigm should not be chosen on an ideological basis but rather with respect to the actual problem at hand. We argue that the involvement of a range of diverse stakeholders might be more favourable to a claimant-based approach to stakeholder management. Meanwhile overcoming the challenges of reverse logistic processes, i.e. returning the transport items, is more suitable to a resource-based view, as the relevant stakeholders must involve their respective resources in the cooperation project. Due to the lack of a specific company that could serve as the basis for gathering actual data, the chapter can offer but a blueprint for other companies aiming to introduce a circular economy solution in their respective industry. The proposed framework and case study are able to assist practitioners in developing their own specific stakeholder approach.

Acknowledgements The authors would like to thank Andreas Holzleithner and Peter Entenfellner from BOOXit for providing some of the materials. A special recognition goes to Ann Ilaria Maryrhofer for editing this chapter. This work was part of a research project granted by the Kreislaufwirtschaft-funding line of the Austrian Research Promotion Agency (FFG) and sponsored by the Austrian Federal Ministry for Climate Action, Environment, Energy, Mobility, Innovation and Technology.

References

Abdallah, T. (2017). Environmental impacts. In *Sustainable mass transit* (pp. 45–59). Elsevier.

Alhaddi, H. (2015). Triple bottom line and sustainability: A literature review. *Business and Management Studies, 1*(2), 6–10. https://doi.org/10.11114/bms.v1i2.752

Arena, U., Mastellone, M., & Perugini, F. (2003). The environmental performance of alternative solid waste management options: A life cycle assessment study. *Chemical Engineering Journal, 96*(1–3), 207–222. https://doi.org/10.1016/j.cej.2003.08.019

Bansal, P. (2005). Evolving sustainably: A longitudinal study of corporate sustainable development. *Strategic Management Journal, 26*(3), 197–218. https://doi.org/10.1002/smj.441

Brickson, S. L. (2007). Organizational identity orientation: The genesis of the role of the firm and distinct forms of social value. *Academy of Management Review, 32*(3), 864–888. https://doi.org/10.5465/AMR.2007.25275679

Bundesministerium für Nachhaltigkeit und Tourismus. (2019). *MTV 2019 in Österreich: Überblick über den Einsatz von Mehrwegtransportverpackungen in Österreich*. Pulswerk. https://www.pulswerk.at/download.php?f=699

Campbell, J. L. (2007). Why would corporations behave in socially responsible ways? An institutional theory of corporate social responsibility. *Academy of Management Review, 32*(3), 946–967. https://doi.org/10.5465/AMR.2007.25275684

Cochran, J. K., & Ramanujam, B. (2006). Carrier-mode logistics optimization of inbound supply chains for electronics manufacturing. *International Journal of Production Economics, 103*(2), 826–840. https://doi.org/10.1016/j.ijpe.2006.01.005

Coelho, P. M., Corona, B., ten Klooster, R., & Worrell, E. (2020). Sustainability of reusable packaging–Current situation and trends. *Resources, Conservation & Recycling: X, 6*, Article 100037. https://doi.org/10.1016/j.rcrx.2020.100037
de Brito, M. P., Dekker, R., & Flapper, S. D. P. (2005). Reverse logistics: A review of case studies. In G. Fandel, W. Trockel, B. Fleischmann, & A. Klose (Eds.), *Distribution logistics* (pp. 243–281). Springer.
DesJardine, M. R., Zhang, M., & Shi, W. (2023). How shareholders impact stakeholder interests: A review and map for future research. *Journal of Management, 49*(1), 400–429. https://doi.org/10.1177/01492063221126707
Dreyer, L., Hauschild, M., & Schierbeck, J. (2006). A framework for Social Life Cycle Impact Assessment. *The International Journal of Life Cycle Assessment, 11*(2), 88–97. https://doi.org/10.1065/lca2005.08.223
Elkington, J. (1998). Accounting for the triple bottom line. *Measuring Business Excellence, 2*(3), 18–22. https://doi.org/10.1108/eb025539
Elkington, J. (2004). Enter the triple bottom line. In A. Henriques & J. Richardson (Eds.), *The triple bottom line: Does it all add up?* (1st ed., pp. 1–16). Routledge. https://www.johnelkington.com/archive/TBL-elkington-chapter.pdf
EuroBox Logistiksysteme. (2023). *Home*. https://www.eurobox-logistiksysteme.com/eurobox/home/
European Commission. (2022, November 30). *European Green Deal: Putting an end to wasteful packaging, boosting reuse and recycling* [Press Release]. https://ec.europa.eu/commission/presscorner/detail/en/ip_22_7155
Fassin, Y. (2008). Imperfections and shortcomings of the stakeholder model's graphical representation. *Journal of Business Ethics, 80*(4), 879–888. https://doi.org/10.1007/s10551-007-9474-5
Freeman, R. E. (2010). *Strategic management: A stakeholder approach*. Cambridge University Press.
Freeman, R. E., Harrison, J. S., Wicks, A. C., Parmar, B. L., & de Colle, S. (2010). *Stakeholder theory: The state of the art*. Cambridge University Press.
Gersmann, H. (2021). *Das Paket der Zukunft?* Taz. https://taz.de/Mehrwegverpackungen-fuer-Online-Handel/!5791615/
Ghobakhloo, M., Fathi, M., Iranmanesh, M., Maroufkhani, P., & Morales, M. E. (2021). Industry 4.0 ten years on: A bibliometric and systematic review of concepts, sustainability value drivers, and success determinants. *Journal of Cleaner Production, 302*, Article 127052. https://doi.org/10.1016/j.jclepro.2021.127052
Greenwood, S. C., Walker, S., Baird, H. M., Parsons, R., Mehl, S., Webb, T. L., & Rothman, R. H. (2021). Many happy returns: Combining insights from the environmental and behavioural sciences to understand what is required to make reusable packaging mainstream. *Sustainable Production and Consumption, 27*, 1688–1702. https://doi.org/10.1016/j.spc.2021.03.022
GS1 Germany. (2023). *Mehrwegsysteme*. https://www.gs1-germany.de/gs1-solutions/supply-chain-management/mehrwegsystem-cc-mtv/
Hahn, R. (2022). *Sustainability management: Global perspectives on concepts, instruments, and stakeholders*. Rüdiger Hahn.
Harvey, N. (1998). *Environmental impact assessment: Procedures, practice, and prospects in Australia*. Oxford University Press.
IFCO SYSTEMS. (2023). *Gut fürs Geschäft. Gut für die Umwelt*. https://www.ifco.com/de/
Ilic, A., Ng, J. W. P., Bowman, P., & Staake, T. (2009). The value of RFID for RTI management. *Electronic Markets, 19*(2–3), 125–135. https://doi.org/10.1007/s12525-009-0011-5
Ji, G., Gunasekaran, A., & Yang, G. (2014). Constructing sustainable supply chain under double environmental medium regulations. *International Journal of Production Economics, 147*, 211–219. https://doi.org/10.1016/j.ijpe.2013.04.012
Karcz, J., & Ślusarczyk, B. (2021). Criteria of quality requirements deciding on choice of the logistic operator from a perspective of his customer and the end recipient of goods. *Production Engineering Archives, 27*(1), 58–68. https://doi.org/10.30657/pea.2021.27.8

Koskela, S., Dahlbo, H., Judl, J., Korhonen, M.-R., & Niininen, M. (2014). Reusable plastic crate or recyclable cardboard box? A comparison of two delivery systems. *Journal of Cleaner Production, 69*, 83–90. https://doi.org/10.1016/j.jclepro.2014.01.045

Landschützer, C., Ehrentraut, F., & Jodin, D. (2015). Containers for the physical internet: Requirements and engineering design related to FMCG logistics. *Logistics Research, 8*(1), Article 8. https://doi.org/10.1007/s12159-015-0126-3

Lemos Junior, W. J. F., dos Reis, L. P. D. A., de Oliveira, V. S., Lopes, L. O., & Pereira, K. S. (2019). Reuse of refillable PET packaging: Approaches to safety and quality in soft drink processing. *Food Control, 100*, 329–334. https://doi.org/10.1016/j.foodcont.2019.02.008

LivingPackets. (2023). *LivingPackets*. https://livingpackets.com

McVea, J. F., & Freeman, R. E. (2005). A names-and-faces approach to stakeholder management: How focusing on stakeholders as individuals can bring ethics and entrepreneurial strategy together. *Journal of Management Inquiry, 14*(1), 57–69. https://doi.org/10.1177/105649260427079

Mitchell, R. K., Agle, B. R., & Wood, D. J. (1997). Toward a theory of stakeholder identification and salience: Defining the principle of who and what really counts. *Academy of Management Review, 22*(4), 853–886. https://doi.org/10.2307/259247

Nikolaou, I. E., Evangelinos, K. I., & Allan, S. (2013). A reverse logistics social responsibility evaluation framework based on the triple bottom line approach. *Journal of Cleaner Production, 56*, 173–184. https://doi.org/10.1016/j.jclepro.2011.12.009

Noland, J., & Phillips, R. (2010). Stakeholder engagement, discourse ethics and strategic management. *International Journal of Management Reviews, 12*(1), 39–49. https://doi.org/10.1111/j.1468-2370.2009.00279.x

Pladerer, C. (2009). *Mehrweg hat Zukunft*. Österreichisches Ökologie-Institut. https://ecology.at/mehrweg09_modell.htm

Pullman, M. E., Maloni, M. J., & Carter, C. R. (2009). Food for thought: Social versus environmental sustainability practices and performance outcomes. *Journal of Supply Chain Management, 45*(4), 38–54. https://doi.org/10.1111/j.1745-493X.2009.03175.x

Rawhouser, H., Cummings, M., & Newbert, S. L. (2019). Social impact measurement: Current approaches and future directions for social entrepreneurship research. *Entrepreneurship Theory and Practice, 43*(1), 82–115. https://doi.org/10.1177/1042258717727718

Samadhiya, A., Kumar, A., Agrawal, R., & Kazancoglu, Y. (2023). Reinventing reverse logistics through blockchain technology: A comprehensive review and future research propositions. *Supply Chain Forum: An International Journal, 24*(1), 81–102. https://doi.org/10.1080/16258312.2022.2067485

Schneikart, G., Mayrhofer, W., Frysak, J., & Löffler, C. (2023). A returnable transport item to integrate logistics 4.0 and circular economy in pharma supply chains. *Technical Journal, 17*(3), 375–382. https://doi.org/10.31803/tg-20230504144856

Schumpeter, J. A. (1934). *The theory of economic development: An inquiry into profits, capital, credit, interest, and the business cycle*. Transaction Publishers.

Soysal, M., Koç, Ç., Çimen, M., & İbiş, M. (2023). Managing returnable transport items in a vendor managed inventory system. *Socio-Economic Planning Sciences, 86*, Article 101504. https://doi.org/10.1016/j.seps.2022.101504

Svenska Retursystem. (2023). *Together towards a sustainable future*. https://www.retursystem.se/en

Wieland, J. (2011). The firm as a nexus of stakeholders: Stakeholder management and theory of the firm. In A. Brink (Ed.), *Corporate governance and business ethics* (pp. 225–244). Springer.

Wieland, J. (2014). *Governance ethics: Global value creation, economic organization and normativity*. Springer.

Wieland, J. (2020). Stakeholder resources and stakeholder interests. In *Relational economics: A political economy* (pp. 69–83). Springer.

Wieland, J., & Schmiedeknecht, M. (2010). *Corporate Social Responsibility (CSR), Stakeholder Management und Netzwerkgovernance* (KIeM Working Paper No. 31/2010). Konstanz Institut für Wertemanagement

Windsor, D. (2006). Corporate social responsibility: Three key approaches. *Journal of Management Studies, 43*(1), 93–114. https://doi.org/10.1111/j.1467-6486.2006.00584.x

Winn, M. I. (2001). Building stakeholder theory with a decision modeling methodology. *Business and Society, 40*(2), 133–166. https://doi.org/10.1177/000765030104000202

WKO. (2022). *KEP-Branchenreport 2022*. https://www.wko.at/service/w/verkehr-betriebsstandort/KEP-Branchenreport-2022.pdf

Xu, X., Chung, S.-H., Lo, C. K., & Yeung, A. C. (2022). Sustainable supply chain management with NGOs, NPOs, and charity organizations: A systematic review and research agenda. *Transportation Research Part E: Logistics and Transportation Review, 164*, Article 102822. https://doi.org/10.1016/j.tre.2022.102822

Xu, X., Lu, Y., Vogel-Heuser, B., & Wang, L. (2021). Industry 4.0 and Industry 5.0—Inception, conception and perception. *Journal of Manufacturing Systems, 61*, 530–535. https://doi.org/10.1016/j.jmsy.2021.10.006

Zamagni, A., Pesonen, H.-L., & Swarr, T. (2013). From LCA to Life Cycle Sustainability Assessment: Concept, practice and future directions. *The International Journal of Life Cycle Assessment, 18*(9), 1637–1641. https://doi.org/10.1007/s11367-013-0648-3

Nils Kruse is ESG Manager at Samsung SDI Europe and Visiting Scholar at the TU Dresden at the Chair of Business Administration, especially Responsible Management. He studied Management and Asian Studies at the HTWG Konstanz, International Management at the Korea University (Seoul, South Korea), International Relations at the American University (Washington D.C., USA) and received his PhD in Business Administration at the Martin Luther University Halle-Wittenberg. Previously, he served as Head of the Institute for Business Ethics and Sustainable Strategy (IBES) at the FHWien der WKW, Senior Consultant and Executive Assistant at PwC Germany, as well as researcher at the Martin Luther University and EWHA University in Seoul.

Melanie Rainer (M.Sc.) is a doctoral candidate, a managing director of Junior Management Science (JUMS) and a Research Associate at the University of Sustainability (CFPU) in Vienna. Besides serving as an adjunct lecturer in sustainable and responsible business, her expertise converges on change management with a primary focus on environmental and social sustainability. Additionally, her research focuses on supply chains, entrepreneurship as well as business and human rights, demonstrating a comprehensive commitment to advancing ethical practices and sustainability across various facets of business.

Gerald Schneikart is Senior Researcher and Project Manager at the University of Applied Science and Communication in Vienna, Austria. Dr. Schneikart has a broad scientific background, which extends to digital transformation and management. After studies in Molecular Biology at the University of Vienna, he completed an industrial PhD programme of GlaxoSmithKline (former Novartis Vaccines) in Siena, Italy, and he received a Doctoral Degree in Life Sciences at the Università di Siena. He then decided to further broaden his expertise with an MBA programme in General Management at the Continuing Education Centre of the TU Wien, including a semester at the Baruch College in New York City.

Walter Mayrhofer is Head of Research at the University of Applied Sciences Vienna and Professour at its Institute of Digital Transformation & Strategy. He studied Electrical Engineering, Business, Law and Economics and Industrial Engineering at the TU Wien, Engineering Management at the University of Colorado at Boulder and completed an MBA at the Danube University Krems. Previously, he served as a researcher at the TU Wien and Danube University Krems, Research Director at the FH Burgenland and Fraunhofer Austria, and Managing

Director of FTI Burgenland GmbH, Forschung Burgenland GmbH and "die angewandte" Continuing Education GmbH. He also has extensive experience with national and international research and industry projects.

Made in the USA
Las Vegas, NV
05 November 2026

57552597R00199